Introduction to Space Flight

 # Introduction to Space Flight

Francis J. Hale

North Carolina State University

PRENTICE HALL
Englewood Cliffs, New Jersey 07632

Library of Congress Cataloging-in-Publication Data

Hale, Francis J.
  Introduction to Space Flight / Francis Joseph Hale
    p. cm.
  Includes bibliological references and index.
  ISBN: 0-13-481912-8
  1. Space Flight. 2. Orbital Mechanics. I. Title.
TL791.H35 1994
629.4'1--dc20
                       93-22436
                         CIP

Acquisitions editor: *DOUG HUMPHREY*
Editorial/production supervision and
  interior design: *RICHARD DeLORENZO*
Copy editor: *BARBARA ZEIDERS*
Cover design: *WENDY ALLING JUDY*
Prepress buyer: *LINDA BEHRENS*
Manufacturing buyer: *DAVID DICKEY*
Editorial assistant: *SUSAN HANDY*
Supplements editor: *ALICE DWORKIN*

©1994 by Prentice-Hall, Inc.
A Simon & Schuster Company
Englewood Cliffs, New Jersey 07632

All rights reserved. No part of this book may be
reproduced, in any form or by any means,
without permission in writing from the publisher.

The author and publisher of this book have used their best efforts in preparing this book. These efforts include the development, research, and testing of the theories and programs to determine their effectiveness. The author and publisher make no warranty of any kind, expressed or implied, with regard to these programs or the documentation contained in this book. The author and publisher shall not be liable in any event for incidental or consequential damages in connection with, or arising out of, the furnishing, performance, or use of these programs.

Printed in the United States of America

10 9 8 7 6 5 4 3 2 1

ISBN 0-13-481912-8

Prentice-Hall International (UK) Limited, London
Prentice-Hall of Australia Pty. Limited, Sydney
Prentice-Hall Canada Inc., Toronto
Prentice-Hall Hispanoamericana, S.A., Mexico
Prentice-Hall of India Private Limited, New Delhi
Prentice-Hall of Japan, Inc., Tokyo
Simon & Schuster Asia Pte. Ltd., Singapore
Editora Prentice-Hall do Brasil, Ltda., Rio de Janeiro

*This book is dedicated to
Colonel Edward N. Hall.
Without Ed, there would be no
MINUTEMAN ballistic missile
weapon systems.*

Contents

PREFACE   xi

CHAPTER 1   INTRODUCTION   1

     1-1   Space as a Place   2
     1-2   Scope of Book   4
     1-3   Units and Symbols   4

CHAPTER 2   TWO-BODY ORBITAL MECHANICS   6

     2-1   Introduction   7
     2-2   The Two-Body Problem   8
     2-3   The Equations and Constants of Motion   11
     2-4   The Equation of Conic Sections   14
     2-5   Some Properties of Conic Sections   17
     2-6   The Physical Significance of Two-Body Trajectories   21
     2-7   The Elevation Angle   24
     2-8   Summary of Relevant Two-Body Equations   25
   Problems   26

CHAPTER 3 GEOCENTRIC ORBITS AND TRAJECTORIES 28

    3-1   Introduction   29
    3-2   Circular Orbits   30
    3-3   Elliptical Orbits   33
    3-4   Orbit Shaping and Impulsive Thrusting   36
    3-5   Other Trajectories and Escape   39
    3-6   Orbital Transfers   42
    3-7   Plane Changes   50
    3-8   Closing Remarks   53
    Problems   55

CHAPTER 4 TIME OF FLIGHT 58

    4-1   Introduction   59
    4-2   The Parabola   60
    4-3   The Ellipse   62
    4-4   The Hypberbola   68
    4-5   Lambert's Theorem   72
    4-6   Closing Remarks   79
    Problems   79

CHAPTER 5 INTERPLANETARY TRANSFERS 81

    5-1   Introduction   82
    5-2   The Solar System   83
    5-3   The Sphere of Influence and Impact Parameter   85
    5-4   The Patched Conic   91
    5-5   Planetary Capture   104
    5-6   Planetary Passage and Gravity Assist   112
    5-7   The Intercept Problem and Fast Transfers   124
    5-8   Canonical Units   134
    5-9   Closing Remarks   137
    Problems   138

CHAPTER 6 VEHICLE AND BOOSTER PERFORMANCE 141

    6-1   Introduction   142
    6-2   Rocket Engines   144
    6-3   The Performance Equations   157

Contents ix

    6-4   Single-Stage Performance   160
    6-5   Restricted Staging   169
    6-6   Generalized Staging   179
    6-7   Powered Boost   185
    6-8   Closing Remarks   192
    Problems   193

CHAPTER 7   ATMOSPHERIC ENTRY   196

    7-1   Introduction   197
    7-2   The Entry Problem   198
    7-3   Ballistic Entry   209
    7-4   Lifting Entry   225
    7-5   The Heating Problem   234
    7-6   Closing Remarks   242
    Problems   244

CHAPTER 8   ORBITAL ELEMENTS AND EARTH TRACKS   246

    8-1   Introduction   247
    8-2   The Orbital Elements   247
    8-3   Orbit Determination   252
    8-4   Satellite Ground Tracks   257
    8-5   Closing Remarks   276
    Problems   277

CHAPTER 9   THE BALLISTIC MISSILE   279

    9-1   Introduction   280
    9-2   The Range Equations   281
    9-3   Trajectories and Sensitivities   292
    9-4   The Rotating Earth   302
    Problems   307

CHAPTER 10   ATTITUDE DYNAMICS AND CONTROL   309

    10-1   Introduction   310
    10-2   The Equations of Motion   311
    10-3   Single-Spin Stabilization   317
    10-4   Dual-Spin Stabilization   325
    10-5   Gravity-Gradient Stabilization   326
    10-6   Active Control Stabilization   333

APPENDIX A   SOME USEFUL VECTOR OPERATIONS   339

APPENDIX B   PLANETARY VALUES *342*

    Table B-1   Physical Data   343
    Table B-2   Orbital Data   343

APPENDIX C   ADDITIONAL ILLUSTRATIONS *344*

    SELECTED REFERENCES *356*

    INDEX *359*

Preface

This introductory book was developed as a teaching text for a one-semester course that has been taught at North Carolina State University for many years. The course is an elective for students in all curricula: engineering and non-engineering. Since the book strives to be self-contained, there are no prerequisites *per se* although an understanding of elementary physics, calculus, and vector operations would be helpful.

The major objectives of this book are to give the reader an understanding of how to get into space, how to get around in space, and how to return to Earth or land on another planet. These objectives might be thought of as the performance aspects of space flight as contrasted with the more specialized aspects such as life support, guidance and control, and communications.

The emphasis of this book is on fundamental concepts and analytic expressions and not on "cook book" relationships that have been derived for a special application and require only the insertion of numbers. For example, there are no expressions nor charts for specified orbital transfers in which only the values of the initial and final orbits are inserted to find the total velocity increment and time of flight. Instead the reader is asked to find the characteristics of the desired transfer trajectory and then determine the individual velocity increments and time of flight.

Since one of the secondary objectives is to give the reader a feel for the magnitude of the numbers involved, there are numerous examples in the text as well as problems at the end of each working chapter. The illustrative examples support the concepts and applications of the text and the problems serve to strengthen and, in some cases, extend the reader's understanding of the textual material. Only a hand calculator is required although the reader may feel free to use computer solutions or a spreadsheet if he or she writes the programs or constructs the spreadsheet. Spreadsheets can be very useful if there are enough trajectories and orbits of interest to warrant the expenditure of time in their construction.

For an introductory book such as this, any claim to merit is based on organization, coverage, and style. The first chapter is short and strictly introductory with a brief look at space and current attitudes toward space along with a few words about the scope of the book and the use of SI units. The second chapter develops the two-body planar equations of motion and the important determination that the trajectories are indeed conic sections. Chapter 3 then applies the results of Chapter 2 to Earth-centered (geocentric) trajectories in near space and extends them to include orbit transformations, orbital transfers, and escape from Earth's gravitational attraction. The concept of the velocity increment, ΔV, to change the energy of a trajectory is first introduced in this chapter. Chapter 4 interrupts the application of the two-body mechanics to develop expressions for determining the time required to move from one point to another along a trajectory, the time of flight (TOF). The TOF is of special interest in determining orbital transfer times and lead angles for the intercept problem. The section on Lambert's theorem can be discussed briefly and passed by without any loss of continuity.

Chapter 5 moves on to outer space and interplanetary trajectories, still limited to two-body mechanics. In this chapter, the important concepts of the patched conic and the velocity budget are developed along with a treatment of planetary capture and planetary passage, the latter being the interesting case of modifying a heliocentric trajectory without the expenditure of propulsive energy (the gravity-assist maneuver). This chapter ends with a description and discussion of canonical units, which may be encountered in practice. They are not used in this book since they are essentially dimensionless and do not give a feel for the realistic magnitude of numbers.

Chapter 6 deals with the problem of generating the velocity increment(s) required to place a spacecraft into desired orbits and trajectories along with an estimation of the space vehicle weights and number of stages required. Propulsion is limited to the high-thrust chemical rocket engines currently in use. Chapter 7 then moves on to take a look at the problems and advantages associated with entering an atmosphere, specifically that of Earth. The ballistic (no lift) entry is treated empirically but in reasonable detail with respect to the decelerations encountered. Lifting entry and the heating problem, on the other hand, are treated only in a qualitative manner since either topic alone could be the focus of a one-semester course.

Prior to Chapter 8, other than discussing the velocity requirements for changing the plane of a trajectory, the motion of the smaller body is restricted to a single plane and is described only in terms of its relationship to the center of the inertial reference, the center of mass of the larger body. In Chapter 8, the remaining orbital elements are introduced along with the concept of the ecliptic, a combination that relates the position of the smaller body to other trajectories and orbits; i.e., to the rest of the system under consideration. It is now possible to discuss the motion of the smaller body (a satellite, for example) over the surface of a rotating Earth (or any other planet) as well as the amount of the Earth's surface that can be observed as a function of the orbital altitude. The ground-track plots in Chapter 8 were provided by the Department of Astronautics at the U.S. Air Force Academy.

The treatment of the ballistic missile in Chapter 9 is included as an interesting application of two-body orbital mechanics to a specific sub-orbital problem. It is not essential to meeting the objectives of the book and may be omitted or not covered in detail. In teaching the course, there have been times when I have omitted the section on optimum trajectories and have merely laid out the ballistic missile problem and its solutions.

Chapter 10 is included at the end of the book to remind the reader that a spacecraft is a three-dimensional object and not a point mass. Consequently, its attitude and dynamic stability must be considered in carrying out a mission. This chapter discusses the major aspects of the stabilization and pointing problems and provides a basis for further study if desired.

This sequence, as laid out above and in the Contents, is my choice and may be changed to suit the interests of the reader or of another instructor. The chapters, and many of the sections, are sufficiently self-contained so that different sequences and topic emphasis are possible. For example, orbital elements and ground tracks could easily be introduced much earlier if desired, say after Chapter 3.

Let me emphasize that this is a first book and does not attempt to cover even the performance aspects of space flight completely and exhaustively. For example, there is no treatment of n-body mechanics, not even a restricted three-body problem as exemplified by the Earth-Moon combination. The goals are modest but by the end of the book the reader should be able to determine the approximate velocity budget for a geocentric or interplanetary mission, the appropximate weight (mass) and number of stages required to accomplish the mission, and the general problems associated with the terminal phase of the mission. This ability implies an understanding of the major factors involved in getting into and out of space along with a feel for the magnitude of the relevant parameters.

In conclusion, I should like to express my appreciation to the following reviewers whose excellent comments and suggestions were both helpful and welcome: Wiley J. Larson, U.S. Air Force Academy; John L. Junkins, Texas A & M University; and a special thanks to Kenneth D. Mease, Princeton University.

Francis J. Hale

Introduction to Space Flight

CHAPTER 1

Introduction

The launch of an Atlas-Centaur two-stage booster with a geocentric satellite payload to study the Earth's ionosphere and magnetosphere. (Courtesy of General Dynamics.)

1-1 SPACE AS A PLACE

Space is a vast place. It starts at the edge of our atmosphere and extends beyond our Moon, beyond the planets in our solar system, beyond our galaxy, beyond the far galaxies; no one knows where it ends. Space is so vast that distances are measured in light-seconds and light-years, where a light-second is 299,800 km (162,900 nmi) and a light-year is on the order of 9.46×10^{12} km (5.14×10^{12} nmi or 5.14 million million nautical miles). The immensity of the universe boggles one's mind; imagine 100 billion stars in our galaxy alone. The Earth is but a grain of sand in this universe.

Space has fascinated human beings for thousands of years and has been the object of observations and the subject of theories as to its creation, composition, and behavior. Although meteors and meteorites have been entering our atmosphere throughout history, it has only been in recent years that objects have left the sensible atmosphere. Ballistic missile warheads were probably the first to leave, followed by unmanned satellites and then by manned vehicles.

Since the first satellite launch (Sputnik) in 1957, public interest and governmental support in this country have undergone some oscillations; there have been highs and lows. In the wake of Sputnik, enthusiasm was high, imaginations were strong, and space beckoned. Building on its experience with aircraft and ballistic missiles, the U.S. Air Force began funding various activities to develop a general space capability. Since it was not obvious exactly what should be done in space and what areas of research and development should be emphasized, the U.S. Air Force decided that a program to land people on the Moon would provide a focus for the development of appropriate space technology as well as stimulate public interest and support. After all, the Moon is the nearest and most ubiquitous celestial body and has long been the destination of imaginary expeditions. As examples of early interest, Cyrano de Bergerac's *Voyage to the Moon* was published in 1657 and Jules Verne's *From the Earth to the Moon* in 1865.

The National Aeronautics and Space Administration (NASA) was created in 1958 to develop and emphasize the civilian and peaceful aspects of space. NASA expanded the Moon landing concept of the Air Force into the Apollo program, which indeed became the center of activity for many years. Then on April 12, 1961, Yuri Gagarian made the first orbital manned flight, and Neil Armstrong and Buzz Aldrin walked on the moon on July 20, 1969.

Concurrent with the Apollo program were unmanned explorations, primarily of the neighboring planets, and a series of terrestial satellites for weather observations, surface mapping, navigation, and communications as well as military reconnaissance satellites. During this period there was informal and formal debate, that is still ongoing, as to the relative merits and benefits of manned and unmanned flight.

With the end of the Apollo program, there was a hiatus in manned operations until the advent of the Space Transportation System (STS), popularly known as the Space Shuttle. It has the objective of demonstrating the ability to get into and out of near-Earth space using a manned reusable space vehicle with the capability of inserting and retrieving payloads. Reusability and launch frequency were to be the keys to the reduction in the cost of placing payloads into various orbits.

There is no apparent consensus among the manned-space supporters as to which course of action to pursue next. The space station groups advocate a station orbiting 552 km (300 nmi) above the surface of the Earth but have different ideas as to its size and function(s) (and cost). For example, one group wants a station devoted to scientific and manufacturing experiments and another group visualizes the space station as a staging base for a manned mission to Mars. Other groups are pushing a lunar station, which as with the space station might be an end in itself, perhaps mining the Moon, or it might serve as a way station on the way to Mars. Incidentally, even the most ardent supporters of a manned Mars mission have trouble justifying the trip, other than the fact that Mars is there.

At the same time, the unmanned advocates are pursuing their claims that the money for manned space stations can better be used to increase the number of unmanned explorations to the planets and their moons, to the asteroids, and possibly into deep space. Their justification cites a search for knowledge and an increased understanding of our solar system.

In the meantime, the military is having its own problems in determining a proper role in space. The original euphoria over "seizing the high ground" and placing warhead launchers on the Moon or on satellites dwindled with the realization that there is no high ground in space and that the cost-effectiveness of such space-based systems would not be competitive with current and advanced terrestrial systems. Even the appeal of the aerospace plane with the promise of space combat or use as a reusable booster was a victim of reality. The support of the Space Defense Initiative (SDI), or Star Wars, which represents the latest effort to develop a combat capability in space, is far from unanimous.

The purpose of the preceding paragraphs was not historical nor was it to favor any viewpoints or philosophies. The purpose was to show that there is no consensus among those who think about space as to what the potential of space is, what should be done, and how it should be done. Although it may not be obvious as to what should be done in space, this has always been the case with unexplored territory. There is nothing wrong with accepting the challenge to explore the unknown or the not-as-well-known-as-we-would-like-it-to-be in order to learn more about this unknown territory with an eye to determining how space should be used and what the possible benefits might be. It is a form of self-deception, however, to rationalize space exploration with statements and reasons that are presented as self-evident and obvious but that are speculative, certainly unproven.

We close this section with the words of an unnamed source: "Space is a place—not a mission."

1-2 SCOPE OF BOOK

Since the subject of space is as vast as its dimensions, it is necessary to limit the coverage of any one book. Consequently, this is an introductory book on space flight performance. It is a book about getting into space, getting around in space, and getting out of space (landing on a planetary surface); the emphasis is on getting around in space.

This could be considered to be a book about the management of energy. The major source of this energy is chemical propulsion; it is used to boost a payload into space, to establish appropriate trajectories (to include transfer trajectories), and to modify trajectories by adding or reducing energy. Trajectory modification can also be accomplished by an appropriate hyperbolic flyby of a large body, such as a planet or a moon. Landings are executed by dissipating energy by the use of propulsion, gas dynamic drag (if an atmosphere exists), or a combination thereof.

Even with the scope limited to the performance aspects of space flight, the coverage is further restricted by a series of approximations and idealizations. For example, there is no treatment of n-body, or even restricted three-body, orbital mechanics and the two-body problem has even been reduced to a central force field problem by neglecting the gravitational attraction of the spacecraft. This is not an astrodynamics book by any means; fortunately, there are such books.

There are many other areas that are related to the behavior of a space vehicle and are not covered or treated in this book, such as vehicle stability and control, navigation, guidance and control, and communications. Nor is there any coverage of the systems required to maintain, protect, and operate the contents of a spacecraft where the contents might be human beings and scientific equipment or simply equipment in the case of unmanned missions. Among the functions of such systems are life support, environmental control, power, and structural integrity.

1-3 UNITS AND SYMBOLS

In keeping with the push toward the standardization of units through metrication, SI (Systéme International) units will be used consistently in this book. The three base units of mass, length, and time will be the kilogram (kg), the meter (m), and the second (s). The one exception will be the use of nautical miles (nmi) and kilometers (km) to describe the altitude above a planetary surface, as is often done in practice.

Sec. 1-3 Units and Symbols

The derived units will be newtons (N) for force and weight (kg-m/s²) and pascal [Pa (N/m² or kg/m-s²)] for a force or weight per unit ~~mass~~ area. Density will be expressed in kg/m³. It should be noted that it is still the custom, even in metric countries, to express weight in kilograms even though the kilogram is a unit of mass and not force.

Although there will be little need for conversion from SI to customary English units, several relationships that might be of interest are

1 nmi = 1.840 km = 1.1515 mi = 6080 ft
1 ft = 0.305 m
1 ft/s = 0.305 m/s = 0.682 mi/h = ~~0.45~~ 0.59 knots
1 lbf = 4.448 N = 0.4535 kg
1 lb/ft² = 47.87 Pa
1 J/kg = 1 N·m/kg = 1 m³/s²

Every attempt is made to conform with standard and customary engineering symbology. The one deliberate exception is the use of capital, rather than lowercase, letters to denote the specific mechanical energy and the specific angular momentum. E, rather than e, is used for the former since e may be used elsewhere to denote the eccentricity of a trajectory (ϵ is used in this book) and H, rather than h, is used for the latter in order to be consistent with the use of e rather than E. Since there is no mention of the total energy or of the total angular momentum in the book, there should be no occasions for further confusion.

CHAPTER 2

Two-Body Orbital Mechanics

The Space Shuttle Atlantis, carrying a geocentric tracking and relay satellite (TDRS), during first-stage ascent and shortly after lift-off. (Courtesy of NASA.)

2-1 INTRODUCTION

Celestial mechanics is defined in one dictionary[1] as "the branch of mechanics concerned with the motion of natural or man-made celestial bodies under the influence of gravity." From this definition we see that the terms *celestial mechanics* and *orbital mechanics* are synonymous and interchangeable. However, *orbital mechanics* is often used to describe the trajectories and orbits of human-made vehicles (spacecraft) only.

Celestial (orbital) mechanics is essentially an application of Sir Isaac Newton's (1642–1727) law of universal gravitation and of his three laws of motion. It is interesting to note that prior to Newton, Johannes Kepler (1571–1630) used his own observations and data from earlier astronomers, such as Tycho Brahe (1546–1601), to develop the three Keplerian laws of planetary motion, which can be stated as follows:

1. The orbits of the planets are ellipses with the Sun at one focus.
2. The line joining a planet to the Sun sweeps out equal areas in equal intervals of time.
3. The square of the period of a planet is proportional to the cube of the major axis of its elliptical orbit.

In the course of developing the two-body equations of motion, we shall demonstrate and verify the validity of these laws of Kepler.

In this chapter we develop the vector equation of motion for two bodies subject only to their mutual gravitational attractions and then simplify it for the case where one body is much smaller than the other (the central force field problem). Then the scalar equations and constants of motion will be obtained by vector operations and it will be shown that the trajectory of the smaller body with respect to the larger body can be represented by the polar equation of a conic. Consequently, all of the

[1] *The World Book Dictionary*, Field Enterprises, Chicago, 1971.

well-known properties and characteristics of conic sections can be applied to the trajectory of the smaller body.

2-2 THE TWO-BODY PROBLEM

Although every body (mass) in the universe is attracted to and has an attraction for every other body, it is, fortunately, possible to treat most trajectories of interest as the solution of the much simpler two-body problem. The central problem in two-body mechanics is the determination of the motion of two point masses (with specified initial positions and velocities) that are subject only to their own mutual gravitational forces. Typically, one mass is considerably smaller than the other and it is its trajectory (orbit) that is of interest. Deviations from the two-body solution may arise because the two bodies are not truly point masses or because of the action of other forces, such as drag or the attraction of other bodies. These effects are often sufficiently small as to be treated as corrections to (or perturbations of) the two-body solution or even to be ignored completely.

Before developing the solution to the two-body problem, it is necessary to introduce *Newton's law of universal gravitation*, which in essence states that *two bodies will exert a force on each other that acts along the line joining the bodies and that is directly proportional to the product of their masses and inversely proportional to the square of the distance between them*. Remember that a "law" is an empirical relationship that is based on observations only and *cannot* be proved.

Let us now consider two homogeneous spheres of any relative size, m_1 and m_2, that can be treated as point masses, that are subject only to their mutual gravitational forces, and that are located in an inertial reference system as shown in Fig. 2-2-1, where $\mathbf{R}_1$ and $\mathbf{R}_2$ are the inertial position vectors of the two masses and $\mathbf{r}$ is the vector joining the two masses. The gravitational force acting on each body can be written in vector form as

$$\mathbf{F} = \frac{-Gm_1 m_2}{r^2} \mathbf{1}_r \qquad (2\text{-}2\text{-}1)$$

where G is the *universal gravitational constant* and $\mathbf{1}_r$ is a unit vector of $\mathbf{r}$ and positive in the direction of m_2.

Newton's second law for a point mass can be written as $\mathbf{F} = m\mathbf{a}$, where $\mathbf{a}$ is the acceleration of the mass with respect to inertial space. Applying Newton's second law to m_1, $\mathbf{a}$ is $\ddot{\mathbf{R}}_1$ and Eq. (2-2-1) can be written, keeping the direction of the force in mind, as

$$m_1 \ddot{\mathbf{R}}_1 = \frac{Gm_1 m_2}{r^2} \mathbf{1}_r \qquad (2\text{-}2\text{-}2)$$

If we cancel the m_1 on both sides of the equation, we obtain an expression for $\ddot{\mathbf{R}}_1$, namely,

$$\ddot{\mathbf{R}}_1 = \frac{Gm_2}{r^2} \mathbf{1}_r \qquad (2\text{-}2\text{-}3)$$

Sec. 2-2   The Two-Body Problem

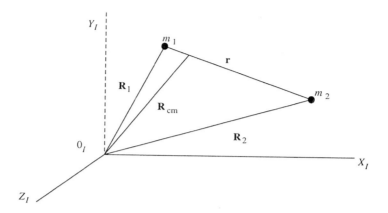

In a similar manner, we obtain a similar expression for $\ddot{\mathbf{R}}_2$,

$$\ddot{\mathbf{R}}_2 = \frac{-Gm_1}{r^2}\mathbf{1}_r \qquad (2\text{-}2\text{-}4)$$

From Fig. 2-2-1 we see that $\mathbf{r} = \mathbf{R}_2 - \mathbf{R}_1$, so that $\ddot{\mathbf{r}} = \ddot{\mathbf{R}}_2 - \ddot{\mathbf{R}}_1$. Subtracting Eq. (2-2-3) from Eq. (2-2-4) and using the relationship that $\mathbf{1}_r = \mathbf{r}/r$ yields

$$\boxed{\ddot{\mathbf{r}} = \frac{-G(m_1 + m_2)}{r^3}\mathbf{r}} \qquad (2\text{-}2\text{-}5)$$

Equation (2-2-5) is the vector differential equation of the relative motion of the two bodies, where the expression $G(m_1 + m_2)$ is generally referred to as the *gravitational parameter* (of the combined masses) and is given the symbol μ.

Let us next look at the relationship of the center of mass (cm) of the system to the center of the larger body. Referring to Fig. 2-2-2 and applying the definition of the center of mass that $\int \mathbf{r}\, dm = 0$, we see that for our two-body system, $m_1 r_1 - m_2 r_2 = 0$. Since $r_2 = r - r_1$, then $r_1 = m_2 r/(m_1 + m_2)$. With the assumption that $m_1 \gg m_2$, r_1 becomes very small and the center of the principal attracting (the larger) body can be taken to be the cm of the system and thus serve as the origin of an inertial reference system. For example, for the Earth–Sun combination with the ratio of the masses $\cong 1/332{,}900$, the cm of the system is on the order of 455 km from the center of the Sun, whereas the distance between the Sun and the Earth is on the order of 150 million kilometers. Furthermore, since the angular velocity of line joining the two bodies is equal to the tangential component of the velocity of each body divided by its distance from the cm, $\mathbf{V}_1 \ll \mathbf{V}_2$. For example, the orbital velocity of the Sun about the cm of the system is on the order of 0.1 m/s, whereas that of the Earth is on the order of 30,000 m/s. It should be pointed out that although this central body approximation is valid for many purposes, there are times when it is not.

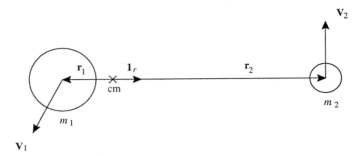

Figure 2-2-2. Two-body system.

Consequently, the two-body system can be represented by Fig. 2-2-3, where the cm of the system is located at the larger point mass and the velocity of the larger mass with respect to the cm is neglected. Remember that the larger mass, m_1, is *not* fixed in space; however, it is the motion of the smaller body with respect to the larger (attracting) body that interests us. It should be pointed out that this central-body approximation is sufficiently accurate for many purposes, but not all.

With $m_1 \gg m_2$, $\mu \cong Gm_1$ and is redefined as the gravitational parameter of the larger body. Equation (2-2-1) can now be rewritten as

$$\mathbf{F} = \frac{-\mu m}{r^2} \mathbf{1}_r \tag{2-2-6}$$

where the subscripts have been dropped and m is the mass of the smaller body. The magnitude of the force in Eq. (2-2-6) is called the *weight* of the smaller body and accelerates the smaller body toward the cm of the attracting body. This acceleration is commonly referred to as the *acceleration due to gravity*, is given the symbol g, and can be found from the expressions

$$F = \frac{\mu m}{r^2} = mg \tag{2-2-7}$$

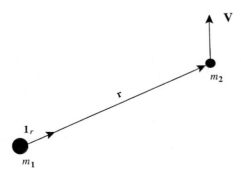

Figure 2-2-3. Simplified two-body system.

Sec. 2-3 The Equations and Constants of Motion

to be

$$g = \frac{\mu}{r^2} \qquad (2\text{-}2\text{-}8)$$

Since μ is a constant for a given attracting body, g is not a constant, varying inversely as the square of the distance of the smaller body from the cm of the attracting body.

In summary, the assumptions used in the development of these restricted two-body equations are that:

1. The larger body is a homogeneous sphere that may be treated as a point mass with its attractive force concentrated at its center.
2. The smaller body is also a point mass.
3. The mass of the larger body is so much greater than that of the smaller body that the cm of the system can be located at the cm of the larger body, and the attractive force of the smaller body can be neglected. The gravitational attraction of the larger body is sometimes referred to as a *central force field*.
4. The gravitational attraction of any other bodies can be ignored.
5. There are no forces other than gravitation acting on the smaller body (i.e., no lift, drag, thrust, solar pressure, etc.).

2-3 THE EQUATIONS AND CONSTANTS OF MOTION

Let us now define an inertial reference system using both Cartesian and polar coordinates, as shown in Fig. 2-3-1, with the origin at the cm of the larger body and where the *position angle v* is measured from the positive x-axis and is positive in the counterclockwise direction.

Using either definition of μ, Eq. (2-2-5) can be written as

$$\ddot{\mathbf{r}} + \frac{\mu}{r^3}\mathbf{r} = 0 \qquad (2\text{-}3\text{-}1)$$

Equation (2-3-1) is the *Newtonian vector equation of motion* that describes the relative motion of the smaller body with respect to the larger. Since the mass of the smaller body has disappeared from the equation, the motion of the smaller body is independent of its mass; this is fortuitous in that the mass of the smaller body (e.g., a spacecraft) does not have to be known. The mass of the larger body manifests itself in the gravitational parameter.

By means of vector operations and manipulations Eq. (2-3-1) can be transformed into perfect differentials that can be integrated to obtain constants of integration that are also constants of the orbital motion and that show that both energy and angular momentum are conserved.

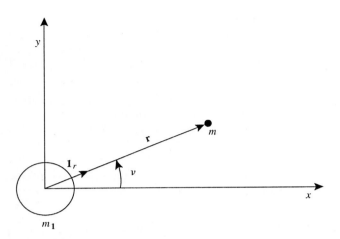

Figure 2-3-1. Inertial reference system.

To show that mechanical energy is conserved, we find the dot product of each term with $\dot{\mathbf{r}}$ so that Eq. (2-3-1) becomes

$$\dot{\mathbf{r}} \cdot \ddot{\mathbf{r}} + \frac{\mu}{r^3}(\dot{\mathbf{r}} \cdot \mathbf{r}) = 0 \qquad (2\text{-}3\text{-}2)$$

since $\mathbf{r} = r\mathbf{1}_r$,

$$\dot{\mathbf{r}} = \frac{d}{dt}(r\mathbf{1}_r) = \dot{r}\mathbf{1}_r + r\frac{d\mathbf{1}_r}{dt}$$

Therefore, the term $(\dot{\mathbf{r}} \cdot \mathbf{r})$ can be written as

$$\dot{\mathbf{r}} \cdot \mathbf{r} = r\dot{r}(\mathbf{1}_r \cdot \mathbf{1}_r) + r^2\left(\mathbf{1}_r \cdot \frac{d\mathbf{1}_r}{dt}\right)$$

However, $(\mathbf{1}_r \cdot \mathbf{1}_r) = 1$ and $(\mathbf{1}_r \cdot r\, d\mathbf{1}_r/dt) = 0$, so that $(\dot{\mathbf{r}} \cdot \mathbf{r}) = r\dot{r}$. Similarly, $(\dot{\mathbf{r}} \cdot \ddot{\mathbf{r}}) = (\mathbf{V} \cdot \dot{\mathbf{V}}) = V\dot{V}$. With these substitutions, Eq. (2-3-2) is a scalar equation, as was expected, and can be written as

$$V\dot{V} + \frac{\mu \dot{r}}{r^2} = 0 \qquad (2\text{-}3\text{-}3)$$

This scalar equation can be written in expanded form as

$$V\frac{dV}{dt} = \frac{-\mu}{r^2}\frac{dr}{dt} \quad \text{or} \quad V\, dV = \frac{-\mu}{r^2}\, dr \qquad (2\text{-}3\text{-}4)$$

With the variables separated, Eq. (2-3-4) can be integrated to obtain

$$\boxed{\frac{V^2}{2} - \frac{\mu}{r} = E} \qquad (2\text{-}3\text{-}5)$$

where E, the constant of integration, is a *constant of the orbital motion*. We recognize it as the *total specific mechanical energy* (energy per unit mass) since it is the sum

Sec. 2-3   The Equations and Constants of Motion

of the specific kinetic energy, $V^2/2$, and the specific potential energy, $-\mu/r$. (Although specific (per unit mass) properties are conventionally assigned lower case letters, E is used to avoid any confusion with the eccentricity, which is often denoted elsewhere by e.) Equation (2-3-5), sometimes referred to as the *vis-viva integral*, shows that the total mechanical energy per unit mass of the smaller body is conserved, which should not surprise us inasmuch as there are no dissipative mechanisms, such as drag, within the closed system.

A few words are in order with regard to the sign and reference datum for the specific potential energy $(-\mu/r)$. We are accustomed to a datum at the surface of the Earth where the potential energy is zero and to a potential energy that increases with altitude. In this case, however, the datum is the center of the larger (the attracting) body, and when r is zero, the potential energy is negatively infinite. As r increases, the potential energy becomes less negative (more positive) and goes to zero when r becomes infinite. At this point the gravitational attraction g is also zero and the smaller body is no longer subject to the attraction of the larger body. We speak of this as the *escape* of the smaller body from the attraction of the larger body. The physical significance of an infinite r will be addressed in a subsequent section.

To show that the angular momentum is conserved, let us return to the vector equation of motion, Eq. (2-3-1), and rewrite it as

$$\ddot{\mathbf{r}} = \frac{-\mu}{r^3} \mathbf{r} \qquad (2\text{-}3\text{-}6)$$

This time we shall find the product of each term crossed with $\mathbf{r}$; that is,

$$\mathbf{r} \times \ddot{\mathbf{r}} = \frac{-\mu}{r^3} (\mathbf{r} \times \mathbf{r}) \qquad (2\text{-}3\text{-}7)$$

But $\mathbf{r} \times \mathbf{r} = 0$, so that Eq. (2-3-9) reduces to $\mathbf{r} \times \ddot{\mathbf{r}} = 0$, which can be integrated with the help of hindsight and backward calculations; for example,

$$\frac{d}{dt}(\mathbf{r} \times \dot{\mathbf{r}}) = (\dot{\mathbf{r}} \times \dot{\mathbf{r}}) + (\mathbf{r} \times \ddot{\mathbf{r}}) \qquad (2\text{-}3\text{-}8)$$

but $\dot{\mathbf{r}} \times \dot{\mathbf{r}} = 0$, so that

$$\mathbf{r} \times \ddot{\mathbf{r}} = \frac{d}{dt}(\mathbf{r} \times \dot{\mathbf{r}}) = 0 \qquad (2\text{-}3\text{-}9)$$

which means that $\mathbf{r} \times \dot{\mathbf{r}}$ is another constant of motion. Since $\dot{\mathbf{r}} = \mathbf{V}$ and $\mathbf{r} \times \mathbf{V}$ is $\mathbf{H}$, *the specific angular momentum* of the smaller body, $d\mathbf{H}/dt = 0$ and $\mathbf{H}$ *is a vector constant of integration that provides two additional constants of motion*. ($\mathbf{H}$ is used rather than the conventional $\mathbf{h}$ to be consistent with the use of E for specific energy and to avoid confusion with h to designate altitude.)

Since $\mathbf{H} = H\mathbf{1}_H$, both its magnitude H and its direction $\mathbf{1}_H$ are constant. Furthermore, since $\mathbf{H}$ is the cross product of $\mathbf{r}$ and $\mathbf{V}$, the plane containing r and V is always perpendicular to $\mathbf{H}$ indicating that the motion of the smaller mass relative to the larger mass (the trajectory or orbit) is restricted to a plane that is fixed in an inertial reference frame (i.e., two-body trajectories are planar). Remember that an

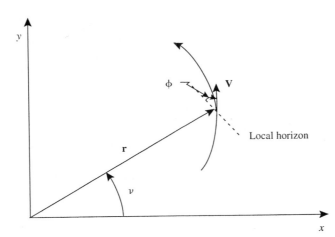

Figure 2-3-2. Plane of motion perpendicular to **H**.

inertial reference frame cannot accelerate or rotate but can have a constant translational velocity.

The relationship of **H** to a trajectory can be seen in Fig. 2-3-2. The *local horizon* is the perpendicular to the position vector **r**, and ϕ, the *elevation angle*,[2] is measured from the local horizon to the velocity vector and is positive in a clockwise direction (above the horizon). The right-hand rule for obtaining the cross product of two vectors shows that the direction of **H** is out of the plane of the trajectory and is fixed in space. The magnitude of the angular momentum vector is

$$H = rV\sin(90° - \phi) = rV\cos\phi \tag{2-3-10}$$

where $(90° - \phi)$ is the angle between **r** and **V**.

At this point in our development, we have a vector equation of motion and two integral relationships, the vis-viva integral [Eq. (2-3-5)] and Eq. (2-3-10), which provide two scalar constants of motion (the mechanical energy E and the magnitude of the angular momentum H) and simple relationships between the time-varying position and velocity of the smaller body at any point on the trajectory (orbit), and we have shown that the direction of **H** is also constant, thus proving that the trajectories are planar and fixed with respect to an inertial reference frame.

2-4 THE EQUATION OF CONIC SECTIONS

Let us return once more to the vector equation of motion, Eq. (2-3-1), and this time cross-multiply each term by **H** to obtain

$$\mathbf{H} \times \ddot{\mathbf{r}} + \frac{\mu}{r^3}(\mathbf{H} \times \mathbf{r}) = 0 \tag{2-4-1}$$

---

[2] The elevation angle may also be called the flight path angle γ.

Sec. 2-4 The Equation of Conic Sections

Using hindsight to evaluate the first term yields

$$\frac{d}{dt}(\mathbf{H} \times \dot{\mathbf{r}}) = \frac{d\mathbf{H}}{dt} \times \dot{\mathbf{r}} + (\mathbf{H} \times \ddot{\mathbf{r}})$$

But $d\mathbf{H}/dt = 0$, so that

$$\mathbf{H} \times \ddot{\mathbf{r}} = \frac{d}{dt}(\mathbf{H} \times \dot{\mathbf{r}}) = \frac{d}{dt}(\mathbf{H} \times \mathbf{V}) \qquad (2\text{-}4\text{-}2a)$$

Expanding the second term in Eq. (2-4-1) with $\mathbf{H} = \mathbf{r} \times \dot{\mathbf{r}}$ yields

$$\frac{\mu}{r^3}[(\mathbf{r} \cdot \mathbf{r})\dot{\mathbf{r}} - (\mathbf{r} \cdot \dot{\mathbf{r}})\mathbf{r}] = \frac{\mu}{r^3}(r^2\dot{\mathbf{r}} - r\dot{r}\mathbf{r}) = \mu\left(\frac{\dot{\mathbf{r}}}{r} - \frac{\dot{r}}{r^2}\mathbf{r}\right)$$

But

$$\mu\left(\frac{\dot{\mathbf{r}}}{r} - \frac{\dot{r}}{r^2}\mathbf{r}\right) = \frac{d}{dt}\left(\frac{\mu}{r}\mathbf{r}\right) = \frac{\mu}{r^3}(\mathbf{H} \times \mathbf{r}) \qquad (2\text{-}4\text{-}2b)$$

Now substitute Eqs. (2-4-2a) and (2-4-2b) into Eq. (2-4-1) to obtain a perfect differential,

$$\frac{d}{dt}\left[(\mathbf{H} \times \mathbf{V}) + \frac{\mu}{r}\mathbf{r}\right] = 0 \qquad (2\text{-}4\text{-}3)$$

Therefore, the bracketed term in Eq. (2-4-3) must be another constant of integration and constant of motion. Let us give it the symbol $-\mathbf{B}$ (the reason for the minus sign will become apparent later), where

$$(\mathbf{H} \times \mathbf{V}) + \frac{\mu}{r}\mathbf{r} = -\mathbf{B} \qquad (2\text{-}4\text{-}4)$$

Now we shall dot-multiply Eq. (2-4-4) by $\mathbf{r}$ to obtain the scalar equation

$$\mathbf{r} \cdot (\mathbf{H} \times \mathbf{V}) + \frac{\mu}{r}(\mathbf{r} \cdot \mathbf{r}) = -\mathbf{B} \cdot \mathbf{r} \qquad (2\text{-}4\text{-}5)$$

The first term expands as

$$\mathbf{r} \cdot (\mathbf{H} \times \mathbf{V}) = -\mathbf{r} \cdot (\mathbf{V} \times \mathbf{H}) = -(\mathbf{r} \times \mathbf{V}) \cdot \mathbf{H} = -\mathbf{H} \cdot \mathbf{H} = -H^2$$

and the second term is simply μr. We see that Eq. (2-4-5) is a scalar equation that can be written as

$$-H^2 + \mu r = -Br \cos\beta \qquad (2\text{-}4\text{-}6)$$

where β is the angle between $\mathbf{B}$ and $\mathbf{r}$. Solving Eq. (2-4-6) for the magnitude of the position vector yields

$$r = \frac{H^2/\mu}{1 + (B/\mu)\cos\beta} \qquad (2\text{-}4\text{-}7)$$

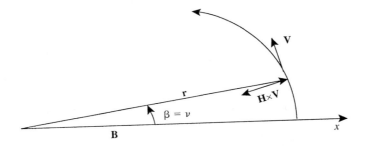

Figure 2-4-1. Relationship of **B** to the trajectory.

We remember that the general equation, in polar coordinates, of a conic with its primary focus at the center of the larger body can be written as

$$r = \frac{p}{1 + \epsilon \cos \nu} \tag{2-4-8}$$

where p, the *semilatus rectum*,[3] is the value of r when the position angle ν is 90° and ϵ, the eccentricity, determines the type and shape of the conic. In comparing Eq. (2-4-8) with Eq. (2-4-7), we see that they are identical in form with $p = H^2/\mu$ and $\epsilon = B/\mu$ and if $\beta = \nu$. Equation (2-4-3) shows that **B**, which is constant in both magnitude and direction, is aligned with both **r** and the cross product of **H** and **V** *only* when **B** lies along the major axis of the conic (the x-axis of the Cartesian coordinate system); therefore, β and ν are identical, as shown in Fig. 2-4-1.

It would be nice to have an expression for B, and thus ϵ, in terms of the other constants of the trajectory. Returning to Eq. (2-4-4) and dotting each side by itself maintains the identity and results in the expression

$$B^2 = H^2\left(V^2 - \frac{2\mu}{r}\right) + \mu^2 \tag{2-4-9}$$

But from Eq. (2-3-8), $V^2 - 2\mu/r = 2E$, so that $B^2 = \mu^2 + 2EH^2$ and

$$\epsilon = \frac{B}{\mu} = \sqrt{1 + \frac{2EH^2}{\mu^2}} \tag{2-4-10}$$

We conclude this section with the observation that the planar trajectory of the smaller body with respect to the center of the attracting (larger) body can be represented by the polar equation of a conic in the form

$$\boxed{r = \frac{H^2/\mu}{1 + \epsilon \cos \nu}} \tag{2-4-11}$$

where the value of ϵ can be found from Eq. (2-4-10).

---

[3] Sometimes p is described as the *parameter* of the conic section.

2-5 SOME PROPERTIES OF CONIC SECTIONS

It is significant and very useful to find that the scalar expression for r, the distance of the smaller body from the center of the larger body, is in the form of the general polar equation of a conic section with the origin at the principal focus. As a consequence, all of the known and familiar properties and characteristics of conic sections (hyperbolas, parabolas, ellipses, and circles) can be applied directly to the planar two-body trajectory. The polar equation of a conic section is repeated here for convenience.

$$r = \frac{p}{1 + \epsilon \cos \nu} \qquad (2\text{-}5\text{-}1)$$

The value of ϵ, the *eccentricity*, is an important parameter; it determines the shape and type of the conic section and thus the trajectory of the smaller body. If $\epsilon > 1$, the trajectory is a hyperbola (an open trajectory). If $\epsilon = 1$, the trajectory is usually a parabola[4] (also an open trajectory). If $0 < \epsilon < 1$, the trajectory is an ellipse (a closed trajectory, an orbit). Finally, if $\epsilon = 0$, the trajectory is a circle, which may be thought of as a degenerate ellipse.

For ballistic missile and planetary satellite trajectories, ellipses and circles are of primary interest. Therefore, let us look first at the major characteristics of an *ellipse*, as shown in Fig. 2-5-1, where the point P represents the location of the smaller body and F_1 (the principal focus) is the center of the larger (attracting) body.

Referring to Fig. 2-5-1, a is the *semimajor axis*, b the *semiminor axis*, and a, b, and c are related by the expression $a^2 = b^2 + c^2$ (see Fig. 2-5-2a). Also, $c/a = \epsilon$, the eccentricity. Points A and B represent the *apsides* of the orbit. Point A, the closest point to the primary focus, is the *periapsis*, and point B, the farthest point from the primary focus, is the *apoapsis*. If the center of the Earth is located at the primary focus, point A is usually referred to as the *perigee* and point B as the *apogee* of the orbit. If, on the other hand, the Sun is located at the primary focus, points A and B are the *perihelion* and *aphelion*, respectively.

Referring this time to Fig. 2-5-2b, we see that at the apsides, $r' + r = (a - c) + (a + c)$, so that

$$r + r' = 2a, \quad \text{a constant} \qquad (2\text{-}5\text{-}2)$$

Equation (2-5-2) is one definition of an ellipse; for example, an ellipse can be drawn by taking a piece of string with a length of $2a$, fastening the ends at the two foci, and then placing the pencil inside the string and drawing.

At periapsis (where $\nu = 0$), r, denoted by r_p, is given by

$$r_p = \frac{p}{1 + \epsilon} \qquad (2\text{-}5\text{-}3a)$$

---
[4] It is also necessary that the semimajor axis, a, be infinite.

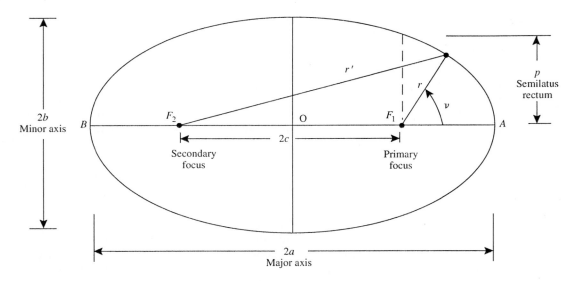

Figure 2-5-1. Major characteristics of an ellipse.

and at apoapsis (where $v = \pi$ radians), r_a is given by

$$r_a = \frac{p}{1 - \epsilon} \qquad (2\text{-}5\text{-}3b)$$

Since $r_a + r_p = 2a$, substituting the two equations above results in the relationship that

$$p = a(1 - \epsilon^2) \qquad (2\text{-}5\text{-}4)$$

and Eq. (2-5-1) can be written as

$$r = \frac{a(1 - \epsilon^2)}{1 + \epsilon \cos v} \qquad (2\text{-}5\text{-}5)$$

Since $p = H^2/\mu$ and $\epsilon = \sqrt{1 + 2EH^2/\mu^2}$ and using Eq. (2-5-4), we can solve for the important relationship that

$$\boxed{E = \frac{-\mu}{2a}} \qquad (2\text{-}5\text{-}6)$$

This equation shows that the energy, which is constant, is inversely proportional to the size of the major axis and is directly proportional to the gravitational parameter, which in turn is directly proportional to the mass of the larger (attracting) body. When a is positive, E is negative and the trajectories (orbits) are ellipses or circles; when a is infinite, E is zero and the trajectory is a parabola; and when a is negative, E is positive and the trajectory is a hyperbola.

Sec. 2-5 Some Properties of Conic Sections

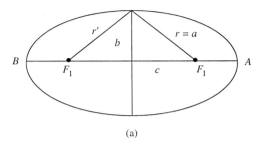

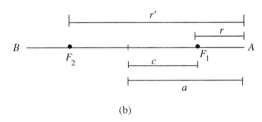

Figure 2-5-2. Additional elliptical relationships.

Returning to the polar equations of the apsides and substituting the relationship that $p = a(1 - \epsilon^2)$ results in

$$r_p = r_{\min} = a(1 - \epsilon)$$
$$r_a = r_{\max} = a(1 + \epsilon) \tag{2-5-7}$$

Finally, it can be shown that

$$\epsilon = \frac{c}{a} = \frac{r_a - r_p}{r_a + r_p} \tag{2-5-8}$$

At the apsides, the elevation angle ϕ is zero and the magnitude of the angular momentum H, which is constant, is simply the product of the distance from the primary focus and the respective velocity at each point. Consequently, the following relationships exist:

$$H = r_p V_p = r_a V_a \tag{2-5-9a}$$

$$V_p = \frac{H}{r_p} = \frac{r_a V_a}{r_p} = V_{\max} \tag{2-5-9b}$$

$$V_a = \frac{H}{r_a} = \frac{r_p V_p}{r_a} = V_{\min} \tag{2-5-9c}$$

Since r_p is the minimum value of the position vector, V_p, the velocity at periapsis, is the highest velocity in the orbit. Correspondingly, V_a, the velocity at apoapsis, is the lowest velocity in the orbit.

The position angle v at any point on a trajectory can be found by solving the polar equation [Eq. (2-5-2)] to obtain the expression

$$v = \cos^{-1}\left[\frac{1}{\epsilon}\left(\frac{p}{r} - 1\right)\right] \qquad (2\text{-}5\text{-}10)$$

Equation (2-5-10) shows that when the smaller body is at the semilatus rectum where $r = p$, $v = 90°$; the semilatus rectum is defined as the value of r when $v = 90°$.

Moving on, the *hyperbola* is the primary trajectory for the escape of a spacecraft from the gravitational attraction of a larger body, such as a planet or a moon, or even the Sun. It is also the trajectory of interest when a spacecraft enters or reenters a sphere of gravitational attraction.

As is the case with the parabola, the hyperbola is an open trajectory (not an orbit) with a periapsis only. The geometry of the hyperbola is sketched in Fig. 2-5-3 with only the major branch and the primary focus F_1 shown. Since the trajectory is not closed, the magnitude of r can increase without limit, and as it approaches infinity the tip of the position vector will approach one of the asymptotes that intersect at the center O, which is outside the trajectory. The semimajor axis a is measured from O to periapsis and is negative, as are b and c, and $\epsilon = -c/-a$; the distance from O to $F_1 = a\epsilon$.

Since the asymptotes represent the position vectors when r is infinite ($\pm\infty$), the angle between the major axis and an asymptote is v_∞, as shown in Fig. 2-5-3. Furthermore, since the conic relationships developed for the ellipse (with the exception of those pertaining to the apoapsis) can be applied to the hyperbola, Eq. (2-5-10) can be used to show that

$$v_\infty = \cos^{-1} -\frac{1}{\epsilon} \qquad (2\text{-}5\text{-}11)$$

The angle between the intersecting asymptotes is the *turning angle* δ; it is the angle through which the velocity along the trajectory turns as a body travels from $-\infty$ to

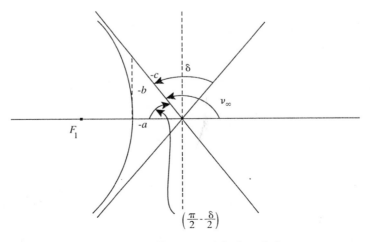

Figure 2-5-3. Geometry of the hyperbola.

$+\infty$. A perpendicular to the major axis at O bisects the turning angle so that the acute angle between the asymptote and the major axis, which is equal to $(\pi - v_\infty)$ radians, is also equal to $(\pi/2 - \delta/2)$ radians. Writing this identity and taking the cosine of both sides yields

$$\cos(\pi - v_\infty) = \cos\left(\frac{\pi}{2} + \frac{\delta}{2}\right) \qquad (2\text{-}5\text{-}12a)$$

Substituting the identities for the cosine of the sum and difference of two angles and using Eq. (2-5-11), Eq. (2-5-12a) can be written as

$$\cos v_\infty = -\frac{1}{\epsilon} = -\sin\frac{\delta}{2} \qquad (2\text{-}5\text{-}12b)$$

The two right-hand terms of Eq. (2-5-12b) can now be solved to obtain an expression for the turning angle as a function of the eccentricity.

$$\boxed{\delta = 2\sin^{-1}\frac{1}{\epsilon}} \qquad (2\text{-}5\text{-}13)$$

The *parabola* represents the boundary between the closed trajectory (orbit) of the ellipse and the open trajectory of the hyperbola. Since the eccentricity of a parabola is unity, $v_\infty = 180°$ and the asymptotes are parallel to and do not intersect the major axis; the turning angle is also 180°, indicating that the trajectory doubles back on itself.

2-6 THE PHYSICAL SIGNIFICANCE OF TWO-BODY TRAJECTORIES

The relationship between ϵ, the eccentricity, and E, the energy, can best be seen from the expression

$$\epsilon = \sqrt{1 + \frac{2EH^2}{\mu^2}} \qquad (2\text{-}6\text{-}1)$$

Since H^2 is always greater than zero, the sign of E determines the value of ϵ and thus the shape of the trajectory. If E is negative ($a > 0$), then $\epsilon < 1$ and the closed trajectory (an orbit) is either an ellipse or a circle. If E is zero ($a = \infty$), then $\epsilon = 1$ and the open trajectory is a parabola. Finally, if E is positive ($a < 0$), then $\epsilon > 1$ and the open trajectory is a hyperbola. Let us look more closely at these four trajectories: the ellipse, the circle, the parabola, and the hyperbola.

The Elliptical Orbit

If the energy E is negative, the semimajor axis a is positive, the eccentricity ϵ is less than unity, and the trajectory is an ellipse. The energy equation [Eq. (2-3-8)] yields the relationship that

$$\frac{V^2}{2} - \frac{\mu}{r} = E < 0 \qquad (2\text{-}6\text{-}2)$$

which shows that the kinetic energy (KE) is always less than the potential energy (PE) if the total energy is to be negative. Since E is constant, and always less than zero, the PE will always be less than zero even when the velocity (and KE) goes to zero. This means that the smaller body *cannot escape* from the gravitational attraction of the larger body but will always be drawn to the larger body. If the radius of the larger body is greater than the periapsis distance, the ellipse will not be able to complete its orbit and will intersect the surface of the larger body; this is the case with a ballistic missile.

The Circular Orbit

If E is negative but equal to $-\mu^2/2H^2$, then $\epsilon = 0$ and the ellipse degenerates into a circle; that is, the smaller body will be in a circular orbit with a constant radius equal to the semimajor axis ($r = a$) and with a constant velocity. With $a = r$ and $E = -\mu/2a$, the energy equation becomes

$$\frac{V^2}{2} - \frac{\mu}{r} = \frac{-\mu}{2r} \tag{2-6-3}$$

which can be solved for the velocity, namely,

$$V_{cs} = \sqrt{\frac{\mu}{r}} = \sqrt{gr} \tag{2-6-4}$$

In Eq. (2-6-4), the subscript "*cs*" denotes a circular satellite (a circular orbit), and g, the acceleration due to gravity, is equal to μ/r^2. Since g varies with r, it is safer to use the expression $(\mu/r)^{1/2}$ for V_{cs}.

The circular satellite is of special interest and can be used to illustrate several salient characteristics that are also applicable to elliptical orbits. First, we see that the orbital velocity V_{cs} decreases as r increases, going to zero in the limit as r goes to infinity. The period (P) of a circular satellite, the time to make one complete revolution about the cm of the attracting body, is easy to calculate because the velocity is constant and the distance traveled is simply the circumference. Therefore,

$$P_{cs} = \frac{2\pi r}{V_{cs}} = \frac{2\pi r}{\sqrt{\mu/r}}$$

so that

$$\boxed{P_{cs} = 2\pi \sqrt{\frac{r^3}{\mu}}} \tag{2-6-5}$$

Notice that the period increases as r increases.

A circular satellite that just skims the surface of the Earth (assumes a spherical Earth with no atmosphere or buildings, etc.) is referred to as an *Earth-surface satellite* and is often used as a baseline or reference orbit. It has the highest velocity

Sec. 2-6 The Physical Significance of Two-Body Trajectories

possible and the shortest period; the period of 84.4 minutes is known as the *Schuler period* and is of interest in inertial navigation. Obviously, for orbits around another planet, such as Mars, the Mars-surface satellite would be the baseline.

The Parabolic Trajectory

If the energy E is zero, the semimajor axis a is infinite, the eccentricity ϵ is unity, and the trajectory is a parabola, an open trajectory. Again using the energy equation, we see that

$$\frac{V^2}{2} - \frac{\mu}{r} = 0 \qquad (2\text{-}6\text{-}6)$$

At any value of r, the KE will be equal to $-\text{PE}$. Consequently, the smaller body has just enough KE to follow the parabolic trajectory to "infinity," where both the velocity and the PE will be zero. The velocity at $r = \infty$ is called the *residual velocity* (or sometimes the *excess velocity*) and will be zero for the parabolic trajectory. By reaching infinity, the smaller body has, in effect, *escaped* from the attraction of the larger body. The residual velocity is given the symbol V_∞ and is often referred to as *V infinity*; for a parabola, $V_\infty = 0$.

With $E = 0$ ($\epsilon = 1$), the velocity at a specified r is called the *escape velocity* V_{esc} for that r and can be found from Eq. (2-6-6) to be

$$V_{esc} = \sqrt{\frac{2\mu}{r}} = \sqrt{2gr} = \sqrt{2}V_{cs} \qquad (2\text{-}6\text{-}7)$$

We see that the magnitude of the escape velocity decreases as r increases, as does V_{cs}, and that V_{esc} is always on the order of 40% greater than V_{cs}.

It should be realized a smaller body launched with an escape velocity will not actually go to infinity. As the gravitational attraction of the larger body decreases, it will eventually become smaller than that of another attracting body (usually the Sun) and the smaller body will phase into an orbit about this second body. For example, although the velocity with respect to the center of the Earth of a spacecraft on a parabolic escape trajectory (from Earth) will become zero (again with respect to the center of the Earth) at infinity, the velocity of the spacecraft with respect to the Sun (the new attracting body) will be the orbital velocity of the Earth about the Sun.

The Hyperbolic Trajectory

If E is greater than zero, the semimajor axis a is negative, the eccentricity ϵ is greater than unity, and the trajectory is a hyperbola, another open trajectory. From the energy equation, we obtain the relationship that

$$\frac{V^2}{2} - \frac{\mu}{r} = E > 0 \qquad (2\text{-}6\text{-}8)$$

which shows that the KE is always greater than the PE. Furthermore, solving the inequality of Eq. (2-6-8) for V shows that the velocity at any point on the trajectory will be greater than the corresponding escape velocity for a parabolic trajectory. Consequently, on a hyperbolic trajectory the smaller body will not only escape the gravitational attraction of the larger body but will also have a finite residual velocity ($V_\infty > 0$) when it reaches infinity and is captured by another larger body (again, usually the Sun). Since V_∞ is greater than zero and is the velocity with respect to the center of the Earth, the spacecraft will have a larger velocity with respect to the Sun than does the Earth. Consequently, the spacecraft will leave the Earth's orbit and proceed on an orbit of its own.

If we let r go to infinity in Eq. (2-6-8), we see that $E = V_\infty^2/2$. With E known, the velocity at any other point on the trajectory, such as at r_1 in Fig. 2-6-1, V_1 can easily be calculated from the energy equation. Conversely, if V_1 and r_1 are known, the energy and V_∞ can be determined.

It should be pointed out that once the energy is known (no matter how determined) for any of the trajectories above, the velocity at any other point on the trajectory can be found from the useful expression

$$\boxed{V = \sqrt{2\left(E + \frac{\mu}{r}\right)}} \tag{2-6-9}$$

2-7 THE ELEVATION ANGLE

The elevation angle ϕ, the angle between the local horizon and the velocity vector, is an important parameter. For example, if V, r, and ϕ are known at any point in a trajectory, the trajectory is completely defined. If H is known, along with V and r, then ϕ at that point is simply $\cos^{-1}(H/rv)$. There may be occasions, however, when it would be nice to have an expression for ϕ in terms of ϵ and ν.

Consider Fig. 2-7-1, in which the velocity vector V is decomposed into a radial

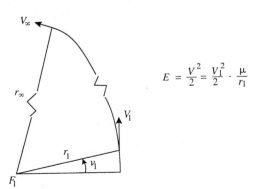

$$E = \frac{V^2}{2} = \frac{V_1^2}{2} - \frac{\mu}{r_1}$$

Figure 2-6-1. Hyperbolic velocity relationships.

Sec. 2-8 Summary of Relevant Two-Body Equations

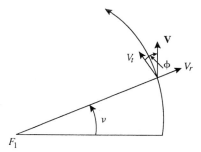

Figure 2-7-1. Velocity vector and its components.

component V_r and a tangential component V_t, where the magnitude of $V_r = \dot{r}$ and the magnitude of $V_t = \dot{\nu} r$. From Fig. 2-7-1 we see that

$$\tan \phi = \frac{V_r}{V_t} = \frac{\dot{r}}{\dot{\nu} r} \tag{2-7-1}$$

We now need an expression for $\dot{r}$. Since

$$r = \frac{p}{1 + \epsilon \cos \nu} \tag{2-7-2}$$

and since both p and ϵ are constants of the motion,

$$\dot{r} = \frac{p \epsilon \dot{\nu} \sin \nu}{(1 + \epsilon \cos \nu)^2} \tag{2-7-3}$$

Substituting Eqs. (2-7-2) and (2-7-3) into Eq. (2-7-1) and solving for ϕ yields the expression

$$\phi = \tan^{-1} \frac{\epsilon \sin \nu}{1 + \epsilon \cos \nu} \tag{2-7-4}$$

2-8 SUMMARY OF RELEVANT TWO-BODY EQUATIONS

At this point, let us summarize, and renumber for convenience, the principal equations and relationships of conic two-body trajectories.

$$r = \frac{p}{1 + \epsilon \cos \nu} \tag{2-8-1}$$

$$p = a(1 - \epsilon^2) = \frac{H^2}{\mu} \tag{2-8-2}$$

$$\epsilon = \sqrt{1 + \frac{2EH^2}{\mu^2}} = \frac{r_a - r_p}{r_a + r_p} \tag{2-8-3}$$

$$E = \frac{V^2}{2} - \frac{\mu}{r} = \frac{-\mu}{2a} \tag{2-8-4}$$

$$H = rV \cos \phi \qquad (2\text{-}8\text{-}5)$$

$$\phi = \cos^{-1} \frac{H}{rV} = \tan^{-1} \frac{\epsilon \sin \nu}{1 + \epsilon \cos \nu} \qquad (2\text{-}8\text{-}6)$$

$$r_a + r_p = 2a \qquad (2\text{-}8\text{-}7)$$

$$r_a = r_{max} = a(1 + \epsilon) \qquad (2\text{-}8\text{-}8)$$

$$r_p = r_{min} = a(1 - \epsilon) \qquad (2\text{-}8\text{-}9)$$

$$V_a = V_{min} = \frac{H}{r_a} = \frac{r_p V_p}{r_a} \qquad (2\text{-}8\text{-}10)$$

$$V_p = V_{max} = \frac{H}{r_p} = \frac{r_a V_a}{r_p} \qquad (2\text{-}8\text{-}11)$$

$$V = \sqrt{2\left(E + \frac{\mu}{r}\right)} \qquad (2\text{-}8\text{-}12)$$

$$\nu = \cos^{-1}\left[\frac{1}{\epsilon}\left(\frac{p}{r} - 1\right)\right] \qquad (2\text{-}8\text{-}13)$$

Remember that the specific energy E and the specific angular momentum H are both constants of motion.

PROBLEMS

2-1. The center of the moon is 0.0026 AU (3.890×10^8 m) from the center of the Earth and has a relative mass of 0.0123.
  (a) Locate the cm of the Earth–Moon system with respect to the center of the Earth. With respect to the center of the Moon.
  (b) Find the ratio of orbital velocities of the Earth and Moon about the cm of the system. If the orbital velocity of the Moon is 1012 m/s, what is the approximate orbital velocity of the Earth about the cm of the Earth–Moon system?

2-2. Using Fig. 2-2-1 and the results of Section 2-2, show that $\ddot{\mathbf{R}}_{cm} = 0$.

2-3. (a) Show that the position angle ν can be found from the expression

$$\nu = \cos^{-1}\left[\frac{1}{\epsilon}\left(\frac{p}{r} - 1\right)\right]$$

  (b) Find ν when $r = p$ (i.e., when the smaller body is at the semilatus rectum).

2-4. Using the expression for ν from Problem 2-3 when r is infinite:
  (a) Write an expression for ν_∞.
  (b) Find ν_∞ for $\epsilon = 1$, a parabola.
  (c) Find ν_∞ for $\epsilon = 3.5$, a hyperbola.
  (d) Find the turning angle δ for parts (b) and (c).

2-5. With $\mu = 3.986 \times 10^{14}$ m³/s², $a = 7.0 \times 10^6$ m, and $\epsilon = 0.35$:
  (a) Find E, H, r_p, r_a, V_p, and V_a.
  (b) Find r, V, and ϕ when $\nu = 110°$.
  (c) Find ν, V, ϕ, and H when $r = 8.813 \times 10^6$ m.

2-6. (a) Show that the semimajor axis of a hyperbola is negative.
(b) Will p, the semilatus rectum, be negative or positive?

2-7. (a) Show that the semimajor axis of a parabola is infinite.
(b) Show that the semilatus rectum of a parabola is equal to twice the periapsis distance (i.e., $p = 2r_p$).

2-8. For a hyperbolic trajectory, show that:
(a) The semimajor axis can be found from $a = -\mu/V_\infty^2$.
(b) The velocity at any point on the trajectory, where the position vector is r, can be expressed as

$$V = \sqrt{V_\infty^2 + V_{esc}^2}$$

where V_{esc} is the escape velocity at r.

2-9. Show that for any two-body trajectory

$$V = \sqrt{\mu\left(\frac{2}{r} - \frac{1}{a}\right)}$$

2-10. Show that the difference in energy between any two trajectories can be expressed in terms of their semimajor axes, namely,

$$E_2 - E_1 = \frac{\mu}{2}\left(\frac{a_2 - a_1}{a_1 a_2}\right)$$

2-11. Show that for a circular satellite, the period can be found from the expression

$$P = \frac{2\pi\mu}{V_{cs}^3}$$

2-12. Use simple differential calculus to show that the position vector is at a maximum or minimum at the apses and write expressions for the position vector at those points.

2-13. Using the polar equation of a conic section [Eq. (2-5-20)], plot five points, sketch, and identify the following trajectories:
(a) $p = 6$ and $\epsilon = 0$
(b) $p = 12$ and $\epsilon = 1$
(c) $p = 8$ and $\epsilon = 0.333$
(d) $p = 18$ and $\epsilon = 2$
(e) $p = 10$ and $\epsilon = 0.666$

2-14. (a) Show that the specific energy E is equal to $-\mu/2a$.
(b) Starting with $E = -\mu/2a$, show that E is also equal to $-V_p V_a/2$.

2-15. (a) Show that the period of a circular satellite can be expressed as a function of its orbital velocity V_{cs}, namely,

$$P = \frac{2\pi\mu}{V_{cs}^3}$$

(b) With $\mu_{Earth} = 3.986 \times 10^{14}$ m³/s², plot a curve for r and P versus V_{cs}.

2-16. Show that at the intersection of an elliptical orbit and the semiminor axis, the position angle v is equal to $\cos^{-1}(-\epsilon)$, the elevation angle ϕ is equal to $\cos^{-1}(\sin v)$, and the specific angular momentum is equal to $(a\mu)^{1/2} \sin v$.

2-17. If in an elliptical orbit, $r_p/r_a = 0.5$, what is the eccentricity of the orbit?

CHAPTER 3

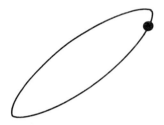

Geocentric Orbits and Trajectories

The Skylab 4 space station in geocentric orbit, as photographed from the command and service modules. (Courtesy of NASA.)

3-1 INTRODUCTION

Now that we have the basic expressions for determining and describing the two-body motion of a point mass, let us apply them to Earth-centered (*geocentric*) trajectories and orbits. We use the following values for the Earth:

$$\mu_E = 3.986 \times 10^{14} \text{ m}^3/\text{s}^2$$

$$r_E = 6.378 \times 10^6 \text{ m } (6378 \text{ km})$$

$$g_o = 9.80 \text{ m/s}^2 \text{ (standard value at sea level)}$$

Remember that at any position on a trajectory, $r = r_E + h$, where h is the altitude (height) above the surface of the Earth. Altitudes are often given in nautical miles; 1 nmi = 1840 m = 1.84 km = 1.1515 mi.

Geocentric orbits and trajectories fall into two basic categories: closed orbits (circular and elliptic) that are involved with the utilization of near-Earth space and open trajectories (primarily hyperbolic) that are used to escape from the Earth's gravitational attraction. The closed orbits have many uses to include investigation of the Earth and its surrounding environment, communication relays, navigational systems (global positioning systems), and military reconnaissance. Another important use of the closed orbit is as a parking orbit from which satellites can be transferred to higher closed orbits (as from the Space Shuttle) and from which interplanetary spacecraft can enter hyperbolic escape trajectories.

We shall first examine circular orbits and then extend our understanding to ellipses. We shall then look at the shaping of orbits (i.e., the transformation of ellipses to circles, and vice versa). Following that, we examine parabolic and hyperbolic trajectories with application to escape from the Earth's gravitational attraction. Next we concern ourselves with transferring from one orbit to another and conclude this chapter with a discussion of plane changing.

It is important to remember that these two-body trajectories and orbits are trajectories in an inertial reference frame that describe the motion of the smaller body (a satellite, spacecraft, meteor, or comet) with respect to the center of the

Earth and not with respect to its rotating surface, where we live and from which we observe such bodies. Ground tracks, which are these trajectories projected onto the surface of a rotating Earth, are described and discussed in Chapter 8.

It should be apparent that the expressions and discussions of this and the following sections are planetocentric in nature and are applicable to trajectories and orbits about any other planet or celestial body (such as a moon).

3-2 CIRCULAR ORBITS

Since the eccentricity ϵ of a circular orbit is zero, the primary focus (the center of the Earth in this case) is also at the center of the circular orbit. Consequently, the semimajor axis a is equal to the position distance r, the orbital velocity V is constant, and the elevation angle ϕ is always zero, as can be seen in Fig. 3-2-1. There are no apsides as such (i.e., no perigee and no apogee).

Since it is often used as a baseline or reference orbit, the first example will be the Earth-surface circular satellite. In this example, as in all trajectory determinations, there usually is more than one method of obtaining a specific answer. One or two alternatives may be shown, but no attempt is made to show all possible solutions. Alternative solutions also serve as a check of a previously determined answer.

Example 3-2-1

For an Earth-surface circular satellite find (a) the orbital velocity, (b) the energy, (c) the angular momentum, and (d) the period.

Solution Since $h = 0$, $r = r_E$. We can use both versions of Eq. (2-6-4) to find V_{cs}, namely:

$$V_{cs} = \sqrt{\frac{3.986 \times 10^{14}}{6.378 \times 10^6}} = 7905 \text{ m/s} = 28{,}460 \text{ km/h}$$

or

$$V_{cs} = \sqrt{9.80 \times (6.378 \times 10^6)} = 7906 \text{ m/s} = 17{,}690 \text{ mph}$$

(b) Using both versions of Eq. (2-8-4) yields

$$E_{cs} = \frac{-3.986 \times 10^{14}}{2 \times (6.378 \times 10^6)} = -3.125 \times 10^7 \text{ m}^2/\text{s}^2$$

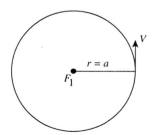

Figure 3-2-1. Circular orbit.

Sec. 3-2 Circular Orbits

or

$$E_{cs} = \frac{(7905)^2}{2} - \frac{3.986 \times 10^{14}}{6.378 \times 10^6} = -3.125 \times 10^7 \text{ m}^2/\text{s}^2$$

(c) From Eq. (2-8-5) with $\phi = 0°$,

$$H_{cs} = (6.378 \times 10^6) \times 7905 = 5.042 \times 10^{10} \text{ m}^2/\text{s}$$

(d) Finally, from Eq. (2-6-5),

$$P_{cs} = 2\pi \sqrt{\frac{(6.378 \times 10^6)^3}{3.986 \times 10^{14}}} = 5069 \text{ s} = 84.4 \text{ min}$$

Of all the geocentric orbits, the Earth-surface circular satellite has the lowest energy, the highest velocity, and the shortest period; it is the minimum-energy minimum-period geocentric orbit. As mentioned in Section 2-6, the period of 84.4 minutes has significance in inertial navigation and is known as the Schuler period.

Example 3-2-2

Let us see what happens when we increase the altitude of the satellite to 552 km (300 nmi), so as to be outside the sensible atmosphere of the Earth, and redo Example 3-2-1.
Solution Since $h = 552$ km, $r = r_E + 552 \times 10^3 = 6.930 \times 10^6$ m.
(a) Using both versions of Eq. (2-6-4) yields

$$V_{cs} = \sqrt{\frac{3.986 \times 10^{14}}{6.930 \times 10^6}} = 7584 \text{ m/s} = 27{,}300 \text{ km/h}$$

If you prefer to use $V_{cs} = (gr)^{1/2}$, you must calculate the new value of g first with $g = \mu/r^2$.

$$g = \frac{3.986 \times 10^{14}}{(6.930 \times 10^6)^2} = 8.30 \text{ m/s}^2$$

and

$$V_{cs} = \sqrt{8.30 \times (6.930 \times 10^6)} = 7584 \text{ m/s}$$

(b) Applying both versions of Eq. (2-8-4) gives

$$E_{cs} = \frac{-3.986 \times 10^{14}}{2 \times (6.930 \times 10^6)} = -2.876 \times 10^7 \text{ m}^2/\text{s}^2$$

or

$$E_{cs} = \frac{(7584)^2}{2} - \frac{3.986 \times 10^{14}}{6.930 \times 10^6} = -2.876 \times 10^7 \text{ m}^2/\text{s}^2$$

(c) Returning to Eq. (2-8-5), since $\phi = 0$,

$$H_{cs} = (6.930 \times 10^6) \times 7584 = 5.256 \times 10^{10} \text{ m}^2/\text{s}$$

(d) Finally, with Eq. (2-6-5),

$$P_{cs} = 2\pi \sqrt{\frac{(6.930 \times 10^6)^3}{3.986 \times 10^{14}}} = 5741 \text{ s} = 95.68 \text{ min}$$

Comparing the two orbits, the earth-surface orbit and the 300-nmi orbit, we see that the higher orbit has a larger (less negative) energy (+8%), a lower orbital velocity (−4%), a higher angular momentum (+4%), and a longer period (+13%). The apparent conflict of the lower orbital velocity and the larger energy requirement of the higher orbit can be resolved by introducing the concept of a pseudo "launch velocity," which can be thought of as the instantaneous velocity required at the surface of a nonrotating Earth to give the satellite the necessary energy for any specified orbit (circular or elliptical). The launch velocity V_L can be obtained from the energy equation [Eq. (2-8-4)] written at the surface of the Earth, namely,

$$E = \frac{V_L^2}{2} - \frac{\mu}{r_E} \qquad (3\text{-}2\text{-}1)$$

where E is the energy of the orbit of interest. If we apply Eq. (3-2-1) to the Earth-surface satellite, we find that the launch and orbital velocities are identical, 7905 m/s. For the 300-nmi orbit, however, using the solution of Eq. (3-2-1) in the form of Eq. (2-8-11) and the value of -2.876×10^7 m²/s² for the energy of the higher orbit, we find V_L to be

$$V_L = \sqrt{2\left(-2.876 \times 10^7 + \frac{3.986 \times 10^{14}}{6.378 \times 10^6}\right)} = 8214 \text{ m/s}$$

We see that the launch velocity of the higher orbit is 4% higher than that for the Earth-surface satellite, which is consistent with the 8% higher energy requirement of the higher orbit (E is proportional to V^2).

These launch (and orbital) velocities are inertial velocities in that they are measured with respect to the center of a nonrotating Earth. The Earth, however, is rotating about the North Pole with an angular velocity ω_E of 7.27×10^{-5} r/s and at its surface has a tangential velocity to the east with a magnitude equal to $\omega_E r_E \cos \lambda$, where λ is the latitude. At the equator, this tangential velocity is a maximum and is equal to 463.7 m/s.

The tangential velocity of the Earth can be used to reduce the launch velocity that must be imparted to a satellite by launching to the east; a westward launch would increase the velocity required. For example, the 300-nmi (552-km) satellite calls for a launch velocity of 8214 m/s. If the launch site is located at 20° north latitude and the launch is to the northeast (an inclination of 45°), the launch component of the Earth's velocity would be 463.1 cos(20°) cos(45°) = 307.7 m/s and the required launch velocity (with respect to the surface of the Earth) would be reduced from 8214 m/s to 7906 m/s, a reduction of 3.75%. A launch due east at the equator would call for a launch velocity of 7751 m/s, a 5.6% reduction, whereas a launch due west from the equator would increase the launch velocity to 8677 m/s, a 5.6% increase.

Returning to our circular satellite, let us increase the altitude, and r, until the period of the satellite matches the period of the Earth, which is 24 h or 86,400 s. The satellite is in what is known as a *geosynchronous Earth orbit* (GEO) and will pass over the same spots on the surface of the Earth every 24 h; the satellite will cross the same spot on the equator every 12 h. If the orbit is equatorial, the satellite will

Sec. 3-3 Elliptical Orbits

remain over one spot on the equator; such a satellite may be referred to as a *geostationary satellite*.

Example 3-2-3

The altitude of a GEO can be determined by first finding r from a rearrangement of Eq. (2-6-5), the expression for the period. Consequently,

$$r_{cs} = \left[\left(\frac{P}{2\pi}\right)^2 \mu\right]^{1/3}$$

so that

$$r_{cs} = \left[\left(\frac{86,400}{2\pi}\right)^2 \times 3.986 \times 10^{14}\right]^{1/3} = 42.24 \times 10^6 \text{ m}$$

Therefore, $h = r_{cs} - r_E = 35.86 \times 10^6$ m = 35,860 km = 19,490 nmi = 22,440 mi above the surface of the Earth. The corresponding energy E is -4.72×10^6 m²/s², the orbital velocity is 3072 m/s, and the launch velocity is 10,750 m/s. Notice the increase in the energy and the decrease in the orbital velocity that accompanied the increase in the period.

Geocentric circular orbits are somewhat loosely classified by altitude and you may find them described by acronyms. An orbit on the order of 300 nmi (552 km) or lower might be referred to as a *low Earth orbit* (LEO), one between an LEO and a GEO might be referred to as a *medium earth orbit* (MEO), and one above a GEO as a *high Earth orbit* (HEO).

3-3 ELLIPTICAL ORBITS

Elliptical orbits may be deliberate, as in the case of certain reconnaissance satellites that come in low over their targets and then make a wide swing out into space to make the tracking and intercept problem more difficult, or they may be unintentional as the result of an error in the injection velocity for a circular orbit. The salient features of an elliptical orbit are sketched in Fig. 3-3-1. Notice that r is no longer equal to a, nor is V constant, nor is ϕ always equal to zero as was the case with the circular orbit. With respect to ϕ, notice that it is positive (above the horizon) when the position angle v is greater than 0° and less than 180° and negative (below the horizon) during the other half of the orbit and is zero (tangent to the trajectory) only at periapsis and apoapsis, which for geocentric orbits are called *perigee* and *apogee*.

Although the velocity in an elliptical orbit varies around the orbit, the period, the time to make a complete revolution, is identical to that for a circular orbit with the same semimajor axis, as will be proved in Chapter 4. For now, without proof, the period of an elliptical orbit can be found from Eq. (2-6-5) by substituting a for r.

$$P = 2\pi\sqrt{\frac{a^3}{\mu_E}} \tag{3-3-1}$$

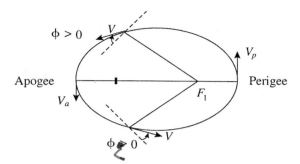

Figure 3-3-1. Elliptical orbit.

No matter what the reason for an elliptical orbit might be, its characteristic values can be determined from a set of given data by applying the appropriate expressions that are summarized in Section 2-8. Let us look at several examples of elliptical orbits and their characteristics. Remember that there is usually more than one way to obtain a particular value. In the examples to follow, some less obvious expressions may be used to remind you that there are alternative expressions.

Example 3-3-1

Determine the characteristics of a geocentric satellite with a perigee altitude of 300 nmi (552 km) and an apogee altitude of 1200 nmi (2208 km).

$$r_p = r_E + 552 \times 10^3 = 6.930 \times 10^6 \text{ m}$$

$$r_a = r_E + 2208 \times 10^3 = 8.586 \times 10^6 \text{ m}$$

$$2a = r_p + r_a = 15.52 \times 10^6 \text{ m and } a = 7.760 \times 10^6 \text{ m}$$

Solution Since $r_p = a(1 + \epsilon)$,

$$\epsilon = 1 - \frac{r_p}{a} = 1 - \frac{6.930 \times 10^6}{7.760 \times 10^6} = 0.1070$$

$$E = \frac{-\mu}{2a} = \frac{-3.986 \times 10^{14}}{15.52 \times 10^6} = -2.568 \times 10^7 \text{ m}^2/\text{s}^2$$

$$p = a(1 - \epsilon^2) = 7.671 \times 10^6 \text{ m}$$

Since $p = H^2/\mu$,

$$H = \sqrt{p\mu} = \sqrt{(7.671 \times 10^6) \times (3.986 \times 10^{14})} = 5.530 \times 10^{10} \text{ m}^2/\text{s}$$

At perigee, $V_p = H/r_p = 7980$ m/s and at apogee, $V_a = H/r_a = 6441$ m/s. Notice that $V_a < V_p$, as it should be. Rather than finding H from p and μ and then using H to find V_p and V_a, we could have found V_p (and V_a) directly from Eq. (2-8-11), which is what we have done before. Using Eq. (3-3-1), the period of the ellipse is

$$P = 2\pi \sqrt{\frac{(7.76 \times 10^6)^3}{3.986 \times 10^{14}}} = 6803 \text{ s} = 113.4 \text{ min}$$

In Example 3-2-2 we examined a circular orbit whose radius r_{cs} is equal to r_p of the ellipse of Example 3-3-1. Since a of the ellipse is greater than r_{cs}, we would

Sec. 3-3 Elliptical Orbits

expect both the energy and the period of the ellipse to be greater than those of the circle, and they are. Since E is greater, V_p is greater than V_{cs} and V_a is less.

Example 3-3-2

For the elliptical orbit of Example 3-3-1, find r, h, V, and ϕ when $\nu = 45°$.
Solution This is a simple problem since we already know the values of p, ϵ, E, and H from Example 3-3-1. From Eq. (2-8-1),

$$r = \frac{7.671 \times 10^6}{1 + 0.107 \cos(45°)} = 7.131 \times 10^6 \text{ m}$$

$$h = r - r_E = 7.53 \times 10^5 \text{ m} = 753 \text{ km} = 409.2 \text{ nmi}$$

Now V should be less than V_p and greater than V_a. V can be found from Eq. (2-8-11) to be

$$V = \sqrt{2\left(-2.568 \times 10^7 + \frac{3.986 \times 10^{14}}{7.131 \times 10^6}\right)} = 7774 \text{ m/s}$$

From Eq. (2-8-5), $\phi = \cos^{-1}(H/rV)$, so that

$$\phi = \cos^{-1}\frac{5.53 \times 10^{10}}{(7.131 \times 10^6) \times 7774} = +4.02°$$

Since the point of interest lies between perigee and apogee ($0° < \nu < 180°$), ϕ is positive (see Fig. 3-3-1) and the velocity vector is above the horizon.

At least two pieces of independent data are required to determine or describe a planar trajectory. For example, if a, the semimajor axis, and ϵ, the eccentricity, are known, the trajectory is defined and the other characteristics can be determined. Similarly, E and H, which are not measurable quantities but must be back-calculated, are sufficient to define a trajectory.

Example 3-3-3

A geocentric satellite is observed to have a velocity of 7620 m/s, an altitude of 2208 km (1200 nmi), and an elevation angle of $+20°$. Find E and H and use these values to find a and ϵ and the other characteristics of the orbit.

$$r = 6.378 \times 10^6 + 2208 \times 10^3 = 8.586 \times 10^6 \text{ m}$$

Solution Since $H = rV \cos\phi$,

$$H = 8.586 \times 10^6 \times 7620 \times \cos(20°) = 6.148 \times 10^{10} \text{ m}^2/\text{s}$$

Using Eq. (2-8-4) yields

$$E = \frac{(7620)^2}{2} - \frac{3.986 \times 10^{14}}{8.586 \times 10^6} = -1.739 \times 10^7 \text{ m}^2/\text{s}^2$$

Now we can find ϵ from Eq. (2-8-3) to be

$$\epsilon = \sqrt{1 + \frac{2 \times (-1.739 \times 10^7) \times (6.148 \times 10^{10})^2}{(3.986 \times 10^{14})^2}} = 0.4154$$

Since $0 < \epsilon < 1$, the satellite is in an elliptical orbit. The semimajor axis can be found from

$$a = \frac{-\mu}{2E} = \frac{-3.986 \times 10^{14}}{2 \times (-1.739 \times 10^7)} = 11.46 \times 10^6 \text{ m}$$

Therefore,

$$r_p = a(1 - \epsilon) = 6.70 \times 10^6 \text{ m} \qquad h_p = 322 \text{ km} = 175 \text{ nmi}$$

and

$$r_a = a(1 + \epsilon) = 16.22 \times 10^6 \text{ m} \qquad h_a = 9842 \text{ km} = 5349 \text{ nmi}$$

As a partial check, you might wish to see if $r_p + r_a = 2a$. Now for the perigee and apogee velocities,

$$V_p = \frac{H}{r_p} = 9176 \text{ m/s}$$

and

$$V_a = \frac{H}{r_a} = 3790 \text{ m/s}$$

Note that V for the observation was 7620 m/s, which is greater than V_a but less than V_p. It is now possible to find r, V, and ϕ at any other point (for a specific ν) on the orbit as well as the period; you may wish to do so to increase your familiarity with the various expressions.

3-4 ORBIT SHAPING AND IMPULSIVE THRUSTING

There are times when it may be necessary or desirable to change the type, shape, or size of an orbit or trajectory. Examples might be the circularization of an elliptical orbit, entering an ellipse, parabola, or hyperbola from a circular orbit, or entering a circular orbit from an ellipse, parabola, or hyperbola.

As an example, let us circularize the elliptical orbit of Example 3-3-1, first at perigee and then at apogee.

Example 3-4-1

The elliptical orbit of Example 3-3-1 has the following characteristics:

$$r_p = 6.930 \times 10^6 \text{ m} \qquad r_a = 8.586 \times 10^6 \text{ m} \qquad a = 7.76 \times 10^6 \text{ m}$$
$$V_p = 7980 \text{ m/s} \qquad V_a = 6441 \text{ m/s} \qquad \epsilon = 0.107$$
$$E = -2.568 \times 10^7 \text{ m}^2/\text{s}^2 \qquad\qquad H = 5.53 \times 10^{10} \text{ m}^2/\text{s}$$

If we are to circularize at perigee, then, as can be seen in Fig. 3-4-1a, $r_{cs} = a_{cs} = r_p$. Since $a_{cs} < a_{\text{ellipse}}$, we should expect the energy of the circular orbit to be less than that of the ellipse. Since $E_{cs} = -\mu/2r_p$,

$$E_{cs} = \frac{-3.986 \times 10^{14}}{2 \times 6.930 \times 10^6} = -2.876 \times 10^7 \text{ m}^2/\text{s}^2 < E_{\text{ell}}$$

Sec. 3-4   Orbit Shaping and Impulsive Thrusting   37

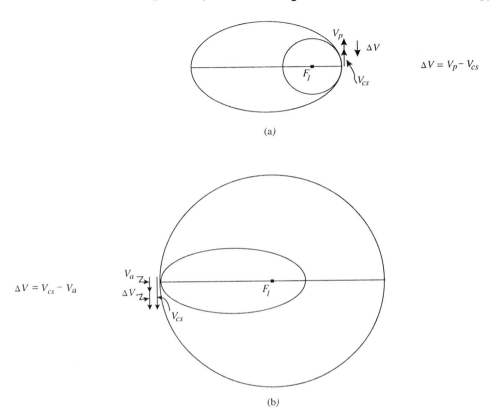

Figure 3-4-1. Circularization of an ellipse: (a) at perigee; (b) at apogee.

Since the energy of the satellite must be reduced by 3.06×10^6 m²/s²$[-2.876 \times 10^7 - (-2.568 \times 10^7)]$ in order to enter the new circular orbit, the question is how to do so. Since r, and thus the potential energy, is the same for both orbits at their point of contact, Eq. (2-8-4) shows that the velocity must be reduced so as to reduce the kinetic energy. The new velocity, V_{cs}, can be found to be [from $V_{cs} = (\mu/r)^{1/2}$]

$$V_{cs} = \sqrt{\frac{3.986 \times 10^{14}}{6.930 \times 10^6}} = 7584 \text{ m/s}$$

which indeed is less than V_p. Since the two velocities are collinear ($\phi = 0$ for both),

$$\Delta V = V_{cs} - V_p = 7584 - 7980 = -396 \text{ m/s}$$

The minus sign indicates the need for a reduction in velocity with an accompanying decrease in the kinetic energy of the spacecraft.

This reduction in the velocity of the spacecraft can be accomplished by the use of an on-board propulsion system to generate a forward thrust to slow the spacecraft down. The application of thrust appears to violate the two-body assumption of force-free flight and it does, for the period of time of thrusting. If, however, the thrust time is kept "small" with respect to the times of the motion, we can neglect the effects of thrust and introduce the concept of *impulsive thrusting* whereby we assume that the ΔV is achieved in an infinitesimal instant of time (i.e., an *instantaneous* ΔV). If the thrust T is the only applied force, then $T = m\, dV/dt$. We can approximate Δt by $\Delta t = \Delta V/(T/m)$, which tells us that Δt can be made "small" by an appropriate choice of the thrust-to-mass ratio (T/m). Throughout this book we assume impulsive thrusting (high thrust) unless otherwise stated.

To go from a circle to an exterior ellipse with perigee at r_{cs}, the procedure is reversed; the kinetic energy of the spacecraft must be increased by increasing the orbital velocity. For the orbits of Example 3-4-1, a transfer from the circular orbit to the elliptical orbit would require a positive ΔV, specifically, $+396.0$ m/s.

Example 3-4-2

Now let us circularize the ellipse at apogee. This time $a_{cs} = r_{cs} = r_a > a_{\text{ellipse}}$ (see Fig. 3-4-1b). Consequently, the energy of the circle will also be greater than the energy of the ellipse, requiring an increase in the orbital velocity.

$$E_{cs} = \frac{-3.986 \times 10^{14}}{2 \times (8.586 \times 10^6)} = -2.321 \times 10^7 \text{ m}^2/\text{s}^2 > E_{\text{ell}}$$

Since the orbital energy of the circular orbit is greater than that of the ellipse by $2.49 \times 10^6 \text{ m}^2/\text{s}^2$, it should be necessary to increase the velocity of the satellite. The new velocity is

$$V_{cs} = \sqrt{\frac{3.986 \times 10^{14}}{8.586 \times 10^6}} = 6814 \text{ m/s}$$

so that

$$\Delta V = V_{cs} - V_a = 6814 - 6441 = +373 \text{ m/s}$$

The plus sign indicates the need to increase the velocity of the spacecraft so as to attain the greater energy required for the new orbit. To return from this circle to the original (interior) ellipse would require a reduction in energy, a decrease in the orbital velocity of -373 m/s.

These two examples could be extended to the circularization of parabolas and hyperbolas and, conversely, to entering parabolas and hyperbolas, either interior or exterior, from a circular orbit. Although the orbit transformations in these two examples were made at perigee and apogee, where $\phi = 0$, the transfers could have been made at any other point along the ellipse. However, at any other point, ϕ of the ellipse would not be equal to zero, although ϕ of the circle would. Consequently, V_{ellipse} and V_{cs} would not be collinear, nor would ΔV; ΔV would be the vector difference, not the scalar difference, between the actual and desired orbital velocities. Discussion of *noncollinear* ΔV's will be deferred to the treatment of plane changes and fast transfers.

3-5 OTHER TRAJECTORIES AND ESCAPE

Parabolas and hyperbolas are open trajectories rather than orbits. When outgoing from the Earth, they represent escape trajectories. When incoming, they represent spacecraft returning from voyages outside the Earth's gravitational field or visitors from space, such as comets and meteors. Let us consider the approach of a large meteor.

Example 3-5-1

A tracking station reports that an approaching meteor has a velocity of 9150 m/s at an altitude of 50,000 nmi (92,000 km) and that the velocity vector makes an angle of 10° with r (see Fig. 3-5-1). Will the meteor hit the Earth? If not, what is the altitude (and velocity) at closest passage?

Solution At the sighting,

$$r = r_E + h = 98.37 \times 10^6 \text{ m} \quad \text{and} \quad \phi = -90° + 10° = -80°$$

$$H = 98.37 \times 10^6 \times 9150 \times \cos(-80°) = 15.6 \times 10^{10} \text{ m}^2/\text{s}$$

Using Eq. (2-8-4) gives

$$E = \frac{(9150)^2}{2} - \frac{3.986 \times 10^{14}}{98.37 \times 10^6} = +3.78 \times 10^7 \text{ m}^2/\text{s}^2$$

The fact that $E > 0$ indicates that the incoming trajectory is a hyperbola. Let us verify that by finding the eccentricity ϵ, using Eq. (2-8-3).

$$\epsilon = \sqrt{1 + \frac{2 \times (3.78 \times 10^7) \times (1.56 \times 10^{11})^2}{(3.98 \times 10^{14})^2}} = 3.55 > 1$$

Since $\epsilon > 1$, the trajectory is definitely a hyperbola. To find r_p and h_p, we shall use the expression that $r_p = a(1 - \epsilon)$. But we need to find a first.

$$a = \frac{-\mu}{2E} = \frac{-3.986 \times 10^{14}}{2 \times (3.78 \times 10^7)} = -5.27 \times 10^6 \text{ m}$$

Notice that the semimajor axis a of a hyperbola is negative. Now,

$$r_p = -5.27 \times 10^6 (1 - 3.55) = 13.44 \times 10^6 \text{ m}$$

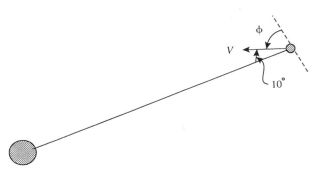

Figure 3-5-1. Incoming meteor.

and

$$h_p = r_p - r_E = 7062 \text{ km} = 3838 \text{ nmi}$$

Finally,

$$V_p = \frac{H}{r_p} = \frac{15.6 \times 10^{10}}{13.44 \times 10^6} = 11{,}610 \text{ m/s}$$

So we see that the incoming meteor will miss the Earth and then escape in an outgoing hyperbola.

Let us now look at an outgoing parabola, which represents the minimum escape velocity and the lowest-energy escape trajectory.

Example 3-5-2

If escape is to be from the surface of a nonrotating Earth, the escape velocity can be found from Eq. (2-6-7) to be

$$V_{\text{esc}} = \sqrt{\frac{2\mu}{r_e}} = \sqrt{\frac{2 \times (3.986 \times 10^{14})}{6.378 \times 10^6}} = 11{,}180 \text{ m/s}$$

Notice that V_{esc} is indeed equal to $(2)^{1/2}$ times the Earth-surface orbital velocity of 7905 m/s of Example 3-2-1. In Section 2-4, V_{esc} was found by setting the energy E and the residual velocity V_∞ equal to zero. Let us check to be sure that both are indeed zero by using Eq. (2-8-4).

$$E = \frac{(11{,}180)^2}{2} - \frac{3.986 \times 10^{14}}{6.378 \times 10^6} = 120 \text{ m}^2/\text{s}^2 \cong 0$$

With $r = \infty$ and $E = 120 \text{ m}^2/\text{s}^2$

$$V_\infty = \sqrt{2\left(E + \frac{\mu}{r_\infty}\right)} = 15.49 \text{ m}^2/\text{s}^2 \cong 0$$

Example 3-5-3

If escape were to be accomplished from some other altitude, such as from a circular orbit at 552 km (300 nmi), V_{esc} can be found simply by substituting the appropriate value of r into Eq. (2-6-7). Since $r = 6.93 \times 10^6$ m, V_{esc} from 552 km would be

$$V_{\text{esc}} = \sqrt{\frac{2 \times (3.986 \times 10^{14})}{6.930 \times 10^6}} = 10{,}720 \text{ m/s}$$

Although V_{esc} from 552 km is 10,720 m/s, the spacecraft already has a circular velocity of 7584 m/s $[(\mu/r)^{1/2}]$. Therefore, the increase in velocity (the ΔV) required to escape is 3136 m/s (10,720 − 7584).

Since V is the velocity of the spacecraft with respect to the Earth and since $V_\infty = 0$ for a parabola, the spacecraft will be motionless with respect to the center of the Earth when it reaches infinity and will travel in the same orbit around the Sun as does the Earth. If the spacecraft is to move away from the Earth after escape and enter into its own *heliocentric* (Sun-centered) orbit, it must have a finite velocity relative to the Earth. If the escape trajectory is a parabola, this finite relative velocity

Sec. 3-5 Other Trajectories and Escape

can be obtained only by impulsive thrusting at r_∞ to achieve an effective $V_\infty > 0$. If, for example, the desired V_∞ is to be 3050 m/s, the total ΔV that must be imparted to the spacecraft in a parabolic escape from the surface of the Earth would simply be the sum of the two velocities:

$$\Delta V_{TOT} = V_{esc} + V_\infty = 11{,}180 + 3050 = 14{,}230 \text{ m/s}$$

To achieve this ΔV_{TOT}, two propulsive firings would be required, one at launch and the other after arrival at infinity. On the other hand, a properly chosen hyperbolic trajectory would require only one firing to achieve both escape and the desired V_∞.

Considering a finite V_∞, at $r_\infty = \infty$, using Eq. (2-8-4),

$$E = \frac{V_\infty^2}{2} + \frac{\mu}{r_\infty} = \frac{V_\infty^2}{2}$$

At r, any other position on the trajectory,

$$\frac{V_r^2}{2} - \frac{\mu}{r} = \frac{V_\infty^2}{2}$$

so that V_r, the velocity at r, is

$$V_r = \sqrt{\frac{2\mu}{r} + V_\infty^2}$$

But $2\mu/r$ is V_{esc}^2 at r. Consequently, with one firing and a single ΔV, the velocity required for escape from any r with a V_∞ that will be greater than zero can be written as

$$\boxed{V_r = \sqrt{V_{esc_r}^2 + V_\infty^2}} \qquad (3\text{-}5\text{-}1)$$

where V_∞ is the desired residual velocity at infinity.

Equation (3-5-1) is the expression for what is known as an *Oberth maneuver*, described by Hermann Oberth (1894–1979) in 1929, which requires only one firing (a single ΔV) for a hyperbolic escape with a specified residual velocity (a $V_\infty > 0$).

Example 3-5-4

To escape from the surface of a nonrotating Earth with an Oberth maneuver and a $V_\infty = 3050$ m/s requires a velocity

$$V = \sqrt{(11{,}180)^2 + (3050)^2} = 11{,}590 \text{ m/s}$$

If we assume that the Earth is nonrotating, the ΔV required for this escape with one ΔV is 11,590 m/s. This is 18.6% less than the ΔV_T of 14,320 m/s required for the comparable two-ΔV parabolic escape that we examined above.

Since ΔV is a measure of the energy that must be supplied by the propulsion system, this is a significant saving. Furthermore, the problems associated with a firing at r_∞ are not trivial. As a consequence, the Oberth maneuver is the accepted escape trajectory but is usually not executed from the surface of the Earth (a *direct escape*) but rather from a circular *parking orbit*. A parking orbit is a transition orbit between the initial launch from the surface of the Earth and the trajectory or orbit of interest; a parking orbit increases operational and launch flexibility and often widens launch windows. A *launch window* is a period of time during which a launch must occur in order to accomplish an assigned mission.

Example 3-5-5

Let us find the ΔV required for the Oberth escape of a spacecraft from a 552-km (300-nmi) circular parking orbit with $V_\infty = 3050$ m/s. At 552 km, $V_{cs} = (\mu/r)^{1/2} = 7584$ m/s and $V_{esc} = 1.414 V_{cs} = 10{,}720$ m/s. The velocity required for the hyperbolic escape from 552 km is

$$V = \sqrt{(10{,}720)^2 + (3050)^2} = 11{,}140 \text{ m/s}$$

Since the spacecraft already has an orbital velocity of 7584 m/s, the increase in velocity required (the ΔV) will be $+3556$ m/s.

On the other hand, a parabolic escape plus a firing at r_∞ would require a total velocity of 13,770 m/s (10,720 + 3050), a 23.6% larger velocity. Although the ΔV required to leave the parking orbit would be $+3136$ m/s (10,720 $-$ 7584), the total ΔV required would be 6186 m/s (3136 + 3050), which is 74% higher than the ΔV of 3556 m/s required for the Oberth escape. This is a significant increase in the amount of fuel that must be boosted to the 552-km parking orbit to be used in the escape maneuver.

3-6 ORBITAL TRANSFERS

In preceding sections we have transferred from one orbit or trajectory to another when there was a point common to both; this is *orbit shaping*. When the orbital changes are small, we may speak of *orbit trimming*. In this section we look at transfers between nonintersecting orbits, starting with transfers between coplanar circular orbits. As was the case with orbit shaping, these transfers involve energy changes in moving from one orbit to the other.

A basic transfer trajectory between two circular orbits is the two-impulse *Hohmann* transfer ellipse, with its perigee at the inner orbit and its apogee at the outer orbit. Since the semimajor axis a is at a minimum, the two-impulse Hohmann transfer, described by Walter Hohmann in 1925, is usually the minimum-energy transfer trajectory between two orbits. (The exception for large separations is discussed at the end of this section.)

In a Hohmann transfer from an inner circular orbit to an outer circular orbit, as shown in Fig. 3-6-1, there are two impulsive ΔV's. The first is ΔV_1, a collinear firing at perigee, which increases the energy of the satellite to match that of the transfer ellipse and the second is ΔV_2, another collinear firing at apogee, which further increases the energy to match that of the outer orbit. Without ΔV_2 the satellite would

Sec. 3-6 Orbital Transfers

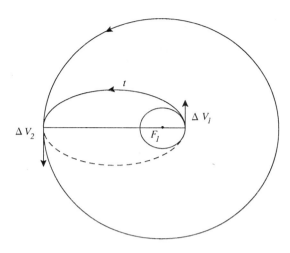

Figure 3-6-1. Hohmann transfer from inner to outer circle.

remain in the transfer ellipse, return to the inner orbit, and continue to orbit the Earth in the elliptical transfer orbit.

Example 3-6-1

A communications satellite was carried by the Space Shuttle into an LEO at an altitude of 322 km (175 nmi) and is to be transferred to a GEO at 35,860 km (19,490 nmi) using a Hohmann transfer. Find the characteristics of the transfer ellipse and the total ΔV required, ΔV_T.

Solution Inner orbit:

$$r_i = r_E + h_i = 6.378 \times 10^6 + 322 \times 10^3 = 6.70 \times 10^6 \text{ m}$$

$$V_i = \left(\frac{\mu}{r_i}\right)^{1/2} = 7713 \text{ m/s}$$

$$E_i = \frac{-\mu}{2r_i} = -2.975 \times 10^7 \text{ m}^2/\text{s}^2$$

Outer orbit:

$$r_o = 42.24 \times 10^6 \text{ m}$$

$$V_o = 3072 \text{ m/s}$$

$$E_o = \frac{-\mu}{2r_o} = -4.718 \times 10^6 \text{ m}^2/\text{s}^2$$

Transfer trajectory:

$$2a_t = r_i + r_o = 48.94 \times 10^6 \text{ m}$$

$$E_t = \frac{-\mu}{2a_t} = -8.144 \times 10^6 \text{ m}^2/\text{s}^2$$

$$E_i < E_t < E_o$$

Making use of Eq. (2-8-11), we find that the velocity at perigee of the transfer trajectory (departure from the inner orbit) is

$$V_{tp} = V_{ti} = \sqrt{2\left(-8.144 \times 10^6 + \frac{3.986 \times 10^{14}}{6.70 \times 10^6}\right)} = 10,130 \text{ m/s}$$

Since ϕ at perigee of the transfer trajectory is zero, Eq. (2-8-4) can be used to find the specific angular momentum and

$$H_t = (6.70 \times 10^6) \times 10,130 = 6.787 \times 10^{10} \text{ m}^2/\text{s}$$

Applying Eq. (2-8-3) to find the eccentricity yields

$$\epsilon_t = \sqrt{1 + \frac{2 \times (-8.144 \times 10^6) \times (6.787 \times 10^{10})^2}{(3.986 \times 10^{14})^2}} = 0.7265$$

The velocity at apogee (arrival at the outer orbit) of the transfer trajectory is

$$V_{ta} = V_{to} = \frac{H}{r_o} = 1607 \text{ m/s}$$

Therefore,

$$\Delta V_1 = V_{ti} - V_i = 10,130 - 7713 = +2417 \text{ m/s}$$

and

$$\Delta V_2 = V_o - V_{to} = 3072 - 1607 = +1465 \text{ m/s}$$

Remember that the sign of the ΔV's indicates the direction of thrusting (whether the energy is to be increased or decreased) and that the total ΔV is the sum of the magnitudes and is not the algebraic sum. In this case, the total ΔV required is

$$\Delta V_T = |\Delta V_1| + |\Delta V_2| = 2417 + 1465 = 3882 \text{ m/s}$$

Since the transfer trajectory is one of the symmetrical halves of an ellipse, the time of flight (TOF) is simply half of the period. Therefore, with $a = 24.47 \times 10^6$ m and Eq. (3-3-1),

$$\text{TOF} = \pi\sqrt{\frac{(24.47 \times 10^6)^3}{3.986 \times 10^{14}}} = 19,050 \text{ s} = 317.4 \text{ min} = 5.29 \text{ h}$$

Example 3-6-2

An orbital transfer vehicle (OTV) has rendezvoused with a communications satellite in a GEO and is to return it to the Space Shuttle, which is in a LEO at 322 km (175 nmi). Find the minimum ΔV_T required.

Solution Since the Hohmann transfer (see Fig. 3-6-2) requires the minimum ΔV_T, this problem is just the reverse of Example 3-6-1 and we can use the data from that example. In this transfer we see that we need to decrease the energy of the OTV–satellite combination to enter the Hohmann transfer and then decrease the energy once more to enter the LEO. Although the energy is decreased in both firings by reducing the velocity, an expenditure of propulsive energy is required to change the energy of the trajectory.

$$\Delta V_1 = V_{to} - V_o = 1607 - 3072 = -1465 \text{ m/s}$$
$$\Delta V_2 = V_i - V_{ti} = 7713 - 10,130 = -2417 \text{ m/s}$$

Sec. 3-6 Orbital Transfers

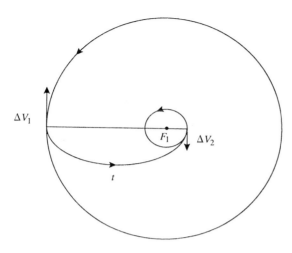

Figure 3-6-2. Hohmann transfer from outer to inner circle.

and

$$\Delta V_T = 1465 + 2417 = 3882 \text{ m/s}$$

Notice that the magnitudes of the total ΔV's are identical, although the directions of thrusting are not. Obviously, the TOF for the return will be identical to that for the ascent.

You may find it interesting to discover that the launch velocity for an orbit, as discussed in Section 3-2, is equal to the ΔV_T of a Hohmann transfer from the surface of a nonrotating Earth to the orbit in question.

Example 3-6-3

Find the ΔV_T for a Hohmann transfer from the surface of a nonrotating Earth to a 322-km (175-nmi) circular orbit and compare it with the launch velocity for that orbit.

Solution $2a_t = r_E + r = 6.378 \times 10^6 + 6.70 \times 10^6 = 13.08 \times 10^6$ m

$$E_t = \frac{-\mu}{2a_t} = -3.047 \times 10^7 \text{ m}^2/\text{s}^2$$

Again, applying Eq. (2-8-11) at $r = r_E$, we have

$$V_{tp} = \sqrt{2\left(-3.047 \times 10^7 + \frac{3.986 \times 10^{14}}{6.378 \times 10^6}\right)} = 8003 \text{ m/s}$$

Since the Earth is assumed to be nonrotating,

$$\Delta V_1 = 8003 - 0 = +8003 \text{ m/s}$$

With $\phi = 0$, $H = rV$, so that

$$H_t = (6.378 \times 10^6) \times 8003 = 5.104 \times 10^{10} \text{ m}^2/\text{s}$$

At 322 km, at the circular orbit,

$$V_{ta} = \frac{H_t}{r} = 7618 \text{ m/s}$$

and

$$V_{cs} = \left(\frac{\mu}{r}\right)^{1/2} = 7713 \text{ m/s}$$

so that

$$\Delta V_2 = 7713 - 7618 = +95 \text{ m/s}$$

Therefore,

$$\Delta V_T = 8003 + 95 = 8098 \text{ m/s}$$

Since E of the circular orbit is $-\mu/2r = -2.975 \times 10^7$ m²/s²,

$$V_L = \sqrt{2\left(-2.975 \times 10^7 + \frac{3.986 \times 10^{14}}{6.378 \times 10^6}\right)} = 8092 \text{ m/s}$$

Allowing for a round-off difference of 6 m/s (0.07%), we see that $\Delta V_T = V_L$ for this transfer from the surface of a nonrotating Earth to a circular orbit. The same relationship holds for Hohmann transfers to elliptical orbits. For example, the launch velocity of the ellipse of Example 3-3-1 is 8581 m/s and the Hohmann ΔV_T value from the Earth to the perigee altitude of 552 km is 8623 m/s, a 0.5% difference, and the Hohmann ΔV_T value to the apogee altitude of 2208 km is 8619 m/s, a 0.4% difference. As the orbital altitude increases, the difference between ΔV_T and V_L increases. For example, for an object in a GEO (at an altitude of 35,600 km), the value of V_L is 1% less than that of ΔV_T for a Hohmann transfer from a nonrotating Earth and for an object at an altitude three times higher (110,400 km), the difference is -9.2%.

In transferring from an inner ellipse to an outer circular orbit, the Hohmann transfer should be initiated at the perigee of the ellipse (see Fig. 3-6-3) so as to keep a_t (and E_t) as small as possible.

Example 3-6-4

Execute a Hohmann transfer from the perigee of the LEO ellipse of Example 3-3-1 ($h_p = 552$ km and $h_a = 2208$ km) to a circular GEO ($h_{cs} = 35,860$ km). Find the individual ΔV's and ΔV_T.

Solution From Example 3-3-1,

$$r_p = 6.930 \times 10^6 \text{ m} \quad \text{and} \quad V_p = 7980 \text{ m/s}$$

$$r_a = 8.586 \times 10^6 \text{ m}$$

For the GEO,

$$r_{cs} = 42.24 \times 10^6 \text{ m} \quad \text{and} \quad V_{cs} = \left(\frac{\mu}{r_{cs}}\right)^{1/2} = 3072 \text{ m/s}$$

From Fig. 3-6-3, the transfer trajectory characteristics are

$$2a_t = r_p + r_{cs} = 49.17 \times 10^6 \text{ m}$$

$$E_t = \frac{-\mu}{2a_t} = -8.106 \times 10^6 \text{ m}^2/\text{s}^2$$

Sec. 3-6   Orbital Transfers

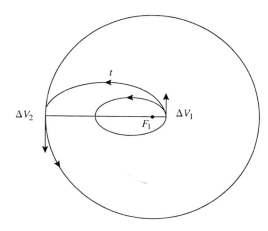

Figure 3-6-3. Hohmann transfer from inner ellipse to outer circle.

Using Eq. (2-8-11) once again yields

$$V_{tp} = \sqrt{2\left(-8.106 \times 10^6 + \frac{3.986 \times 10^{14}}{6.930 \times 10^6}\right)} = 9941 \text{ m/s}$$

$$\Delta V_1 = 9941 - 7980 = +1961 \text{ m/s}$$

$$H_t = 9941 \times (6.930 \times 10^6) = 6.889 \times 10^{10} \text{ m}^2/\text{s}$$

$$V_{ta} = \frac{H_t}{r_{cs}} = 1631 \text{ m/s}$$

$$\Delta V_2 = 3072 - 1631 = 1441 \text{ m/s}$$

$$\Delta V_T = 1961 + 1441 = 3402 \text{ m/s}$$

From Eq. (2-8-3),

$$\epsilon_t = 0.718$$

Since the transfer time is half the period of the transfer ellipse,

$$\text{TOF} = \pi\sqrt{\frac{(24.31 \times 10^6)^3}{3.986 \times 10^{14}}} = 18{,}860 \text{ s} = 314.3 \text{ min} = 5.238 \text{ h}$$

If the transfer trajectory were made from the apogee of the ellipse to the GEO, it could still be classified as a Hohmann transfer, but it should not be the minimum-energy Hohmann since the semimajor axis would be larger than that of Example 3-6-4. Consequently, we would also expect the required ΔV_T to be larger. Problem 3-16, which you may wish to work, shows that $\Delta V_T = 3629$ m/s.

A Hohmann transfer between two elliptical orbits is possible only in the unlikely event that the ellipses share a common major axis; i.e., when the apsides lie on the same line. Otherwise, at least one of the ΔV's will not be collinear. In the case of transfer between elliptical orbits, depending on the eccentricity of the two orbits, it is energy efficient to circularize the departure orbit prior to a Hohmann transfer or to make a Hohmann transfer to a circular orbit that intersects the destination orbit and then transform the circle to the desired ellipse; each of these approaches requires a third ΔV, but all three can be collinear.

By virtue of its smallest possible semimajor axis and two collinear firings, the Hohmann transfer requires the smallest value of ΔV_T; however, it also has the longest time of flight (TOF). There are times, such as in the intercept or rendezvous problems, when a shorter TOF is desired or even required. A *fast transfer* can be elliptical (the usual case), parabolic, or even hyperbolic and is characterized by at least one noncollinear ΔV. Figure 3-6-4 shows a typical fast transfer from an inner circular orbit to an outer circular orbit with a collinear ΔV_1 at the perigee of the transfer trajectory and a noncollinear ΔV_2 at the interception of the outer orbit. Since the semimajor axis of the fast transfer is greater than that of the Hohmann transfer, both the energy and the ΔV_1 of the fast transfer will be greater. Since ΔV_2 of the fast transfer is not collinear, it should be, and is, larger than the ΔV_2 of the Hohmann transfer.

Example 3-6-5

The Hohmann transfer of Example 3-6-1, between a 322-km LEO and a GEO required a $\Delta V_T = 3882$ m/s. Let us arbitrarily double the value of the semimajor axis of the Hohmann transfer ellipse and find the characteristics and ΔV_T of the resulting fast transfer. We use the values from Example 3-6-1 for the inner and outer circular orbits: $V_i = 7713$ m/s, $r_i = 6.70 \times 10^6$ m, $V_o = 3065$ m/s, and $r_o = 42.24 \times 10^6$ m/s.

For the Hohmann transfer, $2a_t = 48.94 \times 10^6$ m. For the fast transfer, set

$$2a_t = 98 \times 10^6 \text{ m}$$

$$E_t = \frac{-\mu}{2a_t} = -4.067 \times 10^6 \text{ m}^2/\text{s}^2 < 0 \quad \text{(elliptical)}$$

Applying Eq. (2-8-11), we find the required transfer velocity at departure to be

$$V_{ti} = \sqrt{2\left(-4.067 \times 10^6 + \frac{3.986 \times 10^{14}}{6.70 \times 10^6}\right)} = 10{,}530 \text{ m/s}$$

Since the elevation angles of V_{csi} and V_{ti} are identically zero, the angle between the two velocities is zero and ΔV_1 and the two velocities are collinear. Consequently,

$$\Delta V_1 = 10{,}530 - 7713 = 2817 \text{ m/s}$$

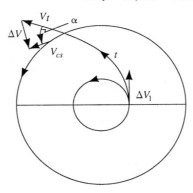

Figure 3-6-4. Fast transfer from an inner to an outer circle.

Sec. 3-6 Orbital Transfers

With $\phi = 0$ at departure,

$$H_t = 10{,}530 \times (6.70 \times 10^6) = 7.055 \times 10^{10} \text{ m}^2/\text{s}$$

The eccentricity can be found from Eq. (2-8-3) to be

$$\epsilon_t = \sqrt{1 + \frac{2 \times (-4.067 \times 10^6)(7.055 \times 10^{10})^2}{(3.986 \times 10^{14})^2}} = 0.863$$

Since $0 < \epsilon < 1$, this fast transfer is elliptical (it might have been hyperbolic) and has an eccentricity that is 19% greater than that of the Hohmann ($\epsilon = 0.7265$). Since the elevation angle ϕ will not be equal to zero at the interception of the outer orbit, V_{to} will not be equal to H_t/r_o and must be found from Eq. (2-8-11).

$$V_{to} = \sqrt{2\left(-4.067 \times 10^6 + \frac{3.986 \times 10^{14}}{42.24 \times 10^6}\right)} = 3277 \text{ m/s}$$

The elevation angle at interception, ϕ_{to}, can be found from Eq. (2-8-5) to be

$$\phi_{to} = \cos^{-1}\frac{H_t}{r_o V_{to}} = 59.36°$$

Since the outer orbit is circular ($\phi_o = 0$), the angle α between the two velocities is equal to 59.36° and the velocities and ΔV_2 are obviously not collinear. ΔV_2 can be found by applying the *law of cosines*, which states that for two noncollinear velocities, V_1 and V_2,

$$\Delta V^2 = V_1^2 + V_2^2 - 2V_1 V_2 \cos \alpha$$

For this example,

$$\Delta V_2 = \sqrt{(3277)^2 + (3065)^2 - 2 \times 3277 \times 3065 \cos(59.36°)} = 3142 \text{ m/s}$$

so that

$$\Delta V_T = 2817 + 3142 = 5959 \text{ m/s}$$

Comparing this value with the Hohmann ΔV_T of 3875 m/s, we see that the fast transfer ΔV_T is 54% higher, a considerable increase that will have a significant effect on the propulsive requirements and size of the spacecraft and the booster. With a Hohmann transfer, ν_0, the position angle at the time of interception of the target orbit, will be 180° in transferring from a lower to a higher orbit. With a fast transfer, ν_0 will be less and can be determined from the conic relationship $r_0 = p/(1 + \epsilon \cos \nu_0)$ with

$$p = \frac{H_t^2}{\mu} = \frac{(7.055 \times 10^{10})^2}{3.986 \times 10^{14}} = 12.49 \times 10^6 \text{ m}$$

(It should be noted in passing that p must always be positive since it is a physical quantity and because H^2 and μ are always positive.) Rearranging the conic relationship to solve for ν_0 and substituting values yields

$$\nu_0 = \cos^{-1}\left[\frac{1}{\epsilon}\left(\frac{p}{r_0} - 1\right)\right] = 144.6°$$

As a partial check, let us use Eq. (2-7-4) to verify ϕ_{to}:

$$\phi_{to} = \tan^{-1}\frac{0.863 \times \sin(144.6°)}{1 + 0.863 \cos(144.6°)} = 59.32°$$

Since the transfer velocities are higher and the distance traveled is less, we would expect the TOF of a fast transfer to be less than that of the Hohmann transfer, which it is. At this point, however, we cannot calculate the TOF between two points on an ellipse and must defer the determination of the new TOF until Chapter 4, which deals with the time of flight along a trajectory.

Returning to consideration of the Hohmann transfer, it can be shown that when the separation between the inner and outer orbits is very large ($r_o > 11.9 r_i$), a three-impulse Hohmann transfer scheme comprising two Hohmann ellipses (often referred to as a *bielliptical transfer*) can be more energy efficient than the two-impulse Hohmann transfer. A three-impulse Hohmann is sketched (not to scale) in Fig. 3-6-5 for transfer between two widely separated circular orbits. The collinear ΔV_1 is positive and takes the satellite on a Hohmann ellipse to an apogee well above the outer orbit where the velocity is small. At this point, a positive ΔV_2 places the satellite on a second Hohmann ellipse whose perigee is tangent to the outer orbit. ΔV_3 is negative and is used to circularize the ellipse. Although the energy requirement may be somewhat less than that for the two-impulse Hohmann, the time of flight will be much greater. For example, in Problem 3-18, there is only a 2% savings in ΔV with the three-impulse transfer but an increase in transfer time from 17.15 h to 39.2 days; from hours to days.

To date, the need for a three-impulse Hohmann transfer has not materialized, nor is it likely to do so in the foreseeable future, at least not until there is a requirement for a satellite or spacecraft in a retrograde orbit at altitudes in excess of 40,000 nmi (73,600 km), which are more than twice the altitude of a GEO.

3-7 PLANE CHANGES

There are times when it is necessary to change the plane of an orbit. For example, it is not possible to enter an equatorial GEO directly unless the launch site is itself

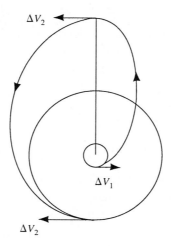

Figure 3-6-5. Bielliptic Hohmann transfer to a high orbit.

Sec. 3-7 Plane Changes

located on the equator; therefore, a plane change is required after the satellite is established in an initial orbit. Since plane changes, as we shall see shortly, require comparatively large ΔV's, every effort is made to avoid the need for large plane changes, and when they are necessary, to execute them so as to minimize the propulsive requirements.

Plane changes call for noncollinear ΔV's, as can be seen in Fig. 3-7-1, and the law of cosines applies here as it did in the case of the fast transfer of the preceding section. In the most general case with both a plane change and an orbit change,

$$\Delta V^2 = V_1^2 + V_2^2 - 2V_1 V_2 \cos \alpha \qquad (3\text{-}7\text{-}1)$$

where V_1 is the velocity of the original orbit, V_2 the velocity of the new orbit, and α the desired plane change.

If the two orbits are identical except for their inclinations (with any orbit changes made prior to or after the plane change), $V_2 = V_1$ and Eq. (3-7-1) becomes

$$\Delta V^2 = 2V_1^2(1 - \cos \alpha)$$

so that

$$\Delta V = \sqrt{2} V_1 \sqrt{(1 - \cos \alpha)} \qquad (3\text{-}7\text{-}2)$$

But

$$\sqrt{1 - \cos \alpha} = \sqrt{2} \sin \frac{\alpha}{2}$$

Therefore, Eq. (3-7-2) becomes

$$\boxed{\Delta V = 2V_1 \sin \frac{\alpha}{2}} \qquad (3\text{-}7\text{-}3)$$

Equation (3-7-3) shows the requirement for a *plane change only*, without any changes in the shape of the orbit. It also shows that to minimize the ΔV required, a plane change should normally be made when V_1 is at a minimum (e.g., at apogee for an ellipse). However, even when V_1 is at a minimum, the ΔV required will be large.

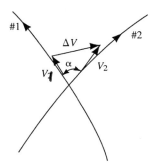

Figure 3-7-1. Plane change.

Example 3-7-1

A communications satellite is in a 322-km (175-nmi) circular parking orbit but not in the desired plane. Find the ΔV required to shift the circular plane by 5°, by 10°, and by 20°.

Solution With $r = 6.70 \times 10^6$ m,

$$V_1 = \left(\frac{\mu}{r}\right)^{1/2} = 7713 \text{ m/s}$$

and

$$\Delta V = 2 \times 7713 \sin\frac{\alpha}{2}$$

α (deg)	ΔV (m/s)
5	672.8
10	1344
20	2679

Example 3-7-2

Rather than making the plane changes while in the LEO, it is decided to make the plane changes after the satellite has been placed into a circular GEO. Find the ΔV's and compare with those of Example 3-7-1.

Solution In the GEO, $r = 42.24 \times 10^6$ m, $V_1 = 3072$ m/s, and

$$\Delta V = 2 \times 3072 \sin\frac{\alpha}{2}$$

α (deg)	ΔV (m/s)
5	268.0
10	535.5
20	1067

By delaying the plane change until the satellite is in the higher orbit where the orbital velocity is lower, the ΔV requirement has been reduced by 60%.

In Example 3-4-1, we found that a collinear ΔV of -396 m/s is required to change the elliptical orbit of Example 3-3-1 with an apogee altitude of 2208 km (1200 nmi) into a circular orbit with an altitude of 552 km (300 nmi) without a plane change. Conversely, a collinear ΔV value of $+396$ m/s is required to go from the circular orbit to the elliptical orbit.

Example 3-7-3

After entering the ellipse of Example 3-3-1 from a circular orbit at 552 km, make a 10° plane change when the spacecraft reaches apogee. Find the total ΔV for the orbit transformation and the plane change.

Solution At r_a, $V_a = 6441$ m/s. Therefore, for the plane change alone,

$$\Delta V = 2 \times 6441 \sin(5°) = 1123 \text{ m/s}$$

The total ΔV required, for the transformation from the circle to the ellipse and for the plane change at apogee, is

$$\Delta V_T = 396 + 1123 = 1519 \text{ m/s}$$

Example 3-7-4

This time, after entering the ellipse, wait until returning to perigee to make the plane change. Find the ΔV for the plane change and the total ΔV to include entering the ellipse.

Solution At r_p, $V_p = 7980$ m/s. Therefore,

$$\Delta V = 2 \times 7980 \sin(5°) = 1391 \text{ m/s}$$

and

$$\Delta V_T = 396 + 1391 = 1787 \text{ m/s}$$

By making the plane change at apogee rather than at perigee, the reduction in ΔV is 15%.

Example 3-7-5

Rather than entering the ellipse and then waiting to make the plane change at apogee, let us see what is required to make the plane change at the time of entering the ellipse at perigee, to use one ΔV rather than two.

Solution At $r_{cs} = r_p = 6.93 \times 10^6$, $V_1 = V_{cs} = 7584$ m/s and $V_2 = V_p = 7980$ m/s. Therefore,

$$\Delta V = \sqrt{(7584)^2 + (7980)^2 - 2 \times 7584 \times 7980 \cos(10°)} = 1413 \text{ m/s}$$

By using a single ΔV to accomplish the orbit modification and plane change simultaneously, there is an overall reduction in the ΔV required of 102.6 m/s (6.7%). Despite this savings, other considerations, such as guidance and control and orbit trimming, may favor the double firing.

There are various schemes and techniques for combining orbital transfers and plane changes so as to optimize the maneuvers that involve mathematical techniques that are beyond the scope of this book.

3-8 CLOSING REMARKS

The specific mechanical energy of each two-body trajectory has a constant value, a value that is inversely proportional to the semimajor axis; if the energy changes, the trajectory changes. At any point on a trajectory, the total energy can be changed by changing the kinetic energy, by increasing or decreasing the velocity (i.e., by introducing an impulsive ΔV to either speed up or slow down the smaller body).

Since an energy source, such as a propulsion system, is needed to generate the desired ΔV, the larger the required ΔV is the larger is the required amount of energy and fuel. The increase in the fuel will increase the weight of the spacecraft, which will be accompanied by an increase in the size and weight of the booster that places the spacecraft into orbit. Consequently, in planning trajectory changes and orbital transfers, every effort should be made to minimize the magnitude of the velocity change.

For trajectory changes within the plane of motion, we have seen that for minimum ΔV, velocity changes should be collinear with the original and final velocity vectors whenever possible. We have not shown, however, other than by example, that velocity changes should be made when the velocity is as high as possible. This can be seen by taking the derivative of the energy, as expressed in Eq. (2-8-4), with respect to the velocity to obtain the simple relationship that

$$dE = V\,dV \quad \text{or} \quad \Delta V \cong \frac{\Delta E}{V} \qquad (3\text{-}8\text{-}1)$$

Equation (3-8-1) shows that for a specified energy change, ΔE, the required velocity change, ΔV, is inversely proportional to the velocity at the time of change; in other words, when you have a choice, make your energy changes when the velocity is at a maximum. For example, elliptical energy changes (velocity changes) should be made at perigee if at all possible.

Plane changes, however, require noncollinear (vector) velocity changes. When only a plane change is involved (no trajectory modification), it has been shown that the plane change should be made when the velocity is at a minimum (i.e., at apogee for an elliptical plane change). If, however, a trajectory modification or an orbital transfer is required as well as a plane change, it is no longer obvious what the velocity changing scheme should be; the correct strategy will depend on the situation. For example, in Example 3-7-5 we saw that a single firing required the lowest ΔV_T of the three possibilities examined.

Let us not forget that our two-body trajectories are ideal trajectories subject to the stated assumptions to include that of impulsive thrusting. Any relaxation of the assumptions leads to perturbations in the orbits. Even though, initially, these perturbations may appear to be small, if they are not corrected by thrusting, their effect over a period of time can be significant and must be considered in the actual operation of spacecraft and satellites. For example, there is enough atmospheric drag at 552 km (300 nmi) to cause an ideal circular orbit to decay, to become an inward-spiraling trajectory that will eventually enter the sensible atmosphere of the Earth. For satellites with long-duration missions, the periodic thrusting required to maintain a desired orbit is known as *station keeping*.

The trajectories and orbits that we have described and discussed in this chapter are simple planar conics with respect to the center of the Earth and would appear as such to an observer located there. However, we do not live at the center of the Earth but rather on its surface, a surface that is rotating about a polar axis. As a consequence, the projection of an orbit on the surface of the rotating Earth, known as a *ground track*, will not be a simple conic section but rather will be a function of orientation of the orbital plane and of the trajectory with respect to a surface reference, principally the equator. The orientation of the plane and the location of the smaller body are described by the orbital elements, which will be defined and discussed along with ground tracks in Chapter 8, after we find out how to get into

Chap. 3   Problems   55

outer space and return to the Earth. In the next chapter we examine the time-of-flight problem.

PROBLEMS

You are encouraged to use sketches to assist you in visualizing the problem under consideration. Keep four significant figures unless told otherwise.

3-1. The Hubble Space Telescope (HST) was placed in a circular orbit at an altitude of 330 nmi (607.2 km).
  (a) Find the orbital characteristics to include the specific kinetic energy, the specific potential energy, the total specific energy, the period, p, ϵ, the orbital velocity, H, and g.
  (b) Find the launch velocity, V_L.

3-2. A satellite has been placed in a circular orbit at an altitude that is three times that of a GEO; $h = 110{,}400$ km.
  (a) Do Problem 3-1(a) and (b).
  (b) Compare results with those of Example 3-2-3 for the GEO.
  (c) Is there anything special about the period and the motion of the satellite with respect to a particular location on the surface of the Earth?

3-3. In inserting a service satellite into the Problem 3-1 orbit of the Hubble Space Telescope, the injection ΔV was larger than necessary. The resulting orbit was elliptical with perigee at 330 nmi (607.2 km) and apogee at 550 nmi (1012 km).
  (a) Find the orbital characteristics: a, ϵ, p, P, E, H, V_p, and V_a.
  (b) Find r, ϕ, and V when $\nu = 60°$; when $\nu = 200°$.

3-4. An object in space is observed to have an altitude of 1200 nmi (2209 km), a velocity of 7000 m/s, and an elevation angle (ϕ) of $+40°$. $a, E, H, r_p, r_a, V_p, V_a, P, \epsilon$
  (a) Give a complete description of the trajectory.
  (b) Is there anything special or different about this trajectory?
  (c) Find V, ν, and ϕ when $r = 6.378 \times 10^6$ m. Sketch the trajectory and locate this point.

3-5. For the elliptical orbit of Example 3-3-3, find r, ϕ, and ν when $V = 6000$ m/s.

3-6. Find the ΔV required to circularize the elliptical orbit of the service satellite of Problem 3-3 along with the period of the new orbit with circularization
  (a) At perigee.
  (b) At apogee.

3-7. The Space Shuttle carried a reconnaissance satellite into a 175-nmi circular parking orbit. Now the satellite is to be placed into an elliptical orbit with perigee at 175 nmi and a period of 24 h.
  (a) Find the ΔV required along with the characteristics of the new orbit.
  (b) Find the ΔV_T required for a Hohmann transfer from the surface of a nonrotating Earth to the parking orbit.
  (c) Find the launch velocity for the parking orbit and compare it with the ΔV_T of part (b).

3-8. The mission objective is to escape from a 175-nmi parking orbit and have a residual velocity (V_∞) of 10,000 m/s.
  (a) Find the ΔV and trajectory characteristics of an Oberth maneuver. a, E, H_e, r_b
  (b) Find the ΔV required for a parabolic escape and compare with part (a).

3-9. A returning spacecraft has a $V_\infty = 6000$ m/s. We want to place it in a 175-nmi parking orbit with a single ΔV.
  (a) What must the value of ϕ_∞ be (in degrees)?
  (b) What are the characteristics of the incoming trajectory?
  (c) Find the ΔV value required to enter the parking orbit.

3-10. Assuming that we can treat the Earth–Moon combination as a two-body closed system, find the orbital velocity of the Moon with respect to the Earth and its period in days. The distance to the Moon is approximately 3.89×10^8 m.

3-11. In Example 3-6-4, the Hohmann transfer was initiated at the perigee of the lower orbit. Initiate the transfer at the apogee of the lower orbit and compare ΔV for this transfer with that of Example 3-6-4.

3-12. Referring to Problem 3-1,
  (a) Find the ΔV_T and time of flight for a Hohmann transfer from the surface of the Earth to the orbit of the Hubble Space Telescope.
  (b) Find the "launch velocity" (V_L) from the energy of the circular orbit and compare with the ΔV_T from part (a).
  (c) If the launch is made from a site at 20° N latitude at an angle (inclination) of 30° with respect to the equator, what will the effect be (the % change) on the ΔV_T of part (a)?

3-13. A communications satellite is in a geosynchronous orbit at an altitude of 35,860 km. Do parts (a) and (b) of Problem 3-12.

3-14. For the satellite of Problem 3-2, do parts (a) and (b) of Problem 3-12.

3-15. A satellite has been carried to a 175-nmi (322-km) circular orbit by a Space Shuttle. The mission objective is to place the satellite into an elliptical orbit with a perigee of 600 nmi (1104 km) and an eccentricity of 0.80.
  (a) Find the characteristics of the ellipse: $a, E, H, r_p, r_a, V_p, V_a, p,$ and P.
  (b) Find $\Delta V_T, \epsilon_t,$ and TOF for a Hohmann transfer from the circular orbit to the perigee of the ellipse. Sketch the transfer trajectory.
  (c) Find $\Delta V_T, \epsilon_t,$ and TOF for a Hohmann transfer to the apogee of the ellipse. Sketch the transfer trajectory.
  (d) Of the two ΔV_T's from parts (b) and (c), which is larger? Can you explain why?

3-16. Do Problem 3-15 for a parking orbit at 300 nmi (522 km) and an elliptical orbit with a perigee of 175 nmi (322 km) and an eccentricity of 0.70.

3-17. Two geocentric elliptical orbits have aligned major axes and their perigees on the same side of the Earth. The first orbit has an $r_p = 7.0 \times 10^6$ m and $\epsilon = 0.3$, whereas the second orbit has an $r_p = 32 \times 10^6$ m and $\epsilon = 0.50$.
  (a) Find the minimum ΔV_T and the TOF and ϵ_t for a transfer from the perigee of the inner orbit to the apogee of the outer orbit. Sketch the transfer.
  (b) Do part (a) for a transfer from the apogee of the inner orbit to the perigee of the outer orbit.
  (c) Compare the ΔV_T's and TOFs and explain any differences.

3-18. The mission objective is to transfer a satellite from a 175-nmi (322-km) circular parking orbit to a high circular orbit at an altitude of 51,150 nmi (94,120 km).
  (a) Find the ΔV_T and the TOF and ϵ_t for a two-impulse Hohmann transfer.
  (b) Find the ΔV_T and the TOF and ϵ_t's for a three-impulse Hohmann transfer with the first transfer to an $r = 9.2 \times 10^8$ m.
  (c) Which transfer do you recommend, and why?

3-19. The mission is to place a satellite into a high circular orbit at 60,000 nmi (110,400 km) with a direct launch from the surface of the Earth (without using an intermediate parking orbit). Do parts (a), (b), and (c) of Problem 3-18.

3-20. Execute a fast transfer from a 175-nmi circular orbit to a GEO ($h = 19{,}490$ nmi) using a trajectory with a departure elevation angle of $0°$ that will cross the outer orbit when $\nu = 100°$. Find ΔV_T and the characteristics of the transfer trajectory. Is it an ellipse?

3-21. Transfer between the orbits of Problem 3-20 using a parabola with a departure angle of $0°$. Find ΔV_T and ν at the crossing of the outer orbit.

3-22. The mission is to transfer from a 300-nmi circular orbit to a geosynchronous orbit at an altitude of 19,490 nmi and at an angle of $15°$ with respect to the circular orbit.
  (a) Use a Hohmann transfer from the circular orbit with a combined orbit and plane change at departure (a single ΔV) with another ΔV at arrival to enter the GEO. Find the ΔV_T for these two ΔV's.
  (b) From the circular orbit enter a coplanar elliptical orbit with $\epsilon = 0.8$ with a single ΔV. Then at apogee use a single ΔV to enter the transfer orbit and simultaneously make the plane change. The third ΔV occurs at arrival at the GEO and is the injection ΔV. Find the ΔV_T for these three ΔV's and compare with part (a).
  (c) Leave the circular parking orbit on a coplanar Hohmann transfer and use a single ΔV at arrival to change the plane and enter the GEO. Find the ΔV_T for these two ΔV's and compare with parts (a) and (b).

3-23. The mission is for a direct launch from the surface of the Earth into a 175-nmi circular orbit that has a $15°$ inclination with respect to the launch plane.
  (a) Find the ΔV_T required for a plane change after injection into the parking orbit (i.e., three $\Delta V_\bullet$'s).
  (b) Find the ΔV_T required if the injection and plane change are accomplished with one ΔV (i.e., two ΔV's).
  (c) Find the ΔV_T required with the plane change made at launch and only an injection ΔV at the parking orbit.

CHAPTER 4

Time of Flight

Artist's conception of the deployed Hubble Space Telescope in orbit after launching from the Space Shuttle. (Courtesy of NASA.)

4-1 INTRODUCTION

In Example 3-6-5 we examined the characteristics of a fast transfer trajectory but were not able to calculate the time required to make the transfer [i.e., the time of flight (TOF)]. In this chapter we develop expressions (with the help of geometry and trigonometry) for determining the time of flight along any trajectory. It is customary to measure orbital (and trajectory) time from the point of closest approach to the primary focus (i.e., from periapsis, as indicated in Fig. 4-1-1). We shall also confirm Kepler's second and third laws of planetary motion.

In this section we develop the general integral equation for determining the time to travel from periapsis to any other point on a trajectory. The following three sections are concerned with developing expressions for determining time of flight along parabolas, ellipses, and hyperbolas, respectively. In the fifth section we use Lambert's theorem to develop expressions for determining the characteristics of an orbit from two measurements of time and position (i.e., when the TOF between two measured positions is known). This orbit determination problem, for which there are no closed-form solutions, is also referred to as the *Gauss problem*.

Returning to the general time-of-flight problem, the expression for the magnitude of the specific angular momentum can be written as

$$H = rV \cos \phi \qquad (4\text{-}1\text{-}1)$$

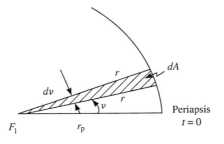

Figure 4-1-1. Differential area dA for an interval of time dt.

But $V \cos \phi$ is the tangential component of the velocity vector (see Fig. 2-7-1), which in turn can be written as $r\dot{v}$, where $\dot{v}$ is the instantaneous angular velocity. With this substitution and with $\dot{v}$ written as dv/dt, Eq. (4-1-1) becomes

$$H = r^2 \frac{dv}{dt} \tag{4-1-2a}$$

Rearranging this equation yields

$$dt = \frac{r^2}{H} dv \tag{4-1-2b}$$

Now it is possible to integrate Eq. (4-1-2b) from $t = 0$, where $v = 0$, to t to obtain

$$t = \frac{1}{H} \int_0^v r^2 \, dv \tag{4-1-3}$$

Returning to Fig. 4-1-1, dA, the differential area swept out during any interval of time dt, is simply the area of a sector, namely,

$$dA = \tfrac{1}{2} r^2 \, dv \tag{4-1-4a}$$

Integrating from $v = 0$, where $A = 0$, to v results in

$$A = \tfrac{1}{2} \int_0^v r^2 \, dv \tag{4-1-4b}$$

or

$$\int_0^v r^2 \, dv = 2A \tag{4-1-5}$$

Substituting Eq. (4-1-5) into Eq. (4-1-3) yields the expressions

$$t = \frac{2A}{H} \quad \text{or} \quad A = \frac{Ht}{2} \tag{4-1-6}$$

Finally, taking the derivative of A with respect to time, we find that

$$\boxed{\frac{dA}{dt} = \frac{H}{2}} \tag{4-1-7}$$

Since H is a constant of motion, dA/dt is also constant, so that $dA = (H/2) \, dt$. Consequently, r sweeps out equal areas in equal intervals of time, *thus confirming Kepler's second law of planetary motion*.

4-2 THE PARABOLA

Let us first look at the time of flight along a parabola, which has an eccentricity of unity ($\epsilon = 1$) and zero specific energy ($E = 0$ and $a = \infty$). Rewriting and renumbering Eq. (4-1-3), for convenience,

Sec. 4-2   The Parabola

$$t = \frac{1}{H}\int_0^v r^2\, dv \qquad (4\text{-}2\text{-}1)$$

we find an expression for r as a function of v, substitute it into Eq. (4-2-1), and integrate.

For a parabola,

$$r = \frac{p}{1 + \cos v} \qquad (4\text{-}2\text{-}2)$$

Since $(1 + \cos v) = 2\cos^2(v/2)$ and $p = H^2/\mu$,

$$r = \frac{H^2}{2\mu \cos^2(v/2)} \qquad (4\text{-}2\text{-}3)$$

Substituting Eq. (4-2-3) into Eq. (4-2-1) and then both changing the variable of integration from v to $v/2$ [$dv = 2d(v/2)$] and replacing $[1/\cos^4(v/2)]$ by $\sec^4(v/2)$ results in

$$t = \frac{H^3}{4\mu^2}\int_0^v \frac{dv}{\cos^4(v/2)} = \frac{H^3}{2\mu^2}\int_0^{v/2} \sec^4\frac{v}{2}\, d\left(\frac{v}{2}\right) \qquad (4\text{-}2\text{-}4)$$

Since $\sec^4(v/2)$ can be written as $\sec^2(v/2)[1 + \tan^2(v/2)]$, it is now possible to integrate the far right-hand integral of Eq. (4-2-4) term by term to obtain an expression for the time of flight from periapsis to any specified value of v. This expression is

$$\boxed{t = \frac{H^3}{2\mu^2}\left(\tan\frac{v}{2} + \frac{1}{3}\tan^3\frac{v}{2}\right)} \qquad (4\text{-}2\text{-}5)$$

Let us use this expression to determine how long it takes to escape from the Earth's surface on a parabolic trajectory.

Example 4-2-1

$$r_E = 6.378 \times 10^6 \text{ m}$$

Assuming launch at perigee,

$$V_{\text{esc}} = \left(\frac{2\mu_E}{r_E}\right)^{1/2} = 11{,}180 \text{ m/s}$$

$$H = r_E V_{\text{esc}} = 7.13 \times 10^{10} \text{ m}^2/\text{s}^2$$

From Eq. (2-8-1) with $\epsilon = 1$, $v = \cos^{-1}(p/r - 1)$. If $r = \infty$, then $v = \cos^{-1}(-1) = 180°$ and $v/2 = 90°$. Since $\tan 90°$ is infinite, Eq. (4-2-5) shows that the time to escape to infinity will also be infinite, which is logical. However, as mentioned earlier, the smaller body will "escape" at some finite distance at which point it is considered to come under the influence of another larger body, specifically the Sun.

Example 4-2-2

Let us, therefore, pick a specific, but large, value for r, such as 920,000 km (500,000 nmi) and see how long it takes to reach that distance.

Solution From Example 4-2-1, $H = 7.13 \times 10^{10}$ m²/s², so that

$$p = \frac{H^2}{\mu} = 12.75 \times 10^6 \text{ m}$$

Rearranging Eq. (2-8-1), with $r = 9.2 \times 10^8$ m, results in

$$\nu = \cos^{-1}\left(\frac{p}{r} - 1\right) = \cos^{-1}\left(\frac{12.75 \times 10^6}{9.2 \times 10^8} - 1\right) = 170.4°$$

With $\nu/2 = 85.2°$, substituting into Eq. (4-2-5) produces

$$t = \frac{(7.13 \times 10^{10})^3}{2\mu^2}(11.91 + 563.0) = 655{,}800 \text{ s} = 182.2 \text{ h}$$

We see that it takes 7.6 days, or slightly more than a week, to "escape" to 920,000 km, which is approximately 144 Earth's radii from the center of the Earth. At that point, $g = \mu/r_\infty^2 = 4.7 \times 10^{-4}$ m/s², which is quite small when compared with the value of 9.80 m/s² at the Earth's surface. Since $E = 0$, Eq. (2-8-4) shows that $V = (2\mu/r)^{1/2}$. When $r = 9.2 \times 10^8$ m, the velocity will be 930.9 m/s, which although obviously not zero is only 8% of the escape velocity of 11,180 m/s and the kinetic energy (KE) at that finite distance is only 0.7% of the KE at the surface of the Earth.

4-3 THE ELLIPSE

Let us now move on to the ellipse and first verify the expression that we have been using for the period of an ellipse. If we rewrite Eq. (4-1-7) as

$$dt = \frac{2}{H} dA \tag{4-3-1}$$

the period is the integral over one complete orbit, from 0 to 2π radians, namely,

$$P = \int_0^{2\pi} dt = \frac{2}{H}\int_0^{2\pi} dA = \frac{2}{H} A_{\text{ellipse}} \tag{4-3-2}$$

But $A_{\text{ellipse}} = \pi ab$, where a is the semimajor axis and b is the semiminor axis. Since $c = a\epsilon$, $a^2 = b^2 + c^2$, and $p = a(1 - \epsilon^2) = H^2/\mu$,

$$b = \sqrt{a^2 - c^2} = \sqrt{a^2(1 - \epsilon^2)} = H\sqrt{\frac{a}{\mu}} \tag{4-3-3}$$

Substituting Eq. (4-3-3) into the expression for the area and further substituting that expression into Eq. (4-3-2) yields

$$\boxed{P = 2\pi\sqrt{\frac{a^3}{\mu}}} \tag{4-3-4}$$

Equation (4-3-4) verifies Kepler's third law of planetary motion that states that the square of the period of a planet is proportional to the cube of the major axis of

Sec. 4-3   The Ellipse

its elliptical orbit. It is interesting to note that the period is independent of the eccentricity.

To find an expression for the time of flight along an ellipse, it is necessary to introduce new orbital parameters, the eccentric anomaly and the mean anomaly. In Fig. 4-3-1, an auxiliary circle has been drawn about the ellipse with its radius r equal to the semimajor axis a and tangent to the ellipse at the apsides. At any point P on the ellipse, a line PR is drawn perpendicular to the major axis and extended to touch the circle at P'. The line OP' and the major axis define a new parameter u, the *eccentric anomaly*. The position angle v will now be renamed and be known henceforth as the *true anomaly*. At this point a third anomaly, the *mean anomaly M*, will be defined arbitrarily by the relationship that

$$M = \frac{2\pi t}{P} \tag{4-3-5}$$

or

$$\boxed{t = \frac{P}{2\pi} M} \tag{4-3-6}$$

where P is the period of the ellipse and t is the time from periapsis. The time of flight (TOF) between any two points on an ellipse, such as A and B, can be found from the expression

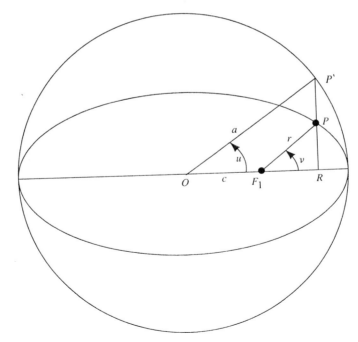

Figure 4-3-1.   Auxiliary circle and the eccentric anomaly.

$$\text{TOF} = t_B - t_A = \frac{P}{2\pi}(M_B - M_A) \qquad (4\text{-}3\text{-}7)$$

Now we find an expression for the mean anomaly M as a function of the orbital parameters by applying trigonometry to assist the integration called for in Eq. (4-1-3), which is rewritten and renumbered for convenience:

$$t = \frac{1}{H}\int_0^v r^2 \, dv \qquad (4\text{-}3\text{-}8)$$

The plan is to find expressions for r as a function of the eccentric anomaly u and to change the variable from v to u.

Referring to Fig. 4-3-1, we see that

$$\cos u = \frac{c + r\cos v}{a}$$

But $c = a\epsilon$ and $r = a(1 - \epsilon^2)/(1 + \epsilon\cos v)$, so that

$$\boxed{\cos u = \frac{\epsilon + \cos v}{1 + \epsilon\cos v}} \qquad (4\text{-}3\text{-}9)$$

This equation can be used to find u when v is known. Solving Eq. (4-3-9) for $\cos v$ yields

$$\cos v = \frac{\cos u - \epsilon}{1 - \epsilon\cos u} \qquad (4\text{-}3\text{-}10)$$

When Eq. (4-3-10) is substituted into the polar equation for r [given in the line preceding Eq. (4-3-9)], the result is an expression for r in terms of u; that is,

$$\boxed{r = a(1 - \epsilon\cos u)} \qquad (4\text{-}3\text{-}11)$$

We now need to change the variable of integration from v to u, to find an expression for dv as a function of u and du. We start by taking the derivative of the $\cos v$, as given in Eq. (4-3-10), with respect to u to obtain

$$\frac{d\cos v}{du} = -\sin v \frac{dv}{du} = \frac{-(1 - \epsilon^2)\sin u}{(1 - \epsilon\cos u)^2} \qquad (4\text{-}3\text{-}12)$$

Now we need an expression for $\sin v$. Figure 4-3-1 shows that

$$\sin v = \frac{PR}{r} \qquad (4\text{-}3\text{-}13)$$

$$PR^2 = r^2 - FR^2$$

and

$$FR = a\cos u - c = a(\cos u - \epsilon)$$

Sec. 4-3 The Ellipse

Since $r = a(1 - \epsilon \cos u)$,
$$PR^2 = a^2(1 - \epsilon^2) \sin^2 u$$
and
$$PR = a \sin u \sqrt{1 - \epsilon^2}$$

With the appropriate substitutions, Eq. (4-3-13) becomes
$$\sin v = \frac{\sin u \sqrt{1 - \epsilon^2}}{1 - \epsilon \cos u}$$

and Eq. (4-3-12) can be rewritten as
$$dv = \frac{\sqrt{1 - \epsilon^2}\, du}{1 - \epsilon \cos u} \tag{4-3-14}$$

Substituting for r and dv in Eq. (4-3-8) and changing the upper limit of integration from v to u ($u = 0$ when $v = 0$) yields
$$t = \frac{a^2\sqrt{1 - \epsilon^2}}{H} \int_0^u (1 - \epsilon \cos u)\, du = \frac{a^2\sqrt{1 - \epsilon^2}}{H} (u - \epsilon \cos u)$$

But
$$H = \sqrt{\mu p} = \sqrt{\mu a(1 - \epsilon^2)}$$

Substituting this expression for H in the preceding equation, we finally obtain an equation for the time of flight along an ellipse, namely,

$$\boxed{t = \sqrt{\frac{a^3}{\mu}} (u - \epsilon \sin u)} \tag{4-3-15}$$

Since the period P is given by
$$P = 2\pi \sqrt{\frac{a^3}{\mu}}$$

Eq. (4-3-15) can also be written as

$$\boxed{t = \frac{P}{2\pi} (u - \epsilon \sin u)} \tag{4-3-16}$$

Comparing Eq. (4-3-16) with Eq. (4-3-6), we see that the *mean anomaly* can be found from the expression

$$\boxed{M = u - \epsilon \sin u} \tag{4-3-17}$$

where u and M must be in radians. Johannes Kepler named the eccentric anomaly and the mean anomaly, and Eq. (4-3-17) is called *Kepler's equation*. The relationships among the three anomalies of an ellipse are as follows:

1. All three are zero at periapsis.
2. All three are equal to π radians at apoapsis.
3. If one anomaly lies between 0 and π radians, so will the other two. Similarly, if one anomaly lies between π and 2π radians, so will the other two. In other words, at any time all three anomalies are in the same half-plane with respect to the major axis.

Note that the mean anomaly is not a physical (i.e., a measurable) quantity, although it can be thought of as the angular position of the projection of the smaller mass as the projection moves along the auxiliary circle with a constant angular velocity.

The time of flight equation may also be written in terms of the average value of $\dot{v}$, which is referred to as the *mean motion*, has the units of radians/second (r/s), and is given the symbol n. Since the mean motion is defined as $2\pi/P$,

$$n = \sqrt{\frac{\mu}{a^3}} \quad \text{r/s}$$

Thus we have a third expression for the time of flight along an elliptical trajectory, to wit,

$$\boxed{t = \frac{u - \epsilon \sin u}{n}} \quad (4\text{-}3\text{-}18)$$

Using the definition of the mean anomaly given in Eq. (4-3-17), the three expressions for the time of flight along an ellipse can be written as

$$\boxed{t = \frac{P}{2\pi} M = \sqrt{\frac{a^3}{\mu}} M = \frac{M}{n}} \quad (4\text{-}3\text{-}19)$$

The procedure for finding the time of flight from periapsis to point A with ϵ and a known and v_A specified is:

Step 1. Find u_A (rad) from $\cos u_A = \dfrac{\epsilon + \cos v_A}{1 + \epsilon \cos v_A}$.

Step 2. Find M_A (rad) from $M_A = u_A - \epsilon \sin u_A$.

Step 3. $t_A = \dfrac{P}{2\pi} M_A = \sqrt{\dfrac{a^3}{\mu}} M_A = \dfrac{M_A}{n}$.

One way to find the time of flight (TOF) between points A and B on an ellipse, after determining M_A and M_B, is to find the respective times from periapsis to A and B and subtract one from the other. A second way is to use the expression of Eq. (4-3-7), which is rewritten and renumbered here.

Sec. 4-3   The Ellipse   67

$$\boxed{\text{TOF} = t_B - t_A = \frac{P}{2\pi}(M_B - M_A)} \qquad (4\text{-}3\text{-}20)$$

Example 4-3-1

Find the TOF from perigee to $\nu = 135°$ (2.36 rad) for a geocentric ellipse with an $\epsilon = 0.9$ and a perigee altitude of 552 km (300 nmi).
Solution $r_p = 6.930 \times 10^6$ m $= a(1 - \epsilon)$, so that $a = 6.930 \times 10^7$ m. Since the period is

$$P = 2\pi\sqrt{\frac{a^3}{\mu}} = 181{,}600 \text{ s} = 3027 \text{ min} = 50.4 \text{ h}$$

$P/2\pi$ and the mean motion n are

$$\sqrt{\frac{a^3}{\mu}} = 28{,}900 \text{ s} \qquad \text{and} \qquad n = 3.460 \times 10^{-5} \text{ r/s}$$

The eccentric anomaly can be found from Eq. (4-3-9),

$$u = \cos^{-1}\frac{0.9 + \cos 135°}{1 + 0.9 \cos 135°} = 1.012 \text{ rad} = 57.96°$$

and the mean anomaly from Eq. (4-3-17):

$$M = 1.012 - 0.9 \sin 1.012 = 0.2489 \text{ rad}$$

We can now calculate the TOF from the appropriate expression in Eq. (4-3-19) to be

$$\text{TOF} = 28{,}900 \times 0.2489 = 7193 \text{ s} = 119.9 \text{ min} = 1.998 \text{ h}$$

Notice that all three anomalies lie between 0 and π radians. Also notice that the TOF (1.998 h) is considerably less than one-half of the period (25.2 h), the time to go from periapsis to apoapsis (where $\nu = 180°$), indicating how much the velocity drops off as apoapsis is approached.

Example 4-3-2

Now find the time to go from A at $\nu = 135°$ (2.36 rad) to B at 220° (3.84 rad) on the ellipse of Example 4-3-1.
Solution

$$u_B = \cos^{-1}\frac{0.9 + \cos(220°)}{1 + 0.9 \cos(220°)} = 1.125 \text{ rad} = 64.45°$$

Since this value of u_B (1.125 rad) does not lie on the same side of the major axis as does ν_B (3.84 rad), *it must be incorrect* and should be replaced by

$$u_B = 2\pi - 1.125 = 5.158 \text{ rad} = 295.6°$$

Consequently,

$$M_B = 5.158 - 0.9 \sin 5.158 = 5.97 \text{ rad} = 342.1°$$

and

$$t_B = 28{,}900 \times 5.97 = 172{,}500 \text{ s} = 2875 \text{ min} = 47.92 \text{ h}$$

Therefore, the TOF between points A and B is

$$\text{TOF} = t_B - t_A = 172{,}500 - 7193 = 165{,}300 \text{ s} = 2755 \text{ min} = 45.9 \text{ h}$$

The TOF between the two points could have been found directly from Eq. (4-3-7) to be

$$\text{TOF} = 28{,}900(5.97 - 0.2489) = 165{,}300 \text{ s}$$

Notice that it takes 45.9 h to travel from point A to point B, whereas it only took 2 h to travel from periapsis to point A. Although the difference in the true anomalies of A and B is only 85°, the two points lie on either side of apoapsis, where the velocity is at its lowest.

Example 4-3-3

Let us find the TOF of the fast transfer trajectory of Example 3-6-5, which had an $\epsilon = 0.863$ and an $a = 49 \times 10^6$ m. This was a transfer from a 322-km LEO to a GEO and the interception of the GEO occurred at $v = 144.6°$ (2.524 rad).

Solution

$$u = \cos^{-1} \frac{0.863 + \cos 2.524}{1 + 0.863 \cos 2.524} = 1.41 \text{ rad} = 80.7°$$

Since this value of u lies on the same side of the major axis as does v, it is acceptable and

$$M = 1.41 - 0.863 \sin 1.41 = 0.558 \text{ rad} = 32°$$

M is also on the same side of the major axis, as it should be. Now to find the period of the transfer ellipse,

$$P = 2\pi \sqrt{\frac{a^3}{\mu}} = 107{,}900 \text{ s} = 1798 \text{ min} = 30 \text{ h}$$

Therefore, the TOF for the fast transfer is

$$\text{TOF} = \frac{P}{2\pi} M = 17{,}170 \times 0.558 = 9581 \text{ s} = 159.7 \text{ min} = 2.66 \text{ h}$$

The Hohmann transfer for this problem would have a semimajor axis $a = 24.47 \times 10^6$ m and a period of 10.6 h. The transfer time would be half of the period or 5.3 h. The fast transfer, therefore, has halved the transfer time but with a 54% increase in the required ΔV_T as calculated in Example 3-6-5.

4-4 THE HYPERBOLA

The hyperbola is characterized by $\epsilon > 1$ and a positive specific energy ($E > 0$). Since $E = -\mu/2a$, a must be negative and since p is a physical distance, it is always positive and may be expressed as

$$p = a(1 - \epsilon^2) = -a(\epsilon^2 - 1) > 0 \quad (4\text{-}4\text{-}1)$$

and r is still

$$r = \frac{p}{1 + \epsilon \cos v}$$

Sec. 4-4 The Hyperbola

Thus the true anomaly is still

$$v = \cos^{-1}\left[\frac{1}{\epsilon}\left(\frac{p}{r} - 1\right)\right] \tag{4-4-2}$$

It is possible to develop a time-of-flight expression for the hyperbola by using geometry and trigonometry as was done for the ellipse. Instead of the auxiliary circle used to define the eccentric anomaly of an ellipse, an equilateral hyperbola, which has an $\epsilon = 1.414$ and the same semimajor axis as the hyperbola of interest, is used to define the hyperbolic eccentric anomaly, which is given the symbol F. Referring to Fig. 4-4-1, if at any point P on a hyperbola, a line is drawn through P perpendicular to the major axis and extended to intersect the equilateral hyperbola at Q, the *hyperbolic eccentric anomaly F* is the angle between the major axis and the line OQ.

Rather than actually deriving the time-of-flight expression, we shall accept and use the relationship that $u = \pm iF$, where $i = \sqrt{-1}$. Since u is defined from 0 to 2π radians and F is defined from minus to plus infinity, the $\pm$ sign is necessary with the choice determined by physical reasoning. In the elliptical relationships, u is replaced by $-F$, $\sin u$ is replaced by $\sinh F$, and $\cos u$ is replaced by $\cosh F$. The resulting expressions are

$$\boxed{\cosh F = \frac{\epsilon + \cos v}{1 + \epsilon \cos v}} \tag{4-4-3}$$

$$\boxed{M = \epsilon \sinh F - F} \tag{4-4-4}$$

It should be noted that with hyperbolas it is not necessary for v, F, and M to be in the same half-plane with respect to the major axis; M can be quite large.

$$\boxed{t = \sqrt{\frac{(-a)^3}{\mu}}\, M} \tag{4-4-5}$$

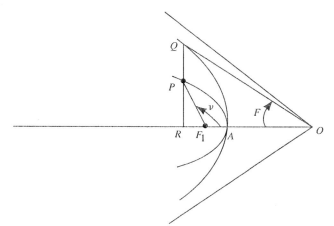

Figure 4-4-1. Equilateral hyperbola and hyperbolic eccentric anomaly F.

In addition,

$$r = a(1 - \epsilon \cosh F) \qquad (4\text{-}4\text{-}6)$$

Example 4-4-1

Example 3-5-1 involved the sighting of a meteor approaching the Earth at an altitude of 92,000 km ($r = 98.37 \times 10^6$ m) with a velocity of 9150 m/s and an elevation angle of $-80°$. We found the trajectory to be a hyperbola with $\epsilon = 3.55$, $E = +3.78 \times 10^7$ m²/s², $H = 15.6 \times 10^{10}$ m²/s², $a = -5.27 \times 10^6$ m, and $r_p = 13.44 \times 10^6$ m. Let us find the time to closest passage.

Solution To find ν, we need p. Using Eq. (4-4-1) gives

$$p = 5.27 \times 10^6 [(3.55)^2 - 1] = 61.1 \times 10^6 \text{ m}$$

or

$$p = \frac{H^2}{\mu} = 61 \times 10^6 \text{ m}$$

From Eq. (4-4-2),

$$\nu = \cos^{-1}\left[\frac{1}{3.55}\left(\frac{p}{r} - 1\right)\right] = 1.68 \text{ rad} = 96.1°$$

Figure 4-4-2 shows that for the incoming meteor, the true anomaly at the sighting is greater than 180° (π radians) and is equal to $(360° - 96.1°)$ or 263.9° (4.6 rad). Since the trajectory is symmetrical, the time from 263.9° to periapsis and the time from periapsis to 96.1° are identical and it does not matter which value of ν we use. Using the smaller value and Eq. (4-4-3), we have

$$F = \cosh^{-1} \frac{3.55 + \cos 1.68}{1 + 3.55 \cos 1.68} = 2.41 \text{ rad}$$

The eccentric anomaly M can be found from Eq. (4-4-4) to be

$$M = 3.55 \sinh 2.41 - 2.41 = 17.19 \text{ rad}$$

The time to closest passage, therefore, is

$$\text{TOF} = \sqrt{\frac{(-a)^3}{\mu}} M = 10{,}420 \text{ s} = 173.7 \text{ min} = 2.9 \text{ h}$$

There would not have been much time for a warning if the meteor had been on a collision course with the Earth.

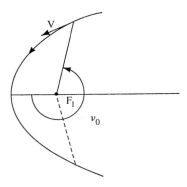

Figure 4-4-2. Incoming meteor of Example 4-4-1.

Sec. 4-4   The Hyperbola

Example 4-4-2

A spacecraft is launched from the surface of the Earth with a booster burnout velocity of 12,190 m/s and a zero burnout elevation angle, $\phi_{bo} = 0°$. Find the time to "escape" to $r = 500{,}000$ nmi (9.2×10^8 m) and the velocity at that distance. Compare the latter with the residual velocity (V_∞) at $r = \infty$.

Solution   The burnout velocity is the velocity at the start of the Keplerian trajectory. A burnout elevation angle of 0° indicates that burnout occurs at perigee. If the burnout altitude is "small" with respect to the radius of the Earth,

$$r_p \cong r_E = 6.378 \times 10^6 \text{ m}$$

$$H = 12{,}190 r_E = 7.775 \times 10^{10} \text{ m}^2/\text{s}$$

$$p = \frac{H^2}{\mu} = 15.16 \times 10^6 \text{ m}$$

The energy of the trajectory at burnout can be found from Eq. (2-8-4) to be

$$E = \frac{(12{,}190)^2}{2} - \frac{3.986 \times 10^{14}}{6.378 \times 10^6} = +11.8 \times 10^6 \text{ m}^2/\text{s}^2$$

Since $E > 0$, the trajectory is a hyperbola and the spacecraft will escape. Knowing E, the semimajor axis, which is negative, can be found to be

$$a = \frac{-\mu}{2E} = -16.89 \times 10^6 \text{ m}$$

and the eccentricity is

$$\epsilon = \sqrt{1 + \frac{2EH^2}{\mu^2}} = 1.38 > 1$$

Rearranging Eq. (2-8-1) to solve for ν at $r = 9.2 \times 10^8$ m yields

$$\nu = \cos^{-1}\left[\frac{1}{1.38}\left(\frac{15.16 \times 10^6}{9.2 \times 10^8} - 1\right)\right] = 2.364 \text{ rad} = 135.5°$$

With ν known, the eccentric anomaly is

$$F = \cosh^{-1}\frac{1.38 + \cos 2.364}{1 + 1.38 \cos 2.364} = 4.387 \text{ rad}$$

and the mean anomaly is

$$M = 1.38 \sinh 4.387 - 4.387 = 51.08 \text{ rad}$$

and

$$\sqrt{\frac{(-a)^3}{\mu}} = 3477 \text{ s}$$

$$\text{TOF} = 3477 \times 51.08 = 177{,}600 \text{ s} = 49.3 \text{ h} = 2.06 \text{ days}$$

When $r = 9.2 \times 10^8$ m, V can be found from Eq. (2-8-11) to be

$$V = \sqrt{2\left(11.8 \times 10^6 + \frac{3.986 \times 10^{14}}{9.2 \times 10^8}\right)} = 4946 \text{ m/s}$$

When r reaches infinity, E will equal V_∞^2, so that

$$V_\infty = \sqrt{2E} = 4858 \text{ m/s}$$

which is only 1.7% less than V at 500,000 nmi. Implicit in this solution is the use of a single-impulse Oberth maneuver for escape. If a two-impulse escape (a parabola plus a second ΔV) were used, V_∞ would be reduced considerably. With $V_{esc} = 11,180$ m/s,

$$V_\infty = 12,190 - 11.180 = 1010 \text{ m/s}$$

4-5 LAMBERT'S THEOREM

The time-of-flight expressions of the preceding sections require a knowledge of the characteristics of the trajectory, characteristics that are often not known for an actual trajectory. In 1761, Johann Heinrich Lambert (1728–1777) postulated a time-of-flight theorem that bears his name. *Lambert's theorem* states that the time to traverse an arc on a conic trajectory is a function only of the semimajor axis (a), the sum of the position radii to the endpoints of the arc traversed ($r_1 + r_2$) and of the chord length of the arc (d). This theorem provides a means of determining the characteristics of a trajectory with data from two observations of a spacecraft or smaller body. Incidentally, the theorem says nothing about the eccentricity of the trajectory, which seems a bit surprising.

Consider Fig. 4-5-1. If A and B are points on an ellipse, the TOF between A and B, using Eq. (4-3-7) and Kepler's equation for the mean anomaly [Eq. (4-3-17)], can be written as

$$\text{TOF} = t_B - t_A = \sqrt{\frac{a^3}{\mu}} \left[(u_B - u_A) - \epsilon (\sin u_B - \sin u_A) \right] \qquad (4\text{-}5\text{-}1)$$

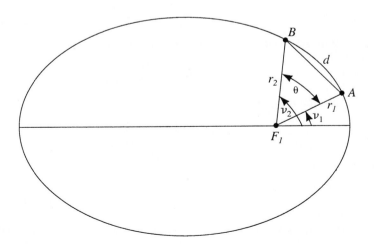

Figure 4-5-1. Ellipse with two observations of a moving body.

Sec. 4-5 Lambert's Theorem

To use Eq. (4-5-1), we need to know ϵ as well as the values of v_B and v_A, which Lambert's theorem does not presuppose. A formulation by Joseph Louis Lagrange (1736–1813) that uses only the conditions of Lambert's theorem can be developed by trigonometric manipulations that are somewhat tedious and will be omitted here. The upshot is the definition of two new auxiliary parameters, α and β, that are defined by the relationships that

$$\cos \alpha = 1 - \frac{r_1 + r_2 + d}{2a} \tag{4-5-2a}$$

$$\cos \beta = 1 - \frac{r_1 + r_2 - d}{2a} \tag{4-5-2b}$$

In these equations, a is the semimajor axis and d is the chord length of the arc between A and B; d can be found by applying the law of cosines,

$$d^2 = r_1^2 + r_2^2 - 2r_1 r_2 \cos \theta \tag{4-5-3}$$

Using the identity that $\cos \alpha = 1 - 2\sin^2(\alpha/2)$, Eqs. (4-5-2) can also be written as

$$\sin \frac{\alpha}{2} = \frac{1}{2}\sqrt{\frac{r_1 + r_2 + d}{a}} \tag{4-5-4a}$$

$$\sin \frac{\beta}{2} = \frac{1}{2}\sqrt{\frac{r_1 + r_2 - d}{a}} \tag{4-5-4b}$$

We see that we do not need to know ϵ in order to find α and β. The arc length d is found from the given data of the problem, usually using the law of cosines.

Using the auxiliary parameters, the TOF expression can be written as

$$\boxed{\text{TOF} = t_B - t_A = \sqrt{\frac{a^3}{\mu}}\left[(\alpha - \sin \alpha) - (\beta - \sin \beta)\right]} \tag{4-5-5}$$

It can also be shown that the auxiliary parameters are related to the eccentric anomalies and the eccentricity by the expressions

$$\alpha - \beta = u_B - u_A \tag{4-5-6a}$$

$$\cos \frac{\alpha + \beta}{2} = \epsilon \cos \frac{u_B + u_A}{2} \tag{4-5-6b}$$

These expressions can also be used to check solutions obtained with Eq. (4-5-5).

Example 4-5-1

Two observations of a spacecraft in a geocentric elliptical orbit were made; the observations were 90° apart. The altitude at the first sighting was 2298 km and at the second sighting was 6476 km. Somehow we know that the semimajor axis is 12×10^6 m. Find the elapsed time between the two observations. Then find the eccentricity, the time of perigee passage with respect to the first observation, the altitude at perigee, and the true anomalies of the two observations.

Solution

$$r_1 = r_E + h_1 = 8.676 \times 10^6 \text{ m}$$

$$r_2 = r_E + h_2 = 12.85 \times 10^6 \text{ m}$$

$$d^2 = r_1^2 + r_2^2 - 2r_1 r_2 \cos(90°) = 2.40 \times 10^{14} \text{ m}^2$$

and

$$d = 15.50 \times 10^6 \text{ m}$$

Using Eqs. (4-5-4) yields

$$\alpha = 2.14 \text{ rad} \quad \text{and} \quad \beta = 0.724 \text{ rad}$$

Since

$$\sqrt{\frac{a^3}{\mu}} = 2082 \text{ s}$$

$$\text{TOF} = 2082[(2.14 - \sin 2.14) - (0.724 - \sin 0.724)]$$

$$= 2573 \text{ s} = 42.9 \text{ min}$$

There is no simple closed-form method for finding the eccentricity, rather there are a series of trigonometric and algebraic relationships to be solved. Accepting the validity of Eq. (4-5-6a) that $u_B - u_A = \alpha - \beta$ and setting $(\alpha - \beta)$, which we now know, equal to ψ, then

$$u_B = \psi + u_A \tag{4-5-7}$$

and

$$\cos u_B = \cos(u_B + \psi) = \cos u_A \cos \psi - \sin u_A \sin \psi \tag{4-5-8}$$

where

$$\psi = 1.416 \text{ rad}$$

Dividing Eq. (4-5-8) by $\cos u_A$ yields the relationship that

$$\frac{\cos u_B}{\cos u_A} = \cos \psi - \sin \psi \tan u_A \tag{4-5-9}$$

Solving Eq. (4-5-9) for $\tan u_A$ yields

$$\tan u_A = \frac{1}{\sin \psi}\left(\cos \psi - \frac{\cos u_B}{\cos u_A}\right) \tag{4-5-10}$$

Since

$$r = a(1 - \epsilon \cos u) \tag{4-5-11}$$

$$\cos u = \frac{a - r}{a\epsilon} \tag{4-5-12}$$

and

$$\frac{\cos u_B}{\cos u_A} = \frac{a - r_2}{a - r_1} \tag{4-5-13}$$

Sec. 4-5   Lambert's Theorem

Using Eq. (4-5-13), Eq. (4-5-10) becomes

$$\tan u_A = \frac{1}{\sin \psi}\left(\cos \psi - \frac{a - r_2}{a - r_1}\right) \quad (4\text{-}5\text{-}14)$$

Substituting the known values into Eq. (4-5-14), yields

$$\tan u_A = 0.4148$$

and

$$u_A = 0.393 \text{ rad} = 22.5°$$

Returning to Eq. (4-5-12) and solving for the eccentricity yields

$$\boxed{\epsilon = \frac{a - r}{a \cos u}} \quad (4\text{-}5\text{-}15)$$

Substituting for a and for r and u at one of the points, say at A, shows that for this example

$$\epsilon = \frac{12 \times 10^6 - 8.676 \times 10^6}{12 \times 10^6 \cos 0.393} = 0.30$$

As a check of our work so far, let us find u_B from Eq. (4-5-13) to be 1.809 rad. Therefore, $M_B = u_B - \epsilon \sin u_B = 1.517$ rad and $M_A = 0.278$ rad. Consequently,

$$\text{TOF} = 2082(1.517 - 0.278) = 2580 \text{ s} = 43 \text{ min}$$

which differs from the previous answer by less than 1%. With M_A known, the time from perigee passage to the first observation (t_A) is

$$t_A = 2082 \times 0.278 = 579 \text{ s} = 9.65 \text{ min}$$

So obviously the spacecraft passed through perigee 9.6 min prior to the first observation.

Since $r_p = a(1 - \epsilon) = 8.4 \times 10^6$ m, $h_p = 2022$ km. The true anomalies can be found from the expression that

$$\nu = \cos^{-1}\left[\frac{1}{\epsilon}\left(\frac{p}{r} - 1\right)\right]$$

Since $p = a(1 - \epsilon^2) = 10.92 \times 10^6$ m, $\nu_A = 0.531$ rad $= 30.4°$, and $\nu_B = 2.095$ rad $= 120°$ and the calculated angle between the two sightings, θ, is 89.6° or approximately 90°.

As a further check using Eqs. (4-5-6), $u_B - u_A = 1.416$ rad and $\alpha - \beta = 1.416$ rad. Furthermore,

$$\cos \frac{\alpha + \beta}{2} = 0.138 \quad \text{and} \quad \epsilon \cos \frac{u_B + u_A}{2} = 0.136$$

Example 4-5-1 was unrealistic in that the value of the semimajor axis was somehow known and used to find the time of flight between the two observations. In practice, the times of the two observations are known (and thus the time of flight) and Lambert's theorem is used to find the semimajor axis and then the other orbital parameters. This is the process of *orbit determination* (the Gauss problem). Unfortunately, there are no closed-form solutions for finding the semimajor axis, and iterative solutions are necessary in which trial values of the semimajor axis are used to find values for the TOF that can be compared with the known value.

Let us redo Example 4-5-1 with the time and angle between observations known as well as the distances to the orbiting object of interest.

Example 4-5-2

Two observations of an orbiting object were made 42.9 min and 90° apart. The altitude of the first observation was 2298 km and of the second was 6476 km. Find a, ϵ, r_p, and the time of perigee passage, as well as the velocity and elevation angle of the orbiter at the first observation.

Solution

$$r_1 = 8.676 \times 10^6 \text{ m}$$

$$r_2 = 12.85 \times 10^6 \text{ m}$$

$$d = 15.5 \times 10^6 \text{ m}$$

$$r_1 + r_2 + d = 37.02 \times 10^6 \text{ m} \quad \text{and} \quad r_1 + r_2 - d = 6.026 \times 10^6 \text{ m}$$

The procedure is to estimate the first trial value of a. One possibility would be $(r_1 + r_2)/2$, which in this case would be a value of 10.8×10^6 m. Using Eqs. (4-5-4), the values of α and β are 2.36 and 0.765 rad, respectively, and the inverse of the mean motion $(1/n)$ is 1778 s. Therefore, from Eq. (4-5-5), the TOF = 2814 s = 46.9 min, which is greater than the observed value of 43.5 min.

For a second trial value, let us choose $a = 13 \times 10^6$ m. The corresponding values of α and β are 2.0 rad and 0.694 rad, respectively, and $1/n = 2348$ s. The resulting TOF is 2433 s or 40.6 min, which is less than the observed value.

Since the correct value of a has been bracketed, it lies somewhere between these two trial values. With great prescience, let us choose a value of 12×10^6 m, which gives values of 2.14 and 0.724 rad for α and β, respectively, and $1/n = 2082$ s, so that the TOF is 2573 s = 42.9 min, the observed time. With the value of the semimajor axis established, the relationships of Example 4-5-1 can be used to find $\epsilon = 0.3$, $r_p = 8.4 \times 10^6$ m, and perigee passage 9.4 min before the first observation.

We need the orbital energy if we are to determine any velocities.

$$E = \frac{-\mu}{2a} = -16.61 \times 10^6 \text{ m}^2/\text{s}^2$$

$$V_p = \sqrt{2\left(E + \frac{\mu}{r_p}\right)} = 7854 \text{ m/s}$$

$$H = r_p V_p = (8.4 \times 10^6) \times 7854 = 6.597 \times 10^{10} \text{ m}^2/\text{s}$$

$$V_1 = \sqrt{2\left(E + \frac{\mu}{r_1}\right)} = 7659 \text{ m/s}$$

$$\phi_1 = \cos^{-1}\frac{H}{V_1 r_1} = 6.89°$$

Sec. 4-5 Lambert's Theorem

Lambert's theorem is true for all conic trajectories, for parabolas and hyperbolas as well as for ellipses. With respect to the parabola, as the semimajor axis a becomes large, α and β become "small." Applying the small angle approximation to Eqs. (4-5-4) yields the approximations that

$$\alpha^2 \cong \frac{r_1 + r_2 + d}{a} \qquad (4\text{-}5\text{-}16)$$

$$\beta^2 \cong \frac{r_1 + r_2 - d}{a} \qquad (4\text{-}5\text{-}17)$$

Without proof, the time of flight between points A and B on a *parabola* can be expressed as

$$\text{TOF} = t_B - t_A = \frac{\alpha^3 - \beta^3}{6n} \qquad (4\text{-}5\text{-}18)$$

Since the mean motion n is $(\mu/a^3)^{1/2}$, Eq. (4-5-18) can be written in expanded form as

$$t_B - t_A = \frac{1}{6\sqrt{\mu}}[(r_1 + r_2 + d)^{3/2} - (r_1 + r_2 - d)^{3/2}] \qquad (4\text{-}5\text{-}19)$$

Example 4-5-3

Two observations of an object in a parabolic trajectory were made 170° apart with the respective distances from the center of the Earth being 6.37×10^6 m and 9.26×10^8 m. Find the time between observations and r_p.

Solution With $\theta = 170° = 2.97$ rad, the law of cosines gives a chord length $d = 9.32 \times 10^8$ m, so that $(r_1 + r_2 + d) = 18.6 \times 10^8$ m and $(r_1 + r_2 - d) = 3.70 \times 10^5$ m. Substituting these values into Eq. (4-5-19) yields

$$\text{TOF} = 6.70 \times 10^5 \text{ s} = 11{,}200 \text{ min} = 186 \text{ h} = 7.75 \text{ days}$$

Since r_p of a parabola is equal to $p/2$, we need a value for the semilatus rectum p in order to find r_p. Since a is infinite, we cannot use the relationship that $p = a(1 - \epsilon^2)$, nor can we use $p = H^2/\mu$ since we have no way to find H. From Battin,[1] however, it is possible to derive an expression for p, *the semilatus rectum of a parabola* as a function of r_1, r_2, and θ; namely,

$$\boxed{p = \frac{r_1 r_2 (1 - \cos\theta)}{r_1 + r_2 - 2\cos(\theta/2)\sqrt{r_1 r_2}}} \qquad (4\text{-}5\text{-}20)$$

Substituting appropriate values into Eq. (4-5-20) gives a value of 12.74×10^6 m for p and an $r_p = 6.37 \times 10^6$ m, which we recognize is the value of r_1, which in turn is the radius of the Earth. You may have recognized this problem as a restatement of Example 4-2-2.

---

[1] Listed in the Selected References at the end of the book.

You may be wondering how we can tell from these two observations that the trajectory is a parabola. The answer is that we cannot. It would be more logical to assume that the trajectory is elliptical and to search for the value of a using the techniques illustrated in Example 4-5-2. If we use 5×10^8 m, approximately equal to $(r_1 + r_2)/2$, for the first trial value of a, $\alpha = 2.6$ rad, $\beta = 0.136$ rad, and TOF $= 1.2 \times 10^6$ s, which is greater than the observed TOF of 6.70×10^5 s. If a is greatly increased to 5×10^{12} m for the second try, we find that α and β become very small and that many significant figures have to be maintained. For this value of a with $\alpha = 0.0192824$ and $\beta = 0.0011267$, the resulting TOF is 6.695×10^5 s, which appears to be identical to that obtained using the parabolic TOF expression. Using Eq. (4-5-14) to find $u_A = -0.0946$ and then using Eq. (4-5-15), we find $\epsilon = 1.004$.

For *hyperbolic orbits*, the sine functions are replaced by the hyperbolic sine functions, so that

$$\sinh \frac{\alpha}{2} = \frac{1}{2}\sqrt{\frac{r_1 + r_2 + d}{-a}} \tag{4-5-21}$$

$$\sinh \frac{\beta}{2} = \frac{1}{2}\sqrt{\frac{r_1 + r_2 - d}{-a}} \tag{4-5-22}$$

$$\text{TOF} = t_B - t_A = \sqrt{\frac{(-a)^3}{\mu}}[(\sinh \alpha - \alpha) - (\sinh \beta - \beta)] \tag{4-5-23}$$

Example 4-5-4

A meteor was sighted at a distance of 98.37×10^6 m from the center of the Earth. The distance at a second sighting was 13.40×10^6 m and the two observations were $96.1°$ apart. Find the time between observations if somehow we know that $a = -5.26 \times 10^6$ m.

Solution With $\theta = 96.1° = 1.68$ rad, the law of cosines shows the arc length to be

$$d = 10^8 \text{ m}$$

Now

$$r_1 + r_2 + d = 2.125 \times 10^8 \text{ m} \quad \text{and} \quad r_1 + r_2 - d = 1.105 \times 10^7 \text{ m}$$

It is now possible to find $\alpha = 3.75$ rad and $\beta = 1.346$. Substituting these values into Eq. (4-5-22), we find that

$$\text{TOF} = 10{,}270 \text{ s} = 171 \text{ min} = 2.85 \text{ h}$$

This example is a rewrite of Example 4-4-1, in which we found the time to travel this distance, using Kepler's equation, to be 10,420 s for a difference of 1.5%.

A more realistic problem would be to know the time between the observations and to find the value of the semimajor axis by trial and error, as was done for the elliptical orbit, until the calculated TOF matches the known TOF.

4-6 CLOSING REMARKS

If the characteristics of a trajectory or orbit are known, it is a simple matter to apply the relationships of Sections 4-2, 4-3, and 4-4 to obtain the time of flight (TOF) between any two points.

In the orbit determination problems of Section 4-5, however, not only are there no closed-form solutions, but also the choice of the formulations to be used is dependent on the type of trajectory, which is unknown at the start of the problem. With respect to the latter, the equations have been reformulated by the use of an auxiliary (universal) variable other than the eccentric anomaly[2] so that a single time-of-flight equation is applicable to all conic orbits. Another consideration in orbit determination is the fact that our equations and relationships are based on position vectors originating at the center of the Earth and on angles between such vectors, whereas the position of an orbiting body is measured with respect to the surface of a rotating Earth. Consequently, coordinate transformations are needed to obtain the inertial position vectors and angular displacements with respect to the center of the Earth, which is the origin of our inertial reference system. Furthermore, there are additional complications such as the actual shape of the Earth, measurement and instrument errors, and atmospheric diffractions.

PROBLEMS

4-1. Using Eq. (4-3-1), develop the expression for the period of a circle.

4-2. A geocentric ellipse has a perigee altitude of 550 nmi (1012 km) and an $\epsilon = 0.560$. What is the period?

4-3. Find the period of a geocentric ellipse with a semilatus rectum $p = 334{,}320$ nmi and an $\epsilon = 0.990$.

4-4. Referring to Example 4-4-2, find the TOF to escape to $r = 1.5 \times 10^6$ nmi and V at that distance. Compare V with V_∞.

4-5. A rescue mission to a space station in a 300-nmi circular orbit starts with a launch with perigee at the surface of the Earth and crosses the space station orbit with $V = 9000$ m/s and an elevation angle of 15°.
(a) Describe the trajectory.
(b) What is the TOF to the space station orbit? How does it compare with the time of a Hohmann transfer?
(c) What launch velocity is required?

4.6. Do the rescue mission of Problem 4-5 with a hyperbolic trajectory with an $\epsilon = 1.8$. Find V and ϕ at the interception of the space station orbit.

4-7. Do Problem 4-6 for a parabolic rescue trajectory.

4-8. A ballistic missile is detected at burnout entering a symmetrical trajectory with $V_{bo} = 6400$ m/s, $\phi_{bo} = +20°$, and $r_{bo} = 6.47 \times 10^6$ m.
(a) Find the characteristics of the trajectory.

[2] For further information, see Bate et al. listed in the Selected References.

(b) What is the apogee altitude and velocity? Locate perigee.
(c) How long does it take the missile to travel from burnout to the corresponding location on the incoming leg? In other words, how much warning is there?
(d) Can you determine the range (nmi) along the surface of the Earth?

4-9. Do Problem 4-8 for $V_{bo} = 6700$ m/s, $\phi_{bo} = +15°$, and $r_{bo} = 6.4 \times 10^6$ m.

4-10. An incoming object is sighted at an altitude of 20,000 nmi with a $V = 8000$ m/s and a $\phi = -65°$.
(a) What type of an object might it be?
(b) Will it impact or fly by?
(c) What is the time to impact or closest passage?

4-11. Do Problem 4-10 for an altitude of 400 nmi, $V = 5500$ m/s, and $\phi = 0°$.

4-12. A geocentric ellipse has a semimajor axis $a = 12 \times 10^6$ m and $\epsilon = 0.450$. Find the time to travel from $\nu = 30°$ to $\nu = 140°$ and the apogee and perigee altitudes.

4-13. A geocentric ellipse has a semimajor axis $a = 15 \times 10^6$ m and $\epsilon = 0.630$. Find the time to travel from $\nu = 200°$ to $\nu = 260°$. Find the altitude at perigee and at apogee.

4-14. Two observations of an object were made 30 min and 60° apart. The first altitude was 2500 nmi (4600 km) and the second was 3500 nmi (6440 km).
(a) Identify the trajectory and determine its approximate characteristics.
(b) Find the approximate velocity and elevation angle of the object at the second observation.

CHAPTER 5

Interplanetary Transfers

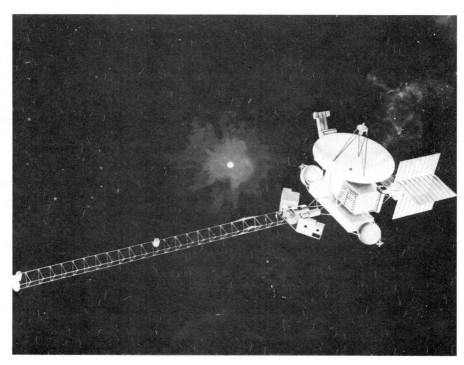

Artist's concept of a solar polar explorer, such as Ulysses, as it approaches the Sun. (Courtesy of NASA.)

5-1 INTRODUCTION

Now that we can move around in near-Earth (inner) space with a certain amount of confidence and dexterity and know how to escape, let us consider traveling outside the gravitational attraction of the Earth. Let us move out into outer space and apply and extend what we have learned to interplanetary travel. This is a busy chapter with many interesting ideas and applications.

The capabilities of current propulsion systems limit travel for the foreseeable future to our solar system. So in the next section we provide a brief description of the solar system and of the simplified model that we use in this book. In the section following we discuss the concept of the sphere of influence of a planet that allows the sequential extension of two-body orbital mechanics to travel between planets; an empirical expression for defining a planetary sphere of influence is also developed. Concomitant with the sphere of influence are the ideas of a collision cross section and an impact parameter that determines whether an incoming space object will hit or miss a planet.

In the next section we describe the patched conic concept, which is applicable to high-thrust (impulsive) propulsion and is the basis of two-body interplanetary travel in which a space object moves from one planetary sphere of influence to another along a heliocentric trajectory. In this section we also introduce the idea of the velocity budget, which is the total ΔV required to accomplish a mission and is a measure of the energy that must be imparted to the spacecraft. It is this velocity budget that determines the choice of propulsion systems and the size and number of stages of the boost vehicles (boosters) that place the spacecraft on its desired trajectories.

In the section following we look at missions whose objective is to examine another planet, either by landing on its surface or by orbiting it. This can be thought of as planetary capture.

When a spacecraft (or other object) enters a planetary sphere of influence without capture, it will leave the sphere of influence with its trajectory modified. This is described as a planetary passage or flyby and the mechanism that modifies the

Sec. 5-2   The Solar System

trajectory is referred to as a gravity assist, a technique whose popularity and use is increasing. A section is devoted to planetary passage and gravity assist.

Since the planets are moving with respect to each other, a destination planet is a moving target and the section following deals with the intercept and rendezvous problems and with the determination of lead angles and launch windows. Although Hohmann transfers are minimum-energy transfers requiring the lowest ΔV_T values, they are also the slowest and can severely restrict the times and conditions for initiating a successful transfer. Consequently, we need to look at fast transfers and their impact on the intercept problem to include the increase in the required ΔV_T, the changes in the time of flight along the intercept (transfer) trajectory, and the effects on lead angles and launch windows.

To give a feeling of the size of actual numbers, we have used and will continue to use throughout the book dimensional numbers such as meters, meters per second, and seconds. Not only are these numbers typically large, as we have seen, but also many of the fundamental distances, such as the gravitational constant, the distance from the Earth to the Sun, and the mass of the Sun are not accurately known and are subject to refinement. Consequently, it is not unusual to run across dimensionless (normalized) units of measure. These dimensionless units, which astronomers call canonical units, are defined and described in the next-to-last section, and the chapter closes with some closing remarks.

5-2 THE SOLAR SYSTEM

The solar system is characterized by great distances and innumerable celestial bodies. Our Sun is one of the estimated 10^{11} (100 billion) stars in our galaxy, the Milky Way, which is approximately 80,000 light-years in diameter. A *light-year* is a large distance; it is the distance traveled in 1 year at the speed of light. Since the speed of light is 2.998×10^5 km/s, a light-year is on the order of 9.460×10^{12} km (5.14×10^{12} nmi), a distance that is over 63,000 times the distance from the Earth to the Sun.

Our Sun is of the order of 30,000 light-years from the center of the Milky Way and has a period of rotation about this center of over 200 million years. When we realize that our galaxy is only one of billions of galaxies, we may be able to visualize the magnitude of the universe. John von Neumann (1903–1957) is paraphrased as having asked "Where are they?", referring to life forms from the universe.

Our solar system comprises the Sun, nine planets, and their satellites (moons), asteroids, comets, meteors and meteorites, dust, and gas particles. The diameter of the solar system is approximately 60×10^8 km (3.2 billion nautical miles), which is on the order of 1/30,000 the distance to the nearest star. Because of the distances involved, the forces exerted by bodies outside our solar system are so small that it is reasonable to treat the solar system as an isolated system with no external forces. As a worst-case example, assume that the nearest star (at 4.3 light-years) has a mass that is 1000 times larger than that of the Sun; the force exerted on the Earth by the Sun will be 74 million times as large as that exerted by this hypothetical dense star.

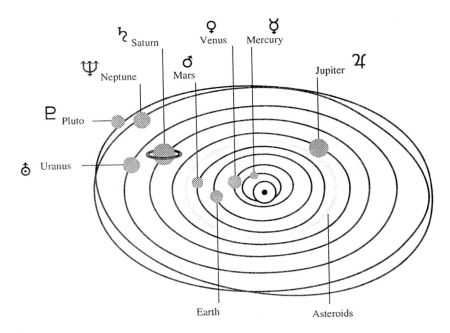

Figure 5-2-1. Solar system.

Physical and orbital data for the nine planets are listed in Tables B-1 and B-2, respectively, in Appendix B. Referring to Fig. 5-2-1 (which obviously is not to scale), the four planets nearest the Sun (Mercury, Venus, Earth, and Mars) comprise the inner solar system and the other five form the outer solar system. The planets can also be divided into two other groups: the terrestrial-type planets (Mercury, Venus, Earth, Mars, and Pluto) and the giant planets (Jupiter, Saturn, Uranus, and Neptune). The asteroids, or minor planets (planetoids), lie principally between Mars and Jupiter. There are thousands of them, with Ceres being the largest, with a diameter on the order of 420 nmi (773 km); the smallest have diameters on the order of 1 nmi (1.84 km). It is interesting to note that the combined planetary mass is on the order of a fifth of a percent (0.2%) of the solar mass and that all the other masses combined only add another 0.02% to that figure; the Sun is the dominant mass.

Examining the orbital data of Table B-2, we see that planetary orbits are elliptical, thus *verifying Kepler's first law of planetary motion*. With the exception of Mercury (the innermost planet) and Pluto[1] (the outermost planet), the eccentricities are small, less than 0.1, leading to the assumption that as a first approximation the planetary orbits are circular. The inclination listed in Table B-2 is the angle that the orbital plane of each planet makes with the *ecliptic*, which is the plane of the Earth's orbit about the Sun. Notice that again with the exception of Mercury and Pluto, the inclinations are small, less than 3.5°, leading to another assumption that

---

[1] Pluto's orbit crosses that of Neptune and has its perihelion inside Neptune's orbit.

as a first approximation the planetary orbits are coplanar. The two approximations of this paragraph lead to the establishment of a *simplified solar system in which the major planets move about the Sun in coplanar circular orbits*.

It is well to keep in mind that our solar system resembles a boulder around which pebbles at large distances revolve; the solar system is mostly space. A few numbers to illustrate the size of the solar system and the distances involved are that the diameter of the Sun is 109 times that of the Earth, its mass is 332,900 times that of the Earth, and the distance from the Earth to the Sun is on the order of 12,000 times the diameter of the Earth. If the Sun were to be represented by a tennis ball with a diameter of 6.35 cm (2.5 in.), the Earth would be a dot with a diameter of only 0.058 cm (0.023 in.) at a distance of 7.62 m (25 ft). An illustration of the flatness of the solar system (the small orbital inclinations) is the fact that if a scale model were to be placed into a circular box 1.284 m (4 ft) in diameter, the height of the box would be 12.7 cm (5 in.); omitting Pluto and Mercury would reduce the height to 7.62 cm (3 in.).

It should be noted that the data in Tables B-1 and B-2 are typical values and that the actual values are more precise and vary somewhat with time. A tabulation of the daily locations of celestial bodies as a function of time (a specific instant of time is an *epoch*) is known as the *ephemeris* and can be found in the Astronomical Almanac for the day of the year of interest. The Almanac is issued by the National Almanac Offices in the United States and Great Britain and is available in some libraries or can be purchased from the U.S. Government Printing Office or H.M. Stationers.

5-3 THE SPHERE OF INFLUENCE AND IMPACT PARAMETER

Since the mass of the Sun is over 1000 times larger than that of the largest planet, its gravitational field is dominant in the solar system except in the regions close to the individual planets where the gravitational attraction of the planet is sufficiently large to be considered the central force field. This region is called the *sphere of influence* (SOI) of the attracting body and is defined in terms of its relationship to a larger body, usually but not necessarily the Sun.

We shall now develop, by less than rigorous methods, an expression for the SOI of an attracting body with respect to another and larger attracting body. Figure 5-3-1 shows a small mass m (a spacecraft or some other object) at the edge of the yet-to-be defined sphere of influence of a larger mass m_2 with both masses in the presence of an even larger mass m_1. Applying Newton's law of gravitational attraction to the force exerted by m_2 on the smallest mass m, we have

$$F_{m,m_2} = \frac{-Gmm_2}{r_2^2} \tag{5-3-1a}$$

where r_2 is the radius of the SOI. Similarly, the force exerted on m by m_1 is

$$F_{m,m_1} = \frac{-Gmm_1}{r_1^2} \tag{5-3-1b}$$

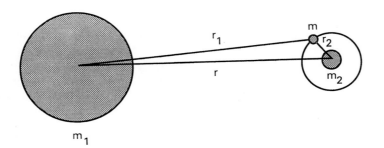

Figure 5-3-1. Sphere of influence (SOI) of m_2 with respect to m_1.

The ratio of the forces can be written as

$$\frac{F_{m_2}}{F_{m_1}} = \left(\frac{r_1}{r_2}\right)^2 \frac{m_2}{m_1} \qquad (5\text{-}3\text{-}2)$$

The magnitude of r_1 will be $\geq (r - r_2)$ and $\leq (r + r_2)$, depending on the location of m with respect to m_1 and m_2.

If we define the *sphere of influence* (sometimes referred to as the *activity sphere*) as the region in which the force exerted by the smaller mass m_2 is much greater than that exerted on m by m_1, we can think of the edge of the SOI at r_2 as being established when the ratio of the forces is approximately a tenth ($F_{m_2}/F_{m_1} \cong 0.1$). Substituting this ratio into Eq. (5-3-2), with the assumption that $r_2 \ll r$ so that $r_1 \cong r$, and then solving for r_2, we obtain an approximate expression for the radius of the SOI, namely,

$$r_2 \cong r\left(\frac{10m_2}{m_1}\right)^{0.5} \qquad (5\text{-}3\text{-}3)$$

Although Eq. (5-3-3) will give satisfactory order of magnitude values for the radius of an SOI, it is customary to use the following empirical relationship instead, whereby

$$\boxed{r_{\text{SOI}} = r\left(\frac{m_2}{m_1}\right)^{0.4}} \qquad (5\text{-}3\text{-}4)$$

where r, the distance between the masses, is the semimajor axis of the planetary orbit in the case of Sun–planet relationships.

Example 5-3-1

Let us determine the radius of the SOI for the Earth with respect to the Sun. Using Eq. (5-3-4) and astronomical units,

$$r_{\text{SOI}} = a_E\left(\frac{m_E}{m_S}\right)^{0.4} = 1 \text{ AU}\left(\frac{1}{332,900}\right)^{0.4} = 0.00618 \text{ AU}$$

where 1 AU is the distance from the Earth to the Sun. Therefore, since 1 AU = 1.496×10^{11} m, the radius of the Earth's SOI with respect to the Sun is 924,500 km

Sec. 5-3 The Sphere of Influence and Impact Parameter

(502,500 nmi), approximately 500,000 nmi. It is customary to take r_{SOI} as a practical value for r_∞, which is why a value of 920,000 km was used in Example 4-3-2 to determine how long it would take to "escape" along a parabolic trajectory. As an aside, if the approximation of Eq. (5-3-3) had been used, the value for the radius of the SOI would have been 0.00548 AU or 820,000 km (446,000 nmi).

Example 5-3-2

Find the radius of the SOI of the Moon with respect to the Earth. The radius of the Moon's orbit about the Earth is 0.0026 AU:

$$r_{SOI} = 0.0026 \text{ AU} \times (0.0123)^{0.4} = 0.000476 \text{ AU}$$

Accordingly, the SOI of the Moon with respect to the Earth extends to 66,970 km (36,400 nmi) from the center of the Moon, and the edge of the Moon's SOI is 323,100 km (175,600 nmi) from the center of the Earth.

The radii of the SOIs of all of the planets are given in Table 5-3-1 in AU, meters, and nmi. These are only order-of-magnitude values, and the so-called spheres are only approximately spherical. With the exception of Pluto, whose mass is in question, notice how large the SOIs of the outer planets are when compared with those of the inner planets, primarily because of the much larger distances from the Sun; the masses of the outer planets (with the exception of Pluto) are also larger.

When an object is leaving a planet, we will assume escape when $r = r_{SOI}$. When an object approaches a planet, we shall assume that it is influenced by the planet's gravitational attraction when $r \leq r_{SOI}$. An object (e.g., a spacecraft) will enter and leave a planetary SOI on either a parabolic trajectory ($V_\infty = 0$) or a hyperbolic trajectory ($V_\infty > 0$), with the hyperbolic trajectory being the more common case.

After entering an SOI, if no action is taken, the spacecraft will either hit (impact) the planet or execute a flyby (parabolic or hyperbolic) and leave the SOI. We can use the *impact parameter* to determine whether there will be an impact or a flyby.

To define the impact parameter, let us look at a hyperbolic trajectory that will just graze the surface of the planet at the periapsis of the trajectory (i.e., $r_p \equiv r_{planet}$),

TABLE 5-3-1 RADIUS OF SPHERE OF INFLUENCE (SOI)

Planet	AU	Meters	Nautical miles
Mercury	0.000747	1.117×10^8	60,710
Venus	0.00411	6.163×10^8	334,900
Earth	0.00618	9.245×10^8	502,500
Mars	0.00386	5.781×10^8	314,200
Jupiter	0.3222	4.820×10^{10}	26,200,000
Saturn	0.3761	5.627×10^{10}	30,580,000
Uranus	0.3457	5.172×10^{10}	28,110,000
Neptune	0.5792	8.664×10^{10}	47,090,000
Pluto	0.0222 ?	3.320×10^{10}	1,804,000

as shown in Fig. 5-3-2. Let us draw a line perpendicular to V_∞ and then a second line that both passes through the center of the planet and is perpendicular to the first line; this second perpendicular will be parallel to V_∞. The segment between V_∞ and this second line through the center of the planet is the impact parameter and is given the symbol b.

If we now draw the radius vector and the local horizon (normal to r), we can define the elevation angle ϕ and see from the figure that the angle between V_∞ and r_∞ is equal to ϕ so that $\cos \phi = b/r_\infty$ or $r_\infty \cos \phi = b$. Therefore, the specific angular momentum of the trajectory can be written as

$$H = V_\infty r_\infty \cos \phi = V_\infty b \tag{5-3-5}$$

Since H is a constant of the trajectory and $r_\infty \cos \phi = b$,

$$V_\infty b = V_p r_p \tag{5-3-6}$$

From the energy equation with $r = r_\infty = \infty$,

$$\frac{V_p^2}{2} - \frac{\mu}{r_p} = \frac{V_\infty^2}{2} \tag{5-3-7a}$$

or

$$V_p^2 = \frac{2\mu}{r_p} + V_\infty^2 \tag{5-3-7b}$$

Recognizing $2\mu/r_p$ as V_{esc}^2 from r_p, the surface of the planet in this case, Eq. (5-3-7b) can be written as

$$V_p = \sqrt{V_{esc}^2 + V_\infty^2} \tag{5-3-8}$$

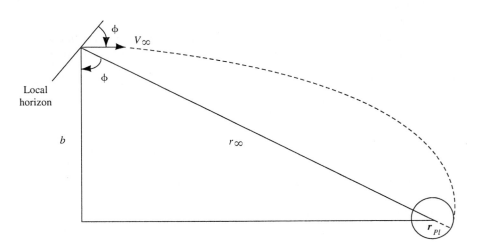

Figure 5-3-2. Grazing hyperbola and the impact parameter b.

Sec. 5-3 The Sphere of Influence and Impact Parameter

We should recognize this as the Oberth escape velocity from the planetary surface, where $r = r_p$.

Substituting Eq. (5-3-8) into Eq. (5-3-6) and solving for b, the impact parameter, yields the expression

$$b = r_{\text{planet}} \sqrt{1 + \frac{V_{\text{esc}}^2}{V_\infty^2}} \quad (5\text{-}3\text{-}9)$$

The *approach distance d* is defined as shown in Fig. 5-3-3 and as with b is the distance between V_∞ (at r_{SOI}, the edge of the SOI) and a parallel line passing through the center of the planet. In the case of the approach distance, however, the periapsis of the incoming trajectory is undefined. From the geometry, the approach distance is

$$d = r_{\text{SOI}} \cos \phi \quad (5\text{-}3\text{-}10)$$

Notice that whereas the impact parameter b is a function of V_∞, the magnitude of the residual velocity $\mathbf{V}_\infty$, the approach distance d is a function of the elevation angle ϕ, the direction of the residual velocity $\mathbf{V}_\infty$. The impact parameter is also proportional to the planetary radius and the approach distance is also proportional to the radius of the SOI.

If no action, propulsive or otherwise, is taken to modify the incoming hyperbolic trajectory, we see that:

1. If $d > b$, there will be a flyby.
2. If $d = b$, there will be a surface graze.
3. If $d < b$, there will be an impact.

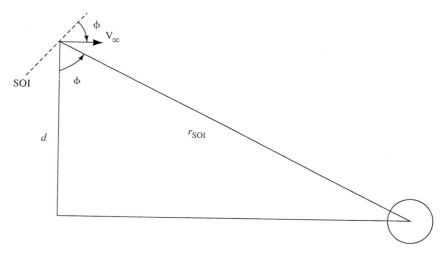

Figure 5-3-3. Approach distance d as a function of r_{SOI} and ϕ.

It is now possible to define a *collision (or capture) cross section* that is presented to the incoming spacecraft (or any other object) as sketched in Fig. 5-3-4, and whose area is a function of V_∞. We can get some idea as to the limiting magnitudes of b and the size of the collision cross section by letting V_∞ go to zero and infinity, respectively, in Eq. (5-3-9):

$$\lim_{V_\infty \to \infty} b = r_{planet}$$

indicating that an approaching spacecraft with an infinite relative velocity would miss the planet unless headed directly toward it $(d = 0)$:

$$\lim_{V_\infty \to 0} b = \infty$$

indicating that any object just reaching the SOI with $V_\infty = 0$ would eventually hit or graze the planet, on a parabolic trajectory in either case.

Example 5-3-3

An approaching spacecraft reaches the SOI of Venus with $V_\infty = 2700$ m/s and $\phi = -85°$.
(a) Find the impact parameter b (m and nmi).
(b) Find the approach distance d (m and nmi).
(c) Using d and b, determine if the spacecraft will impact or fly by.
(d) Find the perigee distance, r_p, and use it to verify part (c).
(e) Find the perigee (or impact) velocity.

Solution

(a) First, we need V_{esc}, from Table B-1 or from

$$V_{esc} = \sqrt{\frac{2 \times 3.248 \times 10^{14}}{6.052 \times 10^6}} = 10{,}360 \text{ m/s}$$

From Eq. (5-3-9),

$$b = 6.052 \times 10^6 \sqrt{1 + \frac{(10{,}360)^2}{(2700)^2}} = 2.40 \times 10^7 \text{ m} = 13{,}040 \text{ nmi}$$

(b) Using Eq. (5-3-10) and the value of the r_{SOI} for Venus from Table 5-3-1,

$$d = 0.00411 \cos(-85°) = 0.000358 \text{ AU} = 5.359 \times 10^7 \text{ m} = 29{,}120 \text{ nmi}$$

(c) Since d is greater than b, the spacecraft will miss the surface of Venus and unless something is done will execute a flyby and leave the SOI of Venus.

(d) $$H = V_\infty d = 1.447 \times 10^{11} \text{ m}^2/\text{s}$$

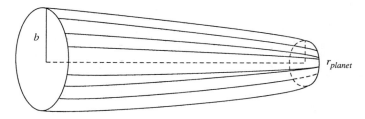

Figure 5-3-4. Collision cross section.

or

$$H = r_{SOI} V_\infty \cos \phi = 1.450 \times 10^{11} \text{ m}^2/\text{s}$$

$$E = \frac{V_\infty^2}{2} = 3.645 \times 10^6 \text{ m}^2/\text{s}^2 > 0 \quad \text{(a hyperbola, as expected)}$$

$$\epsilon = \sqrt{1 + \frac{2EH^2}{\mu^2}} = 1.564 > 1 \quad \text{(a hyperbola)}$$

$$a = \frac{-\mu}{2E} = -4.455 \times 10^7 \text{ m} < 0 \quad \text{(a hyperbola)}$$

$$r_p = a(1 - \epsilon) = +2.513 \times 10^7 \text{ m}$$

$$h_p = r_p - 6.052 \times 10^6 = 1.907 \times 10^7 \text{ m} = 10{,}360 \text{ nmi}$$

Since $r_p > r_{Venus}$ ($h_p > 0$), the spacecraft will not strike the surface of Venus, as was indicated by the fact that d is greater than b.

(e)
$$V_p = \frac{H}{r_p} = 5758 \text{ m/s}$$

As a matter of curiosity, let us find ν at r_{SOI}.

$$p = \frac{H^2}{\mu} = 6.446 \times 10^7 \text{ m}$$

$$\nu = \cos^{-1}\left[\frac{1}{\epsilon}\left(\frac{p}{r} - 1\right)\right] = 125.2°$$

Just for a comparison, ν at $r = \infty$ is $\cos^{-1}(-1/\epsilon) = 129.7°$.

In determining E, r was taken to be infinite. If r_{SOI} is used instead, E becomes 3.118×10^6 m, ϵ becomes 1.496, and h_p becomes 10,750 nmi, a 3.6% difference.

5-4 THE PATCHED CONIC

The *patched conic concept* and approximation avoids the three-body problem in the transition from one central force field to another by using a series of two-body solutions (conic sections) patched at the edge of the relevant sphere of influence. For example, a spacecraft would escape from a departure planet on a planetocentric trajectory that would become a heliocentric trajectory (with the Sun as the attracting force) as the spacecraft leaves the SOI of the departure planet. This change from a planetocentric to a heliocentric trajectory is the *first patch*. When the spacecraft intercepts the SOI of the target (or intermediate) planet, the trajectory will become planetocentric again. This is the *second patch*. It is convenient to divide a mission into phases that are connected by patches.

For example, with a Mars mission the first phase will be a geocentric hyperbola as the spacecraft escapes from the Earth's gravitational attraction. The second phase will be a heliocentric ellipse for the transfer from Earth's orbit to Mars and the third phase will be a hyperbolic approach trajectory with Mars as the attracting force (a planetocentric hyperbola). There will be two patches. At each patch (i.e., at the edge of the planetary SOIs), V_∞ ties the planetocentric and heliocentric trajectories together.

In each of the planetocentric escape trajectories, the residual velocity V_∞ is the scalar velocity of the spacecraft with respect to the planet of interest at "infinity" (the edge of the SOI) and represents the increment of velocity (ΔV) required to enter the heliocentric transfer trajectory of phase II. With a parabolic escape (not the usual case), there is no residual velocity and the magnitude of $V_\infty = 0$; another firing at infinity is needed to generate the ΔV required to enter the transfer trajectory. With a hyperbolic (Oberth) escape and injection at periapsis, the usual case, with the injection velocity V_i specified, V_∞ can be found from

$$V_\infty = \sqrt{V_i^2 - V_{esc}^2} \qquad (5\text{-}4\text{-}1)$$

It should be noted, however, that the value of V_∞ is very sensitive to variations in the injection velocity and much less sensitive to variations in the injection distance from the center of the planet. Considering an Oberth escape, taking the appropriate derivatives of Eq. (5-4-1) yields the following relationships:

$$\frac{dV_\infty}{V_\infty} = \left(\frac{V_i}{V_\infty}\right)^2 \frac{dV_i}{V_i} \qquad (5\text{-}4\text{-}2a)$$

$$\frac{dV_\infty}{V_\infty} = \frac{\mu}{rV_\infty^2} \frac{dr}{r} \qquad (5\text{-}4\text{-}2b)$$

Example 5-4-1

A spacecraft is to escape from the surface of a nonrotating Earth in an Oberth maneuver (with perigee at the Earth's surface) and have a $V_\infty = 2970$ m/s.
(a) Find the nominal V_i.
(b) Find the sensitivity to changes in the nominal V_i and find the new V_∞ if V_i is 10% more than the nominal value; if V_i is 10% less than the nominal value.
(c) Find the sensitivity to errors in r and find V_∞ if the burnout altitude is at 50 nmi (92 km) rather than at the surface ($h = 0$).

Solution

(a) $V_{esc} = (2\mu/r_E)^{1/2} = 11{,}180$ m/s. From Eq. (5-4-1), the nominal $V_i = 11{,}570$ m/s.
(b) Using Eq. (5-4-2a) and the nominal value of V_i yields

$$\frac{dV_\infty}{V_\infty} = \left(\frac{11{,}570}{2970}\right)^2 \frac{dV_i}{V_i} = 15.18 \frac{dV_i}{V_i}$$

which means that for every 1% error in the injection velocity there will be a 15.18% error in V_∞. With a +10% change in V_i, the increase in V_∞ will be +4508 m/s (a 151.8% increase), so that the new $V_\infty = 7478$ m/s. A 10% decrease in V_i, on the other hand, will decrease V_∞ by 4508 m/s so that the resulting V_∞ will be -1538 m/s. Since a negative V_∞ is an impossibility, the implication is that the spacecraft will not have enough energy for a hyperbolic escape but will enter a geocentric elliptical orbit instead.
(c) At the surface of the Earth, $r = 6.378 \times 10^6$ m and

$$\frac{dV_\infty}{V_\infty} = \frac{3.986 \times 10^{14}}{6.378 \times 10^6 (2970)^2} \frac{dr}{r} = 7.085 \frac{dr}{r}$$

Sec. 5-4   The Patched Conic

so that for every 1% error in r at injection, there will be a 7.1% error in V_∞. In this case the error in altitude of +50 nmi (92 km) represents a +1.44% error in r which translates into a 10.2% increase (303.5 m/s) in V_∞, which now will be 3274 m/s.

Since the residual velocity $\mathbf{V}_\infty$ has both a magnitude V_∞ and an elevation angle ϕ, it is definitely a vector. In heliocentric trajectories $\mathbf{V}_\infty$ is the vector difference at the edge of the SOI between the velocity of the spacecraft with respect to the Sun and the velocity of the planet of interest with respect to the Sun, so that

$$\mathbf{V}_\infty = \mathbf{V} - \mathbf{V}_{Pl} \tag{5-4-3a}$$

or

$$\mathbf{V} = \mathbf{V}_{Pl} + \mathbf{V}_\infty \tag{5-4-3b}$$

where $\mathbf{V}$ is the velocity of the spacecraft with respect to the Sun and $\mathbf{V}_{Pl}$ is the velocity of the planet with respect to the Sun.

In examining geocentric orbital transfers, we found that the energy required to enter a new trajectory was a minimum when the velocities were collinear. This is also true for interplanetary transfers and it is to our advantage to plan the escape hyperbola so that V_∞ (the asymptote of the hyperbola) is parallel to the velocity of the planet as is shown in Fig. 5-4-1. There are several items of interest in this figure. First, the heliocentric velocity of the spacecraft after leaving the SOI is simply the linear sum of V_∞ and V_{Pl}, the heliocentric (orbital) velocity of the planet. Also notice that the heliocentric elevation angle at the SOI is zero, although the planetocentric elevation angle is not. Assuming injection into the escape hyperbola at periapsis with a planetocentric elevation angle of zero (a best-case situation), the magnitude of V_i determines the value of V_∞ and the location of the periapsis determines whether or not V_∞ will be parallel to V_{Pl}. Therefore, for the asymptote of the hyperbola to be parallel to the planet's orbital velocity (V_∞ parallel to V_{Pl}), the angle between the planet's velocity and the periapsis vector (the major axis) of the escape hyperbola should be equal to the position angle ν at r_∞ [i.e., $\nu_\infty = \cos^{-1}(-1/\epsilon)$].

Since it is not necessary that the departure trajectory lie in the orbital plane of the planet (the ecliptic in the case of escape from Earth), there is a family of hyperbolas whose asymptotes are parallel to the orbital velocity but who share a common periapsis. Furthermore, since the departure hyperbola of Fig. 5-4-1a is the inverse of the grazing incoming hyperbola of Fig. 5-3-2 that was used to establish the impact parameter, it can easily be shown that when V_∞ is parallel to the heliocentric orbital velocity of the planet, d, the distance between them, is equal to the impact parameter b, as shown in Fig. 5-4-1a, where $d = r_{SOI} \cos \phi$ (ϕ is the planetocentric elevation angle) and b is given by Eq. (5-3-9).

Example 5-4-2

For the spacecraft of Example 5-4-1, which has an Oberth injection velocity of 11,570 m/s, V_∞ is 2970 m/s and is to be parallel to V_E.
   (a) Find the position angle ν and the planetocentric elevation angle ϕ at r_{SOI}, at $r = \infty$. Compare the values.

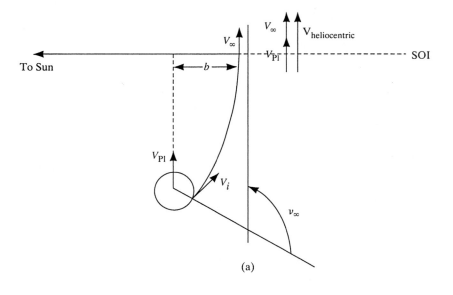

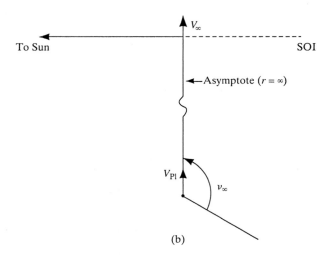

Figure 5-4-1. Escape with V_∞ parallel to V_{Pl}; (a) overview; (b) asymptotic view.

(b) What is the necessary condition for V_∞ to be parallel to the orbital velocity of the Earth about the Sun?

(c) Find the distance between V_∞ and V_E and compare with the impact parameter.

Solution

(a) From Eq. (2-8-1) we find that

$$v = \cos^{-1}\left[\frac{1}{\epsilon}\left(\frac{p}{r} - 1\right)\right]$$

Sec. 5-4   The Patched Conic

We need values for ϵ and p. At injection,

$$E = \frac{V_i^2}{2} - \frac{\mu}{r_p} = 4.436 \times 10^6 \text{ m}^2/\text{s}^2$$

$$a = \frac{-\mu}{2E} = -4.492 \times 10^7 \text{ m}$$

Also at injection,

$$H = r_p V_i = 7.379 \times 10^{10} \text{ m}^2/\text{s}$$

Therefore,

$$p = \frac{H^2}{\mu} = 1.366 \times 10^7 \text{ m}$$

With E and H known, Eq. (2-8-3) can be used to find that $\epsilon = 1.142$. Returning to our equation for ν and using $r_{SOI} = 9.245 \times 10^8$ m, we find that $\nu_\infty = 149.6°$. [With $r = \infty$, ν_∞ is simply $\cos^{-1}(-1/\epsilon) = 151.1°$, a 1% difference.] As for the planetocentric elevation angle, at the edge of the SOI,

$$\phi = \cos^{-1} \frac{H}{rV_\infty}$$

With $r_{SOI} = 9.245 \times 10^8$ m, $\phi = 88.47°$. With $r = \infty$, $\phi = 90°$.

(b) When the angle between the orbital velocity and the major axis is equal to ν_∞ [as found in part (a)], V_∞ will be parallel to the orbital velocity.

(c) $d = 9.245 \times 10^8 \cos(88.47°) = 2.468 \times 10^7 \text{ m}$

$$b = 6.378 \times 10^6 \sqrt{1 + \frac{(11{,}180)^2}{(2970)^2}} = 2.484 \times 10^7$$

There is a difference of only 0.6% between the two values.

Example 5-4-3

For the spacecraft and mission of Example 5-4-2, find the heliocentric velocity of the spacecraft upon escape from the Earth and describe the heliocentric trajectory (orbit).
Solution Since V_∞ and the orbital velocity of the spacecraft are collinear at the edge of the SOI, the heliocentric velocity is $V_\infty + V_E$, so that

$$V = 2970 + 29{,}790 = 32{,}760 \text{ m/s}$$

Since at that point V is perpendicular to the position vector from the Sun, $\phi = 0$, and

$$H = 1.496 \times 10^{11} \times 32{,}760 = 4.90 \times 10^{15} \text{ m}^2/\text{s}$$

and with Eq. (2-8-4),

$$E = -3.504 \times 10^8 \text{ m}^2/\text{s}^2$$

Since E is negative, we see that our geocentric hyperbola has become a heliocentric ellipse. We can verify this by finding the eccentricity ϵ from Eq. (2-8-3) to be 0.211, which indeed is less than 1. The semimajor axis a can be found from $a = -\mu/2E$ to be equal to 1.894×10^{11} m. With these values of a and ϵ, $r_p = a(1 - \epsilon) = 1.494 \times 10^{11}$ m, $r_a = a(1 + \epsilon) = 2.28 \times 10^{11}$ m, and the period is 520.4 days. Since r_p is essentially the distance of Earth from the Sun, r_a is essentially the distance of Mars from the Sun, and the period of a Hohmann transfer is 518 days, we see that this heliocentric orbit is for all practical purposes a Hohmann transfer from Earth to Mars.

Although a departure V_∞ that is collinear with (parallel to) the planet's orbital velocity vector is the normally desired situation, there may be times when V_∞ is not parallel. With the limitations of current propulsion systems and the emphasis on keeping weights low, it is difficult to visualize a deliberate attempt to do so. However, such a situation could arise in practice if the periapsis is not properly located or if the elevation angle at injection were not zero; in the latter case periapsis would not be at injection. In either case, we would have the situation shown in Fig. 5-4-2a, which represents the situation for Example 5-4-4. Although not shown in this figure, d (the offset distance) is no longer perpendicular to the orbital velocity and is greater than the impact parameter b. Furthermore, since V_∞ is no longer parallel, the heliocentric elevation angle is no longer zero but rather is equal to the angular offset of the new periapsis from the periapsis for the parallel V_∞ case. Now the law of cosines is needed to find the heliocentric V using the expression

$$V = \sqrt{v_{Pl}^2 + V_\infty^2 - 2V_{Pl}V_\infty \cos(180° - \phi)} \qquad (5\text{-}4\text{-}4)$$

We would expect V to be less than the value obtained with V_∞ and V_{Pl} collinear.

Example 5-4-4

The spacecraft of Example 5-4-2 has an escape injection velocity of 11,570 m/s, as before, but this time the injection elevation angle $\phi_i = +16°$; V_∞ remains equal to 2970 m/s. Launch (injection) is still from the surface of the Earth and from the same launch site; the angle between the Earth's orbital velocity and a line drawn through the launch site is 149.6°, the v_∞ for a parallel V_∞.
(a) Find the characteristics of the escape hyperbola. Locate the perigee of this trajectory with respect to the launch site and the Earth's orbital velocity.
(b) Find the elevation angle and velocity of the spacecraft with respect to the Sun after escape and determine the characteristics of the heliocentric trajectory. Compare this trajectory with that of Example 5-4-3.

Solution

(a) $\qquad H = 6.378 \times 10^6 \times 11{,}570 \cos 16° = 7.093 \times 10^{10}$ m²/s

Since V and r at injection are unchanged, E of the departure hyperbola and the semimajor axis are also unchanged, so that

$$E = 4.436 \times 10^6 \text{ m}^2/\text{s}^2 \quad \text{and} \quad a = \frac{-\mu}{2E} = -4.492 \times 10^7 \text{ m}$$

$$\epsilon = \sqrt{1 + \frac{2EH^2}{\mu^2}} = 1.132$$

which is slightly less than the eccentricity of 1.142 for the parallel V_∞ escape hyperbola with V_∞ and V_E parallel (Example 5-4-2). With $\epsilon = 1.132$,

$$r_p = a(1 - \epsilon) = 5.929 \times 10^6 \text{ m}$$

and $h_p = -448.6$ km $= -243.8$ nmi below the surface of the Earth; we are obviously

Sec. 5-4   The Patched Conic

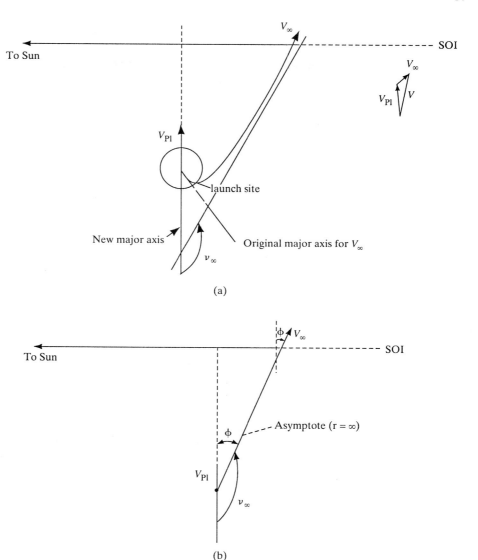

Figure 5-4-2. Escape with V_∞ not parallel to V_{Pl}; (a) overview; (b) asymptotic view.

not launching at perigee in this case. With $p = H^2/\mu = 1.262 \times 10^7$ m, we can find ν_i at injection (launch) to be

$$\nu_i = \cos^{-1}\left[\frac{1}{\epsilon}\left(\frac{p}{r} - 1\right)\right] = 30.17°$$

So we see that perigee is located 30.17° ahead of the injection point (and is still below the surface of the Earth). Since the injection (launch) site on the surface of the Earth is 149.6° in a clockwise direction from the Earth's orbital velocity V_E, the angle between V_E and the major axis of the escape hyperbola is 149.6 + 30.17° = 179.8°, as shown in Fig. 5-4-2.

(b) To locate the asymptote of the escape hyperbola with respect to V_E, we need to find ν_∞. Using the values of p and ϵ obtained in (a) and a value of 9.245×10^8 m for r_{SOI}, $\nu_\infty = 150.6°$, and the heliocentric elevation angle is found to be

$$\phi = 179.8 - 150.6 = 29.2°$$

as shown in Fig. 5-4-2. Consequently, with the application of Eq. (5-4-4), the heliocentric velocity at escape is

$$V = 32{,}410 \text{ m/s}$$

which is 1.0% less than that for the parallel case of Example 5-4-3. Since V has changed, so will the energy and

$$E = \frac{(32{,}410)^2}{2} - \frac{1.327 \times 10^{20}}{1.496 \times 10^{11}} = -3.618 \times 10^8 \text{ m}^2/\text{s}^2$$

which is 3.2% less than the energy of Example 5-4-3. The semimajor axis will be correspondingly less also; $a = 1.833 \times 10^{11}$ m. Since

$$H = 1.496 \times 10^{11} \times 32{,}410 \cos(29.2°) = 4.232 \times 10^{15} \text{ m}^2/\text{s}$$

$$\epsilon = 0.513$$

Now we can find the perihelion and aphelion conditions, namely,

$$r_p = a(1 - \epsilon) = 8.927 \times 10^{10} \text{ m} \qquad V_p = \frac{H}{r_p} = 47{,}410 \text{ m/s}$$

$$r_a = a(1 + \epsilon) = 2.773 \times 10^{11} \text{ m} \qquad V_a = \frac{H}{r_a} = 15{,}260 \text{ m/s}$$

Notice that r_p is less than a_E and r_a is greater than a_M; perihelion and aphelion are no longer at the planetary orbits of Earth and Mars. Since $p = H^2/\mu = 1.350 \times 10^{11}$ m, ν at the transition from the geocentric hyperbola to the heliocentric ellipse is equal to 101.0°, which indicates a large clockwise slewing of the major axis. This is quite a different transfer trajectory than the one we found in Example 5-4-2, which was essentially a Hohmann trajectory.

Although this example shows the effects of V_∞ not being parallel to the orbital velocity when you want it to be, it also indicates that periapsis of the escape trajectory does not need to be at the surface of the planet and that its location with respect to the orbital velocity of the planet can be controlled by the value of the elevation angle at injection. Therefore, if we wish to escape with V_∞ parallel but the injection point is not at the required offset from the orbital velocity, the proper choice of the injection elevation angle (which will be other than zero) will compensate for this deviation and result in a parallel V_∞ at the edge of the SOI. Consequently, *we are no longer constrained to one point in an inertial reference frame for launch into a parallel escape hyperbola*.

In the discussion and examples of this section we have escaped directly from the planetary surface. This is not the best way to escape. As can be seen from Fig.

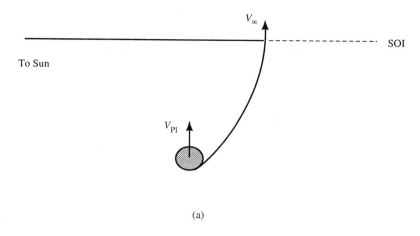

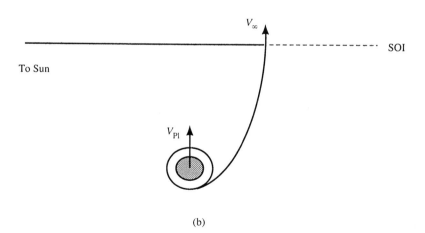

Figure 5-4-3. Planetary escape: (a) from surface; (b) from parking orbit.

5-4-3a, launch is restricted to times when the launch site, which is moving with respect to the inertial reference (the center of the planet), coincides with an acceptable injection point. Entering a circular parking orbit first, however, can provide greater flexibility with respect to acceptable times for injection. For example, a 322-km (175-nmi) parking orbit with a period of 90.96 min and approximately 16 revolutions of the Earth per day can provide more injection opportunities each day than does a 24-h period surface launch site with one launch opportunity per day as well as provide a wider launch envelope (into the parking orbit) and the opportunity to make corrections for deviations in the transfer from the launch site to the parking orbit prior to entering the escape trajectory.

Although there will be an increase in the total ΔV required for escape because of the parking orbit, it is not a large increase. For example, direct injection for escape from the surface of the Earth with a V_∞ of 2970 m/s requires a ΔV of 11,570 m/s. Escape from the 322-km orbit requires $\Delta V = 3587$ m/s and the Hohmann transfer to the 322-km orbit requires an additional ΔV of 8097 m/s ($V_L = 8093$ m/s) for a ΔV_T of 11,680 m/s, which is only a 1% increase in the ΔV requirement. The lower the parking orbit, the smaller the increase in the ΔV and the shorter the period (the greater the number of planetary revolutions and launch opportunities).

If for some reason the spacecraft were to escape from a very high parking orbit where $\mu/r < V_\infty^2/2$, a bielliptic (two-impulse) escape would require less energy than an Oberth escape. The first ΔV would be a collinear firing to slow the spacecraft and place it on an interior ellipse with apogee at the departure point and perigee as low as possible and located so that V_∞ from the second ΔV, which would be the Oberth escape velocity for that altitude, will be parallel to the Earth's orbital velocity. This is not a simple maneuver and it is difficult to visualize an occasion when this escape maneuver would be required.

Before we move on to the interplanetary transfer phase, we should note that we have been looking at escape in order to travel to a planet in an outer orbit (from Earth to Mars, for example) where the heliocentric velocity is the sum of V_∞ and the orbital velocity of the departure planet. For travel to an inner planet, such as from Earth to Venus, the procedure is the same, although now V_∞ is in the opposite direction to the planetary orbital velocity, as shown in Fig. 5-4-4, and the heliocentric velocity is the difference between the planetary velocity and V_∞.

In the preceding examples dealing with the first or escape phase, we have used an arbitrary value for V_∞ without any mention of how it was obtained. The value of V_∞ is determined by the requirements of the second or transfer phase. Let us consider an Earth–Mars mission comprising three phases:

Phase I: geocentric escape (a hyperbola)

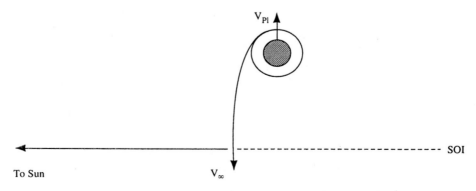

Figure 5-4-4. Escape to an inner planet.

Sec. 5-4 The Patched Conic

Phase II: heliocentric transfer (an ellipse)
Phase III: planetocentric encounter (a hyperbola)

as sketched in Fig. 5-4-5 for a minimum-energy (Hohmann) transfer. In phase III, *planetocentric encounter* implies intercepting and entering the planetary SOI. The objective of our mission analysis is to find the total ΔV required to carry out the mission. This ΔV_T is called the *velocity budget* and is a measure of the energy that must be provided to the spacecraft.

The purpose of phase I is to escape from the Earth and arrive at the start of phase II with a specified V_∞ that is determined by the nature of phase II. So we really start our mission analysis with phase II, which we treat in the next example. Remember that we are using a simplified model of the solar system in which the planetary orbits are circular, time invariant, and coplanar. Planetary and orbital data are provided in Appendix B.

Example 5-4-5

For phase II of an Earth–Mars mission:
(a) Find the V_∞ required for entry into a Hohmann transfer from the orbit of the Earth to the orbit of Mars.
(b) Find the time required to reach the orbit of Mars.

Solution

(a) Let us first collect the relevant data for the two planets and the Sun from Tables B-1 and B-2.

$$\mu_S = 1.327 \times 10^{20} \text{ m}^3/\text{s}^2$$

Mars data:

$$\mu_M = 4.297 \times 10^{13} \text{ m}^3/\text{s}^2$$
$$a_M = 1.524 \text{ AU} = 2.280 \times 10^{11} \text{ m}$$
$$V_M = 24{,}140 \text{ m/s}$$
$$r_M = 3.393 \times 10^6 \text{ m}$$

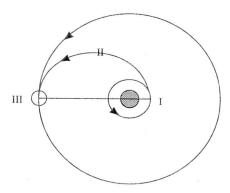

Figure 5-4-5. Phases of a Mars mission.

Earth data:

$$\mu_E = 3.986 \times 10^{14} \text{ m}^3/\text{s}^2$$

$$a_E = 1 \text{ AU} = 1.496 \times 10^{11} \text{ m}$$

$$V_E = 29{,}790 \text{ m/s}$$

$$r_E = 6.378 \times 10^6 \text{ m}$$

Transfer trajectory data:

$$2a_t = a_E + a_M = 2.524 \text{ AU} = 3.776 \times 10^{11} \text{ m}$$

$$E_t = \frac{-\mu_S}{2a_t} = \frac{-1.327 \times 10^{20}}{3.776 \times 10^{11}} = -3.514 \times 10^8 \text{ m}^2/\text{s}^2$$

Since E_t is negative, the transfer trajectory is a heliocentric ellipse, which is what we expected. Since the transfer is from an inner to an outer planet, injection into the transfer ellipse occurs at perihelion. Consequently,

$$V_{tp} = \sqrt{2\left(E_t + \frac{\mu_S}{a_E}\right)} = 32{,}730 \text{ m/s}$$

$$H_t = a_E V_{tp} = 4.896 \times 10^{15} \text{ m}^2/\text{s} \text{ and } \epsilon = 0.2081.$$

$$V_{ta} = \frac{H_t}{a_M} = 21{,}470 \text{ m/s}$$

The ΔV required for injection at perihelion is V_∞. Since V_E and V_{tp} are collinear,

$$V_\infty = \Delta V_1 = V_{tp} - V_E = 32{,}730 - 29{,}790 = 2940 \text{ m/s}$$

(b) The transfer time (TOF) is one-half of the period of the transfer ellipse and is

$$\text{TOF} = \pi \sqrt{\frac{a_t^3}{\mu_S}} = 2.237 \times 10^7 \text{ s} = 6214 \text{ h} = 259 \text{ days} = 8.6 \text{ months}$$

Now that we have a value for V_∞ (2940 m/s), we can return to phase I and calculate the velocity budget for phases I and II, assuming escape from a 175-nmi parking orbit. In the next example we first calculate the ΔV required to place the spacecraft into a 175-nmi parking orbit. Then using the known value of V_∞, the ΔV required to escape from the parking orbit with a one-impulse Oberth (hyperbolic) escape will be determined and added to the ΔV to enter the parking orbit to obtain the total velocity budget for phases I and II. In a similar manner the velocity budget for a two-impulse (parabolic) escape will be calculated and compared with that of the Oberth escape. The ΔV required to enter the parking orbit via a Hohmann transfer is 8097 m/s and the escape velocity from the parking orbit is 10,910 m/s.

Example 5-4-6

The spacecraft of the preceding example is in a 175-nmi parking orbit with a $V_{cs} = 7713$ m/s. The total ΔV to place the spacecraft in this circular orbit, using a Hohmann transfer from the surface of a nonrotating Earth is 8097 m/s. A $V_\infty = 2940$ m/s is needed to enter a heliocentric Hohmann transfer to Mars.

Sec. 5-4 The Patched Conic

(a) Find the ΔV for an Oberth (hyperbolic) escape from the parking orbit as well as the total velocity budget for phases I and II.

(b) Find the ΔV (and velocity budget) for a two-impulse (parabolic) escape from the parking orbit and compare with the velocity budget for the Oberth escape.

Solution

(a) With $V_{cs} = 7113$ m/s in a 175-nmi orbit, $V_{esc} = (2)^{1/2} V_{cs} = 10{,}910$ m/s. If escape from the parking orbit is a one-impulse Oberth hyperbola, the injection velocity is

$$V_{inj} = \sqrt{(10{,}910)^2 + (2940)^2} = 11{,}300 \text{ m/s}$$

and

$$\Delta V_{inj} = 11{,}300 - 7713 = 3587 \text{ m/s}$$

Consequently, the velocity budget for phases I and II is

$$\Delta V_T = 8097 + 3587 = 11{,}680 \text{ m/s}$$

(b) If escape from the parking orbit is a parabola, the escape ΔV is

$$\Delta V_{esc} = 10{,}910 - 7713 = 3197 \text{ m/s}$$

and the total ΔV required for injection into the Hohmann transfer is

$$\Delta V_{inj} = 3197 + 2940 = 6137 \text{ m/s}$$

Therefore, the velocity budget for phases I and II is equal to

$$\Delta V_T = 8097 + 6137 = 14{,}234 \text{ m/s}$$

A comparison of the two velocity budgets shows that the hyperbolic escape (Oberth) trajectory reduces the velocity budget by 2554 m/s (17.9%); it also eliminates the difficulties associated with the additional firing at infinity accompanying a parabolic escape.

Phase III starts when the spacecraft reaches either the sphere of influence (SOI) or the heliocentric orbit of the target planet, Mars in this case. At that point, $V_{ta} = 21{,}470$ m/s; this is the velocity of the spacecraft with respect to the Sun.

The spacecraft will be outside the SOI if the approach distance d remains greater than r_{SOI} upon arrival at the orbit, as sketched in Fig. 5-4-6. (From Table 5-3-1 the r_{SOI} of Mars is equal to 5.781×10^8 m or 314,200 nmi). At this point there are two possibilities:

1. Do nothing. If no (propulsive) action is taken, the spacecraft will continue on the Hohmann ellipse and return to the Earth's orbit; the Earth, however, has moved while the spacecraft has been traveling and will not be awaiting the spacecraft's return.

2. Use propulsion to enter the heliocentric orbit of the target planet. Then at some appropriate time execute an intercept (rendezvous) maneuver to enter the SOI of the planet. (Intercepts are discussed in Section 5-7.) With an error-free Hohmann transfer, the spacecraft and orbital velocities will be

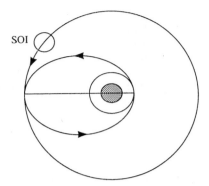

Figure 5-4-6. Failure to intercept the destination SOI.

collinear and the ΔV required by our spacecraft to enter the Mars orbit will simply be the difference between the orbital velocity of Mars (24,140 m/s) and the velocity of the spacecraft at the apogee of the Hohmann transfer (21,470 m/s from Example 5-4-5):

$$\Delta V = V_M - V_{ta} = 24{,}140 - 21{,}470 = +2670 \text{ m/s}$$

If this is done, with an Oberth escape the velocity budget for the entire mission to this point has increased from 11,680 m/s to 14,350 m/s (14.35 km/s). If the spacecraft and orbital velocities are not parallel at the interception of the Mars orbit, a noncollinear (and larger) ΔV will be required.

If, however, the spacecraft enters the SOI at the end of phase II, there are two possible situations:

1. The spacecraft remains within the SOI. This is known as *planetary capture* and is discussed in the next section.
2. The spacecraft executes a hyperbolic flyby and escapes from the SOI and reenters a heliocentric ellipse that has different characteristics than the incoming Hohmann. This hyperbolic flyby is often referred to as *planetary passage* and is discussed in Section 5-6.

In this section we have introduced the concepts of the patched conic and the velocity budget and have examined the first two phases of an interplanetary mission—escape and orbital transfer—in some detail and have introduced the ideas of planetary capture, planetary passage, and intercept.

5-5 PLANETARY CAPTURE

When a spacecraft arrives at an SOI, the approach distance d will either be larger or smaller than or equal to the impact parameter b. In either case the spacecraft will be under the influence of the planet's gravitational attraction. Considering the first two possibilities:

Sec. 5-5 Planetary Capture

1. If $d < b$, the spacecraft will do one of the following:
   a. Impact the planetary surface. Without propulsive or aerodynamic braking, it will be a hard landing. A soft landing with $V = 0$ can be achieved with an approximately collinear propulsive ΔV, where $\Delta V = -V$, V being the impact velocity without braking; or
   b. Enter a parking orbit; to do so requires a noncollinear ΔV and a direct transfer from the incoming hyperbola to the parking orbit.
2. If $d \geq b$:
   a. With no propulsive action (or penetration of any existing atmosphere) the spacecraft will execute a hyperbolic flyby and leave the SOI to enter a modified heliocentric ellipse. This is not planetary capture but rather is the planetary passage mentioned in the preceding section and that is discussed in Section 5-6; or
   b. A ΔV can be used to enter a parking orbit. If the ΔV is to be collinear with the spacecraft's velocity (with respect to the planet), the parking orbit will be at r_p of the incoming hyperbola. Parking orbits at other altitudes can be entered directly with the use of a noncollinear ΔV or by transfer from an intermediate parking orbit at r_p.

In Fig. 5-5-1a we have a spacecraft entering the SOI of an outer planet on the inside of the planetary orbit with $\mathbf{V}$ parallel to $\mathbf{V}_{Pl}$ (a Hohmann transfer), so that the intersection angle is zero and $V_\infty = V - V_{Pl}$. In Fig. 5-5-1b we see the relationships among V_∞, d, and ϕ and the direction of the planetocentric trajectory; ϕ and thus d are determined by the location of the penetration of the SOI. So is the direction of the planetocentric trajectory. In Fig. 5-5-1b the trajectory is counterclockwise; if the penetration of the SOI were to be outside the planet's heliocentric orbit, the planetocentric trajectory would be in the opposite direction (i.e., clockwise).

Since the SOI is small when compared with the other distances of the solar system, changes in the location of SOI penetration and in the magnitude of the

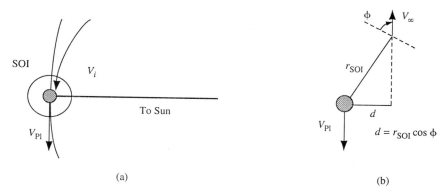

Figure 5-5-1. Entering the SOI of an outer planet on the inside.

approach distance can be made with very small changes in the transfer injection velocity, which is reflected in a change in the escape V_∞.

To obtain some idea as to how small those changes are for a large change in d, let us determine the sensitivity of d to changes in the injection velocity. Taking the derivative of a with respect to the injection velocity with $a = -\mu/2E$, $dE/dV = V$, and $\Delta d = 2\Delta a$, it can easily be shown that

$$\boxed{\frac{\Delta d}{2a} \cong \frac{2aV_i^2}{\mu_S} \frac{\Delta V}{V_i}} \qquad (5\text{-}5\text{-}1)$$

where V_i is equal to V_{tp} of the transfer trajectory. From the data of Example 5-4-5, $V_i = 32{,}730$ m/s and $2a_t = 3.776 \times 10^{11}$, so that

$$\frac{\Delta d}{2a} \cong 3.048 \frac{\Delta V}{V_i}$$

which indicates a very high sensitivity in d for changes in V_i. A difference of ± 1 m/s (out of 32,730 m/s) produces a change in the approach distance on the order of 19,000 nmi; a change of ± 1 ft/s results in a change on the order of 5800 nmi. Small velocity changes can also be made at points along the transfer trajectory and just prior to entering the SOI to change the planetocentric elevation angle and thus the approach distance. Such midcourse and terminal corrections are so commonly used as to be considered standard.

Example 5-5-1

For our Earth–Mars mission of Example 5-4-5, we wish to intercept the SOI of Mars at $d = 150{,}000$ nmi (276,000 km) beyond the Mars orbit.
(a) Find the new transfer injection velocity V_{tp} and the new V_∞.
(b) At the Mars SOI find V, V_∞, and the planetocentric ϕ.
(c) Find the impact parameter and compare with d.
(d) Use Eq. (5-5-1) to determine ΔV_i and the new V_{tp} and compare with the results of part (a). Can you explain the difference?

Solution

(a) In Example 5-4-5, we found the characteristics of the transfer trajectory for $d = 0$; $2a_t = 3.776 \times 10^{11}$ m and $V_{tp} = 32{,}730$ m/s. Now for the new transfer trajectory,

$$2a_t = 3.776 \times 10^{11} + 2.76 \times 10^8 = 3.779 \times 10^{11} \text{ m}$$

$$E_t = \frac{-\mu_S}{2a_t} = -3.512 \times 10^8 \text{ m}^2/\text{s}^2$$

From the energy equation (vis-viva integral),

$$V_{tp} = 32{,}740 \text{ m/s}$$

$$V_\infty = 32{,}740 - 29{,}790 = 2950 \text{ m/s}$$

This new V_∞ is 10 m/s (0.3%) greater than the original V_∞ of 2940 m/s. It takes only a 0.3% increase in the residual velocity and an even smaller increase in the injection velocity to produce this approach distance of 150,000 nmi beyond the Mars orbit. Since

Sec. 5-5 Planetary Capture

the r_{SOI} of Mars is only 314,200 nmi, this approach distance represents 48% of the SOI radius. We see that the approach distance is very sensitive to variations in the injection velocities.

(b) Since the SOI is small, assume that V at the SOI is approximately equal to V_{ta} and that $V_{ta} = H/r_a$, where H is the angular momentum of the Hohmann transfer.

$$H = r_p V_{tp} = 1.496 \times 10^{11} \times 32{,}740 = 4.898 \times 10^{15} \text{ m}^2/\text{s}$$

With H and E known, $\epsilon = 0.2076$.

$$r_a = a(1 + \epsilon) = 2.282 \times 10^{11} \text{ m}$$

and

$$V_{ta} = \frac{H}{r_a} = 21{,}460 \text{ m/s}$$

$$V_\infty = 21{,}460 - 24{,}140 = -2680 \text{ m/s}$$

$$d = r_{SOI} \cos \phi \qquad \phi = \cos^{-1} \frac{d}{r_{SOI}} = +61.48°$$

We could have found ϕ at the SOI from the conditions at the edge of the SOI where the transfer heliocentric ellipse and the planetocentric hyperbola join. H of the hyperbola can be obtained from

$$H = V_\infty d = 2680 \times 2.76 \times 10^8 = 7.397 \times 10^{11} \text{ m}^2/\text{s}$$

But H is also equal to $r_{SOI} V_\infty \cos \phi$, so that

$$\phi = \cos^{-1} \frac{H}{rV} = +61.48°$$

(c) With $V_\infty = 2680$ m/s and $V_{esc} = 5032$ m/s,

$$b = r_M \sqrt{1 + \frac{V_{esc}^2}{V_\infty^2}} = 7.218 \times 10^6 \text{ m} = 3923 \text{ nmi}$$

Since $d > b$, there will be no impact.

(d) Using Eq. (5-5-1), ΔV can be found to be

$$\Delta V = \frac{V_i \Delta d}{2a \times 3.048} = 7.489 \text{ m/s}$$

This is less than the 10 m/s found in part (a). The difference arises from the round-off errors involved in the sum and difference of large numbers of the same order of magnitude with round off to four significant figures, a further indication of the sensitivity of the approach distance to small changes in the injection velocity.

The heliocentric velocity is not always parallel to the planet's orbital velocity at the edge of the SOI but rather may intersect it at some angle β. This is the case with fast transfers and some hyperbolic flyby trajectories used to modify a heliocentric trajectory. In such a case, V_∞ is no longer collinear and must be found from the law of cosines. We defer discussion of finite intersection angles until the next two sections and consider planetocentric trajectories for which d and V_∞ are specified or can be determined.

Let us return to our Hohmann transfer from Earth to Mars and look at several examples dealing with planetary capture. In the first we examine entering parking orbits having specified the approach distance d ($V_{\infty i}$ has been established by the patch between phase II and phase III). In the second we specify the altitude of the parking orbit and determine the requisite conditions at the SOI, specifically the approach distance. Then in the last two examples with $d < b$, we look first at impact and soft landing and then at entering a parking orbit with a noncollinear ΔV.

Planetary capture implies that the spacecraft is somehow constrained to remain within the SOI of Mars, where

$$\mu_M = 4.297 \times 10^{13} \text{ m}^3/\text{s}^2$$

$$r_M = 3.393 \times 10^6 \text{ m}$$

With $V_{ta} = 21{,}470$ m/s (from phase II and Example 5-4-5) and collinear with V_M,

$$V = V_{ta} = V_M + V_\infty$$

where V is the velocity of the spacecraft with respect to the Sun and V_∞ is the velocity of the spacecraft with respect to Mars. Therefore,

$$V_\infty = V_{ta} - V_M = 21{,}470 - 24{,}140 = -2670 \text{ m/s}$$

With the value of V_∞ established, we can find V_{esc} from the surface of Mars and find the corresponding impact parameter b.

$$V_{esc} = \left(\frac{2\mu_M}{r_M}\right)^{1/2} = 5032 \text{ m/s}$$

$$b = r_M\sqrt{1 + \frac{V_{esc}^2}{V_\infty^2}} = 7.239 \times 10^6 \text{ m} = 3934 \text{ nmi}$$

Example 5-5-2

With an approach distance $d = 5000$ nmi (9.2×10^6 m) and a collinear $V_\infty = -2670$ m/s, enter a parking orbit around Mars at periapsis, r_p of the incoming hyperbola. Find the ΔV required for:
(a) A circular orbit.
(b) An elliptical orbit with $\epsilon = 0.5$.
Solution
 (a) At the penetration of the SOI,

$$H = V_\infty d = 2.456 \times 10^{10} \text{ m}^2/\text{s}$$

$$E = \frac{V_\infty^2}{2} = +3.564 \times 10^6 \text{ m}^2/\text{s}^2$$

Note: If $r_{SOI} = 5.781 \times 10^8$ m were used instead of $r = \infty$, E would be 3.490×10^6 m^2/s^2 or 2% less.

$$a = \frac{-\mu_M}{2E} = -6.028 \times 10^6 \text{ m}$$

Sec. 5-5   Planetary Capture

$$\epsilon = \sqrt{1 + \frac{2EH^2}{\mu^2}} = 1.824$$

$$r_p = a(1 - \epsilon) = 4.967 \times 10^6 \text{ m} = 1.464 r_M$$

$$h_p = 0.464 r_M = 1.574 \times 10^6 \text{ m} = 855 \text{ nmi}$$

$$V_p = \frac{H}{r_p} = 4945 \text{ m/s}$$

The velocity of the circular parking orbit is

$$V_{cs} = \left(\frac{\mu_M}{r_p}\right)^{1/2} = 2941 \text{ m/s}$$

Since V_p and V_{cs} are collinear for a transfer at r_p,

$$\Delta V = 2941 - 4945 = -2004 \text{ m/s}$$

Remember that the minus sign merely indicates that the spacecraft must be slowed down; to do so still requires a ΔV (in the opposite direction) and so does *not* reduce the velocity budget, which is the sum of the absolute values of the respective ΔV's. Keeping this in mind and using the lower ΔV of 11,680 m/s for phases I and II, the total velocity budget to put the spacecraft into a circular orbit about Mars is

$$\Delta V_T = 11{,}680 + 2004 = 13{,}684 \text{ m/s} = 13.68 \text{ km/s}$$

As a matter of curiosity, let us find v and ϕ at r_{SOI}.

$$p = \frac{H^2}{\mu} = 1.404 \times 10^7 \text{ m}$$

$$v = \cos^{-1}\left[\frac{1}{\epsilon}\left(\frac{p}{r} - 1\right)\right] = 122.3°$$

$$\phi = \cos^{-1}\frac{d}{r_{SOI}} = 89.08°$$

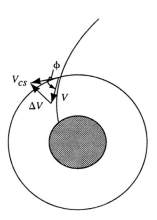

Figure 5-5-2. Noncollinear transfer of Example 5-5-5.

Since V_∞ is parallel to the orbital velocity of Mars, the angle between the orbital velocity and the periapsis position vector is $\nu = 122.3°$. (Note that $\nu_\infty = \cos^{-1}(-1/\epsilon) = 123.2°$.)

(b) With respect to the elliptical orbit with $r_p = 4.967 \times 10^6$ m and $\epsilon = 0.5$,

$$a = \frac{r_p}{1-\epsilon} = 9.993 \times 10^6 \text{ m}$$

$$E = \frac{-\mu_M}{2a} = -2.163 \times 10^6 \text{ m}^2/\text{s}^2$$

$$V_p = \sqrt{2\left(E + \frac{\mu}{r_p}\right)} = 3602 \text{ m/s}$$

$$\Delta V = 3602 - 4945 = -1343 \text{ m/s}$$

Here we see that entering this elliptical parking orbit requires a smaller ΔV (33% smaller) than does the circular orbit, and the total velocity budget is now 13,020 m/s or 4.8% lower. It should be noted that the type and characteristics of the desired parking orbit are determined by the nature of the mission of the spacecraft with respect to Mars.

A higher or lower parking orbit can be attained by using a Hohmann transfer from either of these parking orbits. In addition, a higher orbit could have been entered by a direct transfer from the hyperbolic arrival trajectory using a noncollinear ΔV.

Now let us specify the desired value of r_p and determine the associated value of the approach distance, keeping the value of V_∞ found from phase II.

Example 5-5-3

Let the altitude of the desired parking orbit be 855 nmi (1574 km) with $V_\infty = -2670$ m/s. Find the required approach distance.

Solution

$$r_p = r_M + 1.574 \times 10^6 = 4.967 \times 10^6 \text{ m}$$

$$E = \frac{V_\infty^2}{2} = 3.564 \times 10^6 \text{ m}^2/\text{s}^2$$

$$V_p = \sqrt{2\left(E + \frac{\mu}{r_p}\right)} = 4934 \text{ m/s}$$

$$H = r_p V_p = 2.455 \times 10^{10} \text{ m}^2/\text{s}^2$$

But H is also equal to $V_\infty d$, so that

$$d = \frac{H}{V_\infty} = 9.195 \times 10^6 \text{ m} = 4997 \text{ nmi}$$

which is essentially the 5000-nmi approach distance of Example 5-5-2.

If the approach distance is less than the impact parameter, the arrival hyperbola will intersect the surface of the planet (Mars). Let us examine the possibilities.

Example 5-5-4

The spacecraft arrives at the SOI of Mars with an approach distance of 1500 nmi (2.760×10^6 m), which is less than the impact parameter of 3934 nmi, and with $V_\infty = -2670$ m/s.

Sec. 5-5 Planetary Capture

(a) Find the velocity and elevation angle at impact.
(b) What ΔV is required just prior to impact if the landing velocity is to be zero (a soft landing)? Ignore any atmospheric drag and find the propulsion requirements.

Solution

(a)
$$E = \frac{V_\infty^2}{2} = 3.564 \times 10^6 \text{ m}^2/\text{s}^2$$

$$V_{\text{impact}} = \sqrt{2\left(E + \frac{\mu_M}{r_M}\right)} = 5697 \text{ m/s}$$

$$H = V_\infty d = 7.369 \times 10^9 \text{ m}^2/\text{s}$$

$$\phi = \cos^{-1}\frac{H}{rV} = -67.59°$$

As a check, the impact velocity should be equal to the Oberth escape velocity from the surface of Mars. With V_{esc} from the surface = 5032 m/s,

$$V_i = \sqrt{V_{\text{esc}}^2 + V_\infty^2} = 5696 \text{ m/s}$$

(b) Since the velocity at touchdown, V_L, will be

$$V_L = V_i + \Delta V$$

and since the desired $V_L = 0$,

$$\Delta V = V_L - V = 0 - 5696 = -5696 \text{ m/s} = -5.696 \text{ km/s}$$

When we compare this ΔV with the ΔV_T of 11,680 m/s for phases I and II and the 2004 m/s to enter the circular parking orbit, we see that propulsive braking is quite costly in terms of the energy expenditure required.

The last example in this section on planetary capture deals with entering a parking orbit from an arrival hyperbola that will impact the planet if no propulsive action is taken ($r_p < r_M$).

Example 5-5-5

For the spacecraft on the arrival hyperbola of Example 5-5-4 ($V_\infty = -2670$ m/s and $d = 2.760 \times 10^6$ m), find the ΔV required to enter a circular parking orbit at 855 nmi (1574 km).

Solution From Example 5-5-4,

$$E = 3.564 \times 10^6 \text{ m}^2/\text{s}^2$$

At $r_{cs} = r_M + 1.574 \times 10^6 = 4.967 \times 10^6$ m,

$$V = \sqrt{2\left(E + \frac{\mu}{r}\right)} = 4943 \text{ m/s}$$

$$V_{cs} = \left(\frac{\mu}{r}\right)^{1/2} = 2941 \text{ m/s}$$

But V and V_{cs} are not collinear, as can be seen from Fig. 5-5-2. Using the data from Example 5-5-4 yields

$$H = V_\infty d = 2670 \times (2.760 \times 10^6) = 7.369 \times 10^9 \text{ m}^2/\text{s}$$

but H is also equal to $r_{cs} V \cos \phi$, so that

$$\phi = \cos^{-1} \frac{7.369 \times 10^9}{(4.967 \times 10^6) \times 4943} = -72.53°$$

The sign of the elevation angle is negative because the spacecraft is on an incoming trajectory. Applying the law of cosines to find ΔV gives us

$$\Delta V = \sqrt{V^2 + V_{cs}^2 - 2VV_{cs} \cos \phi} = 4935 \text{ m/s}$$

This is a considerable (146%) increase over the collinear ΔV of 2004 m/s of Example 5-5-2 required to transfer into a circular orbit at r_p at the same altitude. This ΔV of 4935 m/s is approaching the value for a soft all-propulsive landing.

It is time to move on to the next section. In this section we have introduced the patched conic concept that allows us to use simple two-body (central force) mechanics to calculate velocity budgets for interplanetary travel. Even though these ΔV's are approximate, even order-of-magnitude values, they are extremely useful in mission planning and feasibility analyses. Since they represent the energy that must be imparted to the spacecraft, they affect the size and nature of the booster and propulsion systems. Furthermore, they can be used as figures of merit to compare various transfer and capture schemes as was done in this section with several parking orbits around Mars.

In this section we also demonstrated by example how sensitive the trajectories and velocity budgets are to variations in the injection velocities, thus emphasizing the importance of the navigation, guidance, and control systems.

5-6 PLANETARY PASSAGE AND GRAVITY ASSIST

In the preceding section we deferred discussion of the planetocentric arrival hyperbola in which no propulsive action was taken (and there was no atmospheric penetration). A single hyperbolic flyby (planetary passage) might be for the purpose of a one-time look at a planet or a moon. It is also possible that a hyperbolic flyby (swingby) is for the purpose of changing the magnitude and direction of the heliocentric velocity so as to modify the heliocentric trajectory without the need to use propulsion. Both objectives could be, and have been, met in a single flyby.

The use of swingby(s) or flyby(s) is a technique that is referred to as *gravity assist* and that is of interest and of increasing use for journeys deep into space, either inner or outer, to reduce the propulsive velocity budget. Planetary probes generally swing by Venus on the way to Mercury and swing by Mars or Jupiter on the way to the outer planets and asteroids. There is one asteroid rendezvous mission in which the probe first travels inward around Venus and returns for a swing by Earth on its way to its passage of the asteroid(s); this is known as a VEGA (Venus–Earth gravity assist) maneuver. A flyby of Jupiter can be used to execute a large plane change out of the ecliptic and place the spacecraft in a polar orbit about the Sun itself (as with the Ulysses mission). There is no limit to the number and use of such flybys. A

Sec. 5-6 Planetary Passage and Gravity Assist

Mercury swingby produces the largest energy change because of its closeness to the Sun and its correspondingly high heliocentric velocity. A Jupiter swingby gives the largest trajectory deflection for a given V_∞ because of its large mass.

A spacecraft enters an SOI with a $\mathbf{V}_{\infty i}$, its incoming velocity vector with respect to the planet, that is the vector difference between the incoming heliocentric velocity $\mathbf{V}_i$ and the heliocentric orbital velocity of the planet $\mathbf{V}_{\text{Pl}}$. $V_{\infty i}$, the magnitude of $\mathbf{V}_{\infty i}$, determines the specific energy of the incoming hyperbola, and in conjunction with the planetary approach distance d or periapsis r_p, also determines the specific angular momentum. Since the energy of the flyby trajectory is constant, the magnitude of the outgoing residual velocity $V_{\infty o}$ is equal to $V_{\infty i}$ but its direction has been changed, having been rotated through the turning angle δ of Section 2-5. As a consequence, the outgoing heliocentric velocity $\mathbf{V}_o$ is not equal to $\mathbf{V}_i$; both the magnitude and direction are different so that the heliocentric trajectory after planetary passage (flyby) has been modified. In addition to a change in the energy, there will be changes in the length and direction of the major axis, shifting the location of perihelion as well as changing the energy of the trajectory.

Since there are no external energy inputs (no propulsion and no drag) during the flyby and since energy must be conserved, any change in the energy of the trajectory must be accompanied by a corresponding change in the energy of the planet. The conservation of the energy can be expressed as

$$m_{\text{S/C}} \Delta E_{\text{S/C}} + m_{\text{Pl}} \Delta E_{\text{Pl}} = 0 \quad \text{or} \quad \Delta E_{\text{Pl}} = -\frac{m_{\text{S/C}}}{m_{\text{Pl}}} \Delta E_{\text{S/C}}$$

Example 5-6-1

A 2000-kg spacecraft made a hyperbolic flyby of Mars and increased the energy of its trajectory by 4.1×10^8 m^2/s^2.
(a) What is the corresponding decrease in the energy of the orbit of Mars?
(b) What effect did this hyperbolic flyby have on the orbital velocity of Mars?
Solution

(a) From Table B-1, $m_M = 6.441 \times 10^{23}$ kg:

$$\Delta E_M = -\frac{2000}{6.441 \times 10^{23}} \times 4.1 \times 10^8 = -1.273 \times 10^{-12} \text{ m}^2/\text{s}^2$$

which indeed is a small quantity.

(b)

$$\Delta E_M = \frac{V_2^2}{2} - \frac{V_1^2}{2} \quad \text{or} \quad V_2 = \sqrt{V_1^2 + 2\Delta E_M}$$

With V_1, the nominal orbital velocity of Mars, equal to 24,140 m/s, substituting ΔE into the expression above yields a value of 24,140 m/s for V_2. In other words, the effects of this single flyby on the energy and velocity, and thus on the trajectory, of Mars are so small as to be unobservable with our system of four significant figures.

Let us now develop *general* expressions that allow us to describe and define a planetary passage, to find the magnitude and direction of the outgoing heliocentric velocity. Let us start by looking at the geometry at the beginning of the encounter when the spacecraft enters the SOI, as sketched in Fig. 5-6-1. At this point we shall not be concerned with the location of the SOI entry with respect to the planet and Sun. In this figure β_i is the intersection angle between the planetary velocity V_{Pl} and the incoming heliocentric velocity V_i and ζ_i is the angle between the planetary velocity and $V_{\infty i}$. When $\beta_i = 0$, the two velocities are parallel (collinear), as would be the case with a Hohmann interplanetary transfer. (With a circular planetary orbit as shown, β_1 is also the heliocentric elevation angle ϕ_1.)

Applying the law of cosines gives

$$V_{\infty i} = \sqrt{V_{Pl}^2 + V_i^2 - 2V_{Pl} V_i \cos \beta_i} \tag{5-6-1}$$

where V_{Pl}, the orbital velocity of the planet, is obtained from Table B-2, and V_i, the heliocentric velocity of the spacecraft at encounter, is found from the transfer trajectory, as is β_i. Consequently, $V_{\infty i}$ is known and ζ_i can be found from the law of sines, namely,

$$\frac{V_{\infty i}}{\sin \beta_i} = \frac{V_i}{\sin(180° - \zeta_i)}$$

so that

$$\boxed{180° - \zeta_i = \sin^{-1} \frac{V_i \sin \beta_i}{V_{\infty i}}} \tag{5-6-2}$$

Note that *if $V_i \cos \beta_i < V_{Pl}$, then $\zeta_i > 90°$*; otherwise, $\zeta_i < 90°$. This means that when V_{Pl} and V_i are parallel (collinear) and $\beta_i = 0°$ (as with a heliocentric Hohmann transfer), $\zeta_i = 180°$ for an outer-planet passage where $V_{Pl} > V_i$ and $\zeta_i = 0°$ for an inner-planet passage where $V_{Pl} < V_i$.

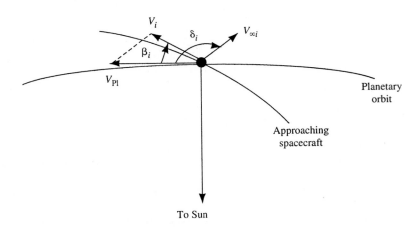

Figure 5-6-1. Initiation of a planetary flyby.

Sec. 5-6 Planetary Passage and Gravity Assist

After the flyby, V_∞ has been turned through the angle δ and $V_{\infty o}$ makes an angle ζ_o with V_{Pl}, as shown in Fig. 5-6-2. The angle δ is the turning angle of Section 2-5. It is the angle between the hyperbolic asymptotes, is a function of the eccentricity ϵ, and can be found from

$$\delta = 2 \sin^{-1}\frac{1}{\epsilon} \qquad (5\text{-}6\text{-}3)$$

The direction of rotation of δ is important inasmuch as it determines the value of ζ_o, the angle between the planetary velocity V_{Pl} and the outgoing heliocentric velocity V_o. In Fig. 5-6-2a, δ rotates in a clockwise (CW) direction; a clockwise rotation can often be associated with a passage in front of the planet. In Fig. 5-6-2b, δ rotates in a counterclockwise (CCW) direction, associated at times with a passage behind the planet. Rather than attempting to memorize relationships between type of passage and direction of rotation, it is easier to use a crude sketch to determine the direction of rotation.

Returning to Fig. 5-6-2, α is the angle opposite V_o and is only of interest in applying the law of cosines to find V_o. Since $\alpha = 180° - \zeta_o$, no matter how δ rotates, $\cos \alpha = -\cos \zeta_o$. Therefore, applying the law of cosines with $V_{\infty o} = V_{\infty i} = V_\infty$ and with $\cos \alpha$ replaced by $-\cos \zeta_o$ yields an expression for V_o.

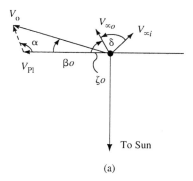

(a)

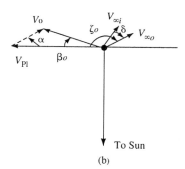

(b)

Figure 5-6-2. Completion of a planetary flyby: (a) CCW; (b) CW.

$$\boxed{V_o = \sqrt{V_{Pl}^2 + V_\infty^2 + 2V_{Pl}V_\infty \cos\zeta_o}} \qquad (5\text{-}6\text{-}4)$$

Now we need an expression for ζ_o. Examination of the angles in the figures shows that with a CCW rotation,

$$180° - \zeta_o = (180° - \zeta_i) + \delta$$

and with a CW rotation

$$180° - \zeta_o = 180° - (\zeta_i + \delta)$$

These two expressions reduce to

$$\boxed{\zeta_o = \zeta_i \pm \delta} \qquad (5\text{-}6\text{-}5)$$

where the plus sign is used when the turning angle δ is clockwise (CW) and the minus sign is used when the turning angle δ is counterclockwise (CCW). Repeating a previous hint, a sketch is helpful in visualizing the flyby trajectory and the sense of δ.

β_o, the intersection of V_o and V_{Pl} after passage, can be found from the law of sines to be

$$\boxed{\beta_o = \sin^{-1}\left(\frac{V_\infty}{V_o}\sin\zeta_o\right)} \qquad (5\text{-}6\text{-}6)$$

where β_o (and β_i) is positive in a clockwise direction from V_{Pl}. If β_o is greater than β_i, the outgoing velocity has been turned away from the Sun. If $\beta_o < \beta_i$, the velocity has been turned toward the Sun. As mentioned previously, a sketch can be helpful.

To find δ, we need the eccentricity ϵ, which can be found from the expression

$$\epsilon = \sqrt{1 + \frac{2EH^2}{\mu_{Pl}^2}}$$

Once $V_{\infty i}$ is known [from Eq. (5-6-1)], E can be found from

$$E = \frac{V_{\infty i}^2}{2} = \frac{V_{\infty o}^2}{2} = \frac{V_\infty^2}{2}, \qquad \text{a constant}$$

Now all we need to find ϵ is a value for H, which requires a knowledge of either the approach distance d or the periapsis distance r_p.

If d is known, then

$$H = dV_\infty$$

If r_p is known, then

$$V_p = \sqrt{2\left(E + \frac{\mu_P}{r_p}\right)}$$

Sec. 5-6 Planetary Passage and Gravity Assist

and

$$H = r_p V_p$$

To summarize the data requirements for a planetary passage (hyperbolic flyby), we need:

1. $\mathbf{V}_i$ and $\mathbf{V}_{Pl}$ (V_i, V_{Pl}, and β_i)

and

2. Either d or r_p.

Let us first look at the special case where V_i and V_{Pl} are parallel ($\beta_o = 0°$); i.e., the incoming spacecraft is on a Hohmann transfer.

Example 5-6-2

A spacecraft on a Hohmann transfer from Earth to Mars is scheduled for a hyperbolic flyby behind Mars (a shady-side passage) intercepting the SOI with $V_i = 21{,}470$ m/s and parallel to V_M. The approach distance d is 5000 nmi (9.20×10^6 m) and is on the far side of Mars, as shown in Fig. 5-6-3.
(a) Find V_o, the ΔV of the passage, and the outgoing intersection angle β_o. Has V_o been turned toward or away from the Sun?
(b) Find the characteristics of the modified trajectory and compare them with those of the original Hohmann transfer from Earth (Example 5-4-5).
(c) Sketch the flyby trajectory from the viewpoint of an observer at the center of Mars. An observer at the center of the Sun.
(d) If the approach distance is on the near side of Mars (toward the Sun) for a sunny-side passage, find V_o and β_o and compare with the results of part (a).

Solution

(a) Since V_i is parallel to V_M, $\beta_i = 0$ and

$$V_\infty = V_i - V_M = 21{,}470 - 24{,}140 = -2670 \text{ m/s}$$

The energy of the hyperbola is

$$E = \frac{(2670)^2}{2} = 3.564 \times 10^6 \text{ m}^2/\text{s}^2 > 0$$

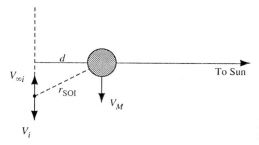

Figure 5-6-3. Initiation of the Mars flyby of Example 5-6-2.

$$H = dV_\infty = 9.20 \times 10^6 \times 2670 = 2.456 \times 10^{10} \text{ m}^2/\text{s}$$

$$\epsilon = \sqrt{1 + \frac{2EH^2}{\mu_M^2}} = 1.824 > 1$$

$$\delta = 2 \sin^{-1} \frac{1}{1.824} = 66.5° = 1.160 \text{ rad}$$

With $\beta_i = 0$, $V_i \cos \beta_i = V_i < V_M$; therefore, ζ_i will be greater than 90° and will be equal to 180°. For tutorial purposes we shall check the value of ζ_i using Eq. (5-6-1). With $\beta_i = 0$,

$$180° - \zeta_i = \sin^{-1} \frac{V_i \sin \beta_i}{V_\infty} = 0° \quad \text{or} \quad 180°$$

Since $\zeta_i > 90°$,

$$\zeta_i = 180° = \pi \text{ radians}$$

Since the rotation of δ is CW, the plus sign will be used in Eq. (5-6-5) to find ζ_o,

$$\zeta_o = \zeta_i + \delta = 180° + 66.5° = 246.5° = 4.30 \text{ rad}$$

Now we can find V_o from Eq. (5-6-4) to be

$$V_o = \sqrt{V_M^2 + V_\infty^2 + 2V_M V_\infty \cos \zeta_o} = 23{,}200 \text{ m/s}$$

$$\Delta V = 23{,}200 - 21{,}470 = +1730 \text{ m/s} \quad (8\% \text{ increase})$$

This increase of 1730 m/s in the heliocentric velocity was obtained without any expenditure of propulsive energy, thus reducing the amount of fuel to be carried aboard the spacecraft and the weight of both the spacecraft and its booster.

Using Eq. (5-6-6), the outgoing intersection angle β_o is

$$\beta_o = \phi_o = \sin^{-1}\left[\frac{2670}{23{,}200} \sin(246.5°)\right] = -6.06°$$

Since β_o is less than β_i ($\beta_i = 0°$), the outgoing heliocentric velocity V_o has been turned toward the Sun.

(b) The energy of the modified heliocentric trajectory can be found from the vis-viva integral to be

$$E = \frac{V_o^2}{2} - \frac{\mu_S}{a_M} = -3.129 \times 10^8 \text{ m}^2/\text{s}^2 < 0 \quad (\text{an ellipse})$$

The energy of the heliocentric trajectory has been increased by 10.9% by virtue of the flyby. The semimajor axis is

$$a = \frac{-\mu_S}{2E} = 2.120 \times 10^{11} \text{ m}$$

which is 12.8% greater than the incoming semimajor axis. Since Mars is assumed to be in a circular orbit, the heliocentric elevation angle at departure from the SOI is equal to the intersection angle; $\phi_o = \beta_o = -6.026°$ and

$$H = a_M V_o \cos \phi_o = 5.260 \times 10^{15} \text{ m}^2/\text{s}$$

Sec. 5-6 Planetary Passage and Gravity Assist

$$p = \frac{H^2}{\mu_s} = 2.085 \times 10^{11} \text{ m}$$

$$\epsilon = \sqrt{1 + \frac{2EH^2}{\mu_s^2}} = 0.1294$$

Since ϵ of the modified trajectory is 37.8% smaller, the modified trajectory will be more circular than the transfer trajectory and the apsides will be located at

$$r_a = a(1 + \epsilon) = 2.743 \times 10^{11} \text{ m}$$

$$r_p = a(1 - \epsilon) = 1.846 \times 10^{11} \text{ m}$$

Aphelion will lie outside the Mars orbit, and perihelion will lie between the orbits of Mars and Earth. The modified trajectory will no longer intercept the orbit of Earth. The location of the new perihelion can be determined by finding the value of the true anomaly v_o at the beginning of the modified trajectory.

$$v_o = \cos^{-1}\left[\frac{1}{\epsilon}\left(\frac{p}{a_M} - 1\right)\right] = -131.3°$$

This is the angle from the new perihelion and since the true anomaly for this point on the original Hohmann transfer trajectory was $-180°$, the major axis of the modified trajectory has been rotated $-48.7°$ $(-180° + 131.3°)$, in a clockwise direction, from the major axis of the original transfer trajectory. The period of the modified trajectory is 616.2 days, which is 19% longer than the 518-day period of the original trajectory.

(c) Figure 5-6-4a is a sketch of the hyperbolic flyby as viewed by an observer at the center of Mars. Figure 5-6-5 is a sketch of the modified heliocentric trajectory with the incoming Hohmann transfer shown as a dashed line.

(d) When the approach distance is on the side toward the Sun (the sunny side), the turning angle will be counterclockwise so that

$$\zeta_o = \zeta_i - \delta = 180 - 66.5 = 113.5°$$

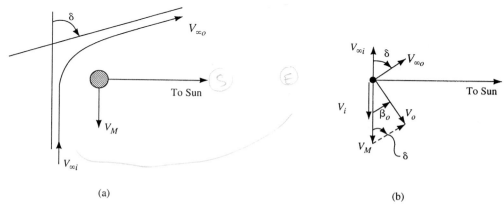

Figure 5-6-4. Clockwise passage of Mars: (a) planetocentric view; (b) vector diagram.

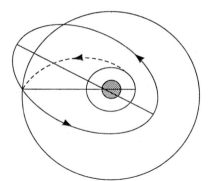

Figure 5-6-5. Modified trajectory of Example 5-6-2 after a Mars flyby.

and

$$V_o = 23{,}200 \text{ m/s}$$

We see that the value of V_o is unchanged, that it is independent of the direction of rotation of δ. On the other hand, β_o will be different, being positive rather than negative.

$$\beta_o = +6.06°$$

In this case $\beta_o > \beta_i$, so that the velocity is turned away from the Sun.

Let us now look at a Hohmann transfer to an inner planet, where the orbital velocity will be less than the incoming velocity, with the flyby periapsis specified this time rather than the approach distance.

Example 5-6-3

A spacecraft is to complete a Hohmann transfer to Venus with a hyperbolic flyby with the altitude of closest passage (periapsis) at 500 km (271.7 nmi). The spacecraft is to pass between the Sun and Venus. $V_i = 37{,}730$ m/s and is parallel to V_V, which is equal to 35,040 m/s.
  (a) Sketch the vector diagram of the flyby, showing the relation to the Sun, and determine the direction of rotation of the turning angle.
  (b) Find the value of the turning angle δ and the magnitude of V_o.
  (c) Find the value of β_o. Has V_o been turned toward or away from the Sun?
Solution
  (a) The sketch in Fig. 5-6-6 shows that δ will be clockwise (CW) and that V_o will be turned away from the Sun. It also shows that $\zeta_i = 0°$.
  (b) To find δ, we need to find ϵ.

$$V_\infty = V_i - V_V = 37{,}730 - 35{,}040 = +2690 \text{ m/s}$$

$$E = \frac{V_\infty^2}{2} = \frac{(2690)^2}{2} = 3.618 \times 10^6 \text{ m}^2/\text{s}^2$$

$$r_p = r_V + h_p = 6.552 \times 10^6 \text{ m}$$

Sec. 5-6 Planetary Passage and Gravity Assist

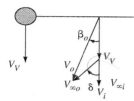

Figure 5-6-6. Clockwise passage of Venus.

$$V_p = \sqrt{2\left(E + \frac{\mu_V}{r_p}\right)} = 10{,}310 \text{ m/s}$$

$$H = r_p V_p = 6.758 \times 10^{10} \text{ m}^2/\text{s}$$

and

$$\epsilon = \sqrt{1 + \frac{2EH^2}{\mu_V^2}} = 1.146$$

Therefore,

$$\delta = 2\sin^{-1}\frac{1}{1.146} = 121.5°$$

Since $V_i > V_V$, $\zeta_i = 0°$, and since δ is CW,

$$\zeta_o = +\delta = 121.5°$$

and

$$V_o = 33{,}710 \text{ m/s} < V_i$$

(c) Using Eq. (5-6-6) gives

$$\beta_o = +3.753°$$

Since $\beta_o > \beta_i$, the heliocentric velocity has been turned away from the Sun, as can be seen in Fig. 5-6-6.

A comparison of the results of Examples 5-6-2 and 5-6-3 indicates that a Hohmann flyby of an outer planet will increase the heliocentric velocity irrespective of the rotation of δ, whereas a flyby of an inner planet will decrease the heliocentric velocity.

These two examples were both Hohmann transfers with the initial intersection angle, β_i, equal to zero. Let us look at an example where $\beta_i \neq 0°$.

Example 5-6-4

The spacecraft of Example 5-6-2 arrives at the far side of Mars with the same approach distance of 5000 nmi (9200 km), but this time with a finite intersection angle, $\beta_i = +20°$. V_i is still equal to 21,470 m/s.
(a) Find V_o and β_o and compare these values with those of Example 5-6-2.
(b) Find the periapsis altitude and compare it with that of Example 5-6-2.
(c) Do part (a) for an approach on the near side of Mars, between Mars and the Sun.

Solution

(a) The turning altitude will still be clockwise and ζ_i will still be greater than 90° since $V_i \cos \beta_i < V_M$, which is still 24,140 m/s. Use Eq. (5-6-1) to find $V_{\infty i}$:

$$V_{\infty i} = +8345 \text{ m/s} = V_{\infty o} = V_\infty$$

which is much larger than the V_∞ of 2670 m/s when $\beta_i = 0°$. Use Eq. (5-6-2) to find ζ_i:

$$180° - \zeta_i = 61.6°$$

so that

$$\zeta_i = 118.4°$$

Now to find ϵ and δ,

$$E = \frac{(8345)^2}{2} = 3.482 \times 10^7 \text{ m}^2/\text{s}^2$$

$$H = dV_\infty = 7.674 \times 10^{10} \text{ m}^2/\text{s}$$

and

$$\epsilon = \sqrt{1 + \frac{2EH^2}{\mu_M^2}} = 14.94$$

so that $\delta = 7.674°$. Since the rotation of δ is clockwise,

$$\zeta_o = \zeta_i + \delta = 118.6 + 7.674 = 126.3$$

With ζ_o known, V_o is found from Eq. (5-6-4) to be

$$V_o = 20{,}350 \text{ m/s}$$

which is 1123 m/s less than V_i; the heliocentric velocity has been decreased by this flyby. The outgoing intersection angle is

$$\beta_o = 19.31°$$

Since β_o is slightly less than β_i, the heliocentric velocity has been turned slightly toward the Sun. This flyby has decreased V_o rather than increased its magnitude and the deflection toward the Sun is slight, much less than the deflection for the Hohmann approach.

(b) With ϵ known, one way to find r_p is to first find $a = -\mu/2E$. For this problem

$$a = -6.170 \times 10^5 \text{ m}$$

$$r_p = a(1 - \epsilon) = 8.60 \times 10^6 \text{ m}$$

and

$$h_p = r_p - r_M = 5208 \text{ km} = 2830 \text{ nmi}$$

For the hyperbolic trajectory of Example 5-6-2,

$$a = -6.028 \times 10^6 \text{ m}$$

and with $\epsilon = 1.824$,

$$r_p = 4.967 \times 10^6 \text{ m}$$

and
$$h_p = 1574 \text{ km} = 855.6 \text{ nmi}$$

(c) If the entry at the SOI is on the near side of Mars, toward the Sun, δ will be counterclockwise. Therefore,

$$\zeta_o = \zeta_i - \delta = 118.4 - 7.764 = 110.7°$$
$$V_o = 22{,}580 \text{ m/s} > V_i$$

and

$$\beta_o = 20.23° > \beta_i$$

With this CCW flyby, V_o has increased by 1060 m/s (4.9%) and has been turned slightly away from the Sun.

A comparison of the results of Examples 5-6-2 and 5-6-4 shows that variations in the initial orientation and location of the entry into the planetary SOI can have large and seemingly contradictory effects on the flyby trajectory and thus on the modified heliocentric trajectory.

Although it may not be wise to attempt to generalize the effects of hyperbolic flybys, these few examples do indicate some rules of thumb. *When $\beta_i = 0°$ (incoming Hohmann transfers) outer-planet passages increase V_o and inner-planet passages decrease V_o regardless of the direction of rotation of the turning angle δ.* The correlation between the direction of rotation of δ and the direction of the outgoing velocity is a function of whether the planet is an outer or inner planet; a CW rotation about an outer planet turns the velocity toward the Sun, whereas a CW passage of an inner planet turns the velocity away from the Sun.

However, *when $\beta_i \neq 0°$, it is the direction of rotation of δ, not the location of the planet, that decides whether V_o increases or decreases;* a CW rotation decreases the velocity and a CCW rotation increases the velocity. Furthermore, the direction of the outgoing velocity vector is a function of both the direction of rotation of δ and of the location of the planet: that is, the relationship between the relative magnitudes of V_i and V_{Pl}, as was the case when $\beta_i = 0°$. A final conclusion is that the larger the turning angle is the greater is the impact on the magnitude and direction of the outgoing velocity.

Keeping these conclusions in mind, it is still a good idea to sketch the vector diagram and general geometry of each flyby. Although the direction of rotation of δ is unambiguous, the type of flyby is often used without specifically mentioning the direction of δ. The CW rotation of δ may be referred to as a front-side or leading-edge flyby, for example, and a CCW rotation as a back-side or trailing-edge flyby or passage.

The value of the incoming intersection angle β_i is a function of the type of interplanetary transfer trajectory. For a Hohmann ellipse, $\beta_i = 0$, and for a fast transfer, $\beta_i \neq 0$. For a specified intersection angle, large changes in the flyby hyperbola and subsequent trajectory can be achieved by small changes in the location of the entry to the planetary SOI, which change the approach distance d. Gravity-assist flybys are not restricted to the planets and can be used to make plane changes. One deep-space probe used a flyby of Titan, Saturn's largest moon, to move out of the plane of the ecliptic. You may run into the terms *pumping* and *cranking* with respect to gravity-assist maneuvers. *Pumping* refers to changing the energy of a trajectory and *cranking* to changing the angular deflection, either planar or out of plane.

Although it appears that intermediate flybys en route to a final destination would increase the time of flight, it is not so theoretically. It should be possible in some special cases, with careful planning, to reduce the propulsive velocity budget without increasing the flight time; for some missions there might even be a decrease in the flight time.

As can be imagined, the computational, navigational, and guidance and control efforts required to plan and execute a flyby or series of flybys so as to exploit successfully the trajectory changes are tremendous. All we could to do in this section was to introduce the principles of planetary passage together with its potential and implications.

5-7 THE INTERCEPT PROBLEM AND FAST TRANSFERS

In our discussions of interplanetary transfers we have ignored the fact that the planets are moving in their orbits with different velocities and thus have relative motion with respect to each other. If the spacecraft is to intercept the SOI of the target planet, the origin and target planets must have the appropriate angular relationship at the time of injection into the transfer trajectory. This angular relationship, called the *lead angle* ψ, is the angle between the origin planet and target planet at the time of departure. The lead angle for a Hohmann transfer from Earth to Mars (from an inner to an outer planet) is shown in Fig. 5-7-1. While the spacecraft is on its transfer trajectory, the target planet (Mars in this case) moves through the angle theta θ, which can be found from the simple relationship

$$\theta = \omega_M T$$

In this expression, ω_M is the orbital angular velocity (mean motion) of Mars:

$$\omega_M = \frac{2\pi}{P_M} = \frac{2\pi}{687.0} = 9.145 \times 10^{-3} \text{ r/day} = 0.5240 \text{ deg/day}$$

and T is the time of flight (TOF), which for a Hohmann transfer to Mars is

$$T = \frac{P_t}{2} = \pi \sqrt{\frac{(a_E + a_M)^3}{8\mu_S}} = 2.237 \times 10^7 \text{ s} = 259.0 \text{ days}$$

Therefore,

$$\theta = 0.5240 \times 259 = 135.7°$$

and the lead angle ψ is

$$\psi = 180° - \theta = 44.3°$$

Sec. 5-7 The Intercept Problem and Fast Transfers

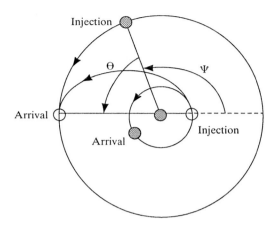

Figure 5-7-1. Intercept problem for an Earth–Mars Hohmann transfer.

The occurrence of this lead angle is a function of chronological time but will recur periodically. Since the relative angular velocity of Mars with respect to Earth is $\omega_M - \omega_E$, this (and any other) lead angle will occur again when the elapsed time Δt (with $\omega_E = 2\pi/365.2 = 0.01720$ r/day $= 0.9858$ deg/day) is

$$\Delta t = \frac{360°}{|\omega_M - \omega_E|} = \frac{360°}{|0.524 - 0.9858|} = 779.6 \text{ days} = 2.135 \text{ yr}$$

and every 2.135 years thereafter. This interval of time is called the *synodic period* of Mars with respect to Earth and is defined as the period of time for any specified lead angle to reappear. The synodic periods of the planets with respect to the Earth are shown in Table 5-7-1.

In addition to the synodic periods with respect to Earth, each planetary pair has a synodic period that can be obtained by using the relevant values of ω_{Pl} in Table 5-7-1. For example, the synodic period of Mercury with respect to Venus is 0.395 year or 144.3 days, whereas the synodic period of Neptune and Pluto is 495 years.

TABLE 5-7-1 PLANETARY SYNODIC PERIODS WITH RESPECT TO EARTH

Planet	ω_{PL} (rad/yr)	Synodic period (yr)
Mercury	26.11	0.1369
Venus	10.21	1.600
Mars	3.340	2.135
Jupiter	0.5297	1.092
Saturn	0.2132	1.035
Uranus	0.0748	1.012
Neptune	0.0381	1.006
Pluto	0.0254	1.004

$\omega_E = 6.283$ rad/yr.

This table of synodic periods shows that the periods are long, generally in excess of a year. This means that if the launch window for a particular transfer is missed, we must wait a long time for the next available window or compute a new trajectory for a different lead angle.

These synodic periods also indicate that round trips to the planets will be longer than the period of the Hohmann transfer since they must include a wait at the target planet for the correct lead angle for the return trip to occur. For example, a round trip to Mars comprises a 259-day Hohmann transfer to Mars, planetary capture and a visit to include a wait, as yet undetermined, for the return lead angle to develop, and a 259-day Hohmann return for a total mission duration in excess of 518 days.

Let us now look at the return from Mars to Earth on the other half of the Hohmann transfer. Since the inner planet, Earth, has a higher angular velocity than the outer planet, Mars, the Earth will be behind Mars at the time of injection; that is, the return lead angle will be negative, as can be seen in Fig. 5-7-2 and from

$$\psi_{ret} = 180° - \theta = 180° - \omega_E T = -75.32°$$

where $\omega_E T = 0.9858 \times 259 = 255.3°$.

It is interesting to note that at the time of arrival of the spacecraft at Mars on the outbound leg, Earth will be 255.3° from the injection point or 75.32° ($-\psi_{ret}$) beyond the major axis, as indicated in Fig. 5-7-1. Therefore, before the spacecraft can enter its homeward leg, Earth must travel through $(360 - 2\psi_{ret})$ degrees; the *wait time* T_w can be found from

$$T_w = \frac{360° - (2 \times 75.32)}{0.9858 - 0.5240} = 453.4 \text{ days} = 1.24 \text{ yr}$$

Consequently, the round trip to Mars will take 971.4 days or 2.66 years, a long trip for a manned flight. As a matter of possible interest, Mars will have traveled through 237.6° of its orbit during the wait period, as shown in Fig. 5-7-2.

Let us now look at the intercept and lead angle problem for any transfer trajectory (not just Hohmann transfers) and for any relative orientation of the two

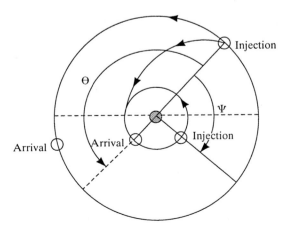

Figure 5-7-2. Intercept problem for a return from Mars.

Sec. 5-7 The Intercept Problem and Fast Transfers

planets. We shall, however, only consider transfer from an inner circular orbit to an outer circular orbit but will remove the limitation for injection at periapsis. In Fig. 5-7-3, a denotes the inner (origin) planet and b the outer (target) planet; the subscripts i and f denote injection and arrival, respectively; and the subscript t refers to the transfer trajectory.

The lead angle ψ can be obtained from

$$\psi = v_{bf} - v_{ai} - \omega_b T \qquad (5\text{-}7\text{-}1)$$

where T is the time of flight. Repeating the equations developed in Section 4-3 for elliptical trajectories,

$$T = \sqrt{\frac{a_t^3}{\mu_S}} (M_{bf} - M_{ai}) \qquad (5\text{-}7\text{-}2)$$

where M_{bf}, the mean anomaly at arrival, is defined as

$$M_{bf} = u_{bf} - \epsilon_t \sin u_{bf} \qquad (5\text{-}7\text{-}3a)$$

and M_{ai}, the mean anomaly at injection, is defined as

$$M_{ai} = u_{ai} - \epsilon_t \sin u_{ai} \qquad (5\text{-}7\text{-}3b)$$

The eccentric anomaly at arrival, u_{bf}, can be found from the expression

$$u_{bf} = \cos^{-1} \frac{\epsilon_t + \cos v_{bf}}{1 + \epsilon_t \cos v_{bf}} \qquad (5\text{-}7\text{-}4)$$

There is a similar expression for u_{ai}, the eccentric anomaly at injection.

Note that r_b, the distance from the Sun at arrival, is equal to a_b, the distance of the outer planet from the Sun, providing an equation with two unknowns that can be useful in obtaining a solution to the intercept problem:

$$r_b = a_b = \frac{a_t(1 - \epsilon_t^2)}{1 + \epsilon_t \cos v_{bf}} \qquad (5\text{-}7\text{-}5)$$

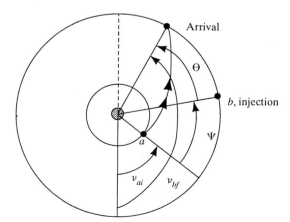

Figure 5-7-3. General intercept problem from inner to outer planet.

There are several classes of intercept problems, two examples of which will be examined next for purposes of showing how such problems might be solved and to illustrate how expensive such trajectories are in terms of the ΔV and energy requirements. Unfortunately, there are no closed solutions for most of the problems and iterative techniques and/or computer solutions are in order. Although Eq. (5-7-1) provides for injection at any point on the intercept trajectory, injection is done at periapsis ($v_{ai} = 0°$) whenever possible to reduce the ΔV requirements. For very small lead angles (very short flight times) it may be necessary or desirable to have injection at a point other than periapsis (i.e., $v_{ai} > 0$).

Let us first look at a problem in which, for some reason or another, we are interested in intercepting the target at a known point. The procedure for solving such a problem will be illustrated in the following example.

Example 5-7-1

For an Earth–Mars transfer trajectory the intercept of Mars is to occur when the angle between the injection point and the arrival point (at Mars) is equal to 120°.
(a) Find the lead angle ψ, the transfer time T, the transfer injection velocity V_i, and V_∞ with respect to Earth.
(b) Find the phase I velocity (i.e., the injection velocity for an Oberth escape from the surface of the Earth) and the velocity budget for phases I and II.
(c) Upon arrival of the spacecraft at Mars, find the velocity V_i, the intersection angle β_i, and $V_{\infty i}$.
(d) Compare this trajectory with the Hohmann transfer of Example 5-4-5.

Solution

(a) To simplify the problem and reduce the injection velocity requirements, we shall inject at periapsis so that $v_{ai} = 0°$ and $v_{bf} = v_f = 120°$. Solving Eq. (5-7-5) for a_t, the semimajor axis of the intercept trajectory, yields

$$a_t = \frac{a_M(1 + \epsilon \cos v_f)}{1 - \epsilon^2}$$

Since $r_p = a_E = a_t(1 - \epsilon)$,

$$a_t = \frac{a_E}{1 - \epsilon}$$

Equating the two expressions produces

$$\frac{1 + \epsilon \cos v_f}{1 + \epsilon} = \frac{a_E}{a_M}$$

which can be solved to find that

$$\epsilon = 0.2974$$

From either of the expressions for a_t,

$$a_t = 2.129 \times 10^{11} \text{ m}$$

From Eq. (5-7-4),

$$u_{bf} = 1.811 \text{ rad}$$

From Eq. (5-7-3a),

$$M_{bf} = 1.522 \text{ rad}$$

Sec. 5-7 The Intercept Problem and Fast Transfers

With $v_{ai} = 0$, $u_{ai} = M_{ai} = 0$, and from Eq. (5-7-2),

$$T = 1.298 \times 10^7 \text{ s} = 150.2 \text{ days}$$

With $\omega_M = 0.5240$ deg/day, Eq. (5-7-1) produces

$$\psi = 120° - (0.524 \times 150.2) = 41.30°$$

With a_t known,

$$E_t = \frac{-\mu_S}{2a_t} = -3.116 \times 10^8 \text{ m}^2/\text{s}^2$$

and V_i (V_{tp}) is found from

$$V_i = V_{tp} = \sqrt{2\left(E_t + \frac{\mu_S}{a_M}\right)} = 33,920 \text{ m/s}$$

$$V_\infty = 33,920 - 29,790 = 4130 \text{ m/s}$$

(b) With $V_{esc} = 11,180$ m/s (from the surface of the Earth with no residual velocity),

$$V_{\text{Oberth}} = \sqrt{(11,180)^2 + (4130)^2} = 11,920 \text{ m/s}$$

the velocity budget for phases I and II is

$$\Delta V_T = 11,920 \text{ m/s}$$

(c) $H_t = a_E V_i = 5.074 \times 10^{15}$ m²/s and $\epsilon_t = 0.2974$. Upon arrival at Mars,

$$V_{bf} = \sqrt{2\left(E_t + \frac{\mu_S}{a_M}\right)} = 23,250 \text{ m/s}$$

and

$$\beta_i = \phi = \cos^{-1}\frac{H}{a_E V_{bf}} = 16.83°$$

so that with the law of cosines

$$V_{\infty i} = \sqrt{V_M^2 + V_{bf}^2 - 2V_M V_{bf} \cos \beta_i} = 6991 \text{ m/s}$$

(d) In comparing this trajectory with the Hohmann transfer, which has an Oberth velocity of 11,560 m/s, we see that we have shortened the trip to Mars by 109 days at the cost of an increase of 360 m/s (a 3% increase). However, we must not forget the increase of the velocity with respect to Mars, $V_{\infty i}$, and its impact on the capture ΔV or on a flyby. We also see that there is little change in the lead angle, from 44.3° to 41.3°; the principal change is the increase of the eccentricity from 0.2081 to 0.2974.

Let us consider one more example, one in which the time of flight is specified.

Example 5-7-2

Find the lead angle for a 100-day elliptical transfer from Earth to Mars.
Solution Assuming injection at periapsis, we have

$$\psi = v_{bf} - \omega_M T$$

Knowing T and ω_M, we need to know v_{bf}, which can be found from

$$v_{bf} = \cos^{-1}\left[\frac{1}{\epsilon}\left(\frac{p}{a_M} - 1\right)\right]$$

where $p = a_t(1 - \epsilon^2)$. So we need to find a_t and ϵ but do not have two separate expressions. Consequently, it is necessary to pick a trial value for one or the other and start iterating. Let us try a value of a_t that is greater than a_M to ensure that the spacecraft at least crosses the orbit of Mars. Let the first value of a_t be equal to $1.4 a_M = 3.192 \times 10^{11}$ m.

$$E_t = \frac{-\mu_S}{2a_t} = -2.079 \times 10^8 \text{ m}^2/\text{s}^2$$

and

$$V_{ai} = V_{pt} = \sqrt{2\left(E + \frac{\mu_S}{a_E}\right)} = 36{,}860 \text{ m/s}$$

Therefore,

$$V_\infty = 36{,}860 - 29{,}790 = 7070 \text{ m/s}$$

This large increase in V_∞, and in the Oberth escape velocity, will affect the size and design of the booster, as we shall see in Chapter 6.

$$H_t = a_E V_{ai} = 5.514 \times 10^{15} \text{ m}^2/\text{s}$$

and

$$\epsilon = \sqrt{1 + \frac{2EH^2}{\mu_S^2}} = 0.5312$$

Now we are in a position to find v_{bf} from

$$v_{bf} = \cos^{-1}\left\{\frac{1}{\epsilon}\left[\frac{a_t(1-\epsilon^2)}{a_M} - 1\right]\right\} = 89.46°$$

[We could have used $p = H^2/\mu_S$ instead of $a_t(1 - \epsilon^2)$.] With v_{bf} known, u_{bf} is found from Eq. (5-7-4) to be 1.024 rad, and with Eq. (5-7-3a), $M_{bf} = 0.570$. Now we can find the time of flight T from Eq. (5-7-2) to be 103.3 days. This is surprisingly close to our goal of 100 days; it may even be close enough for the preliminary planning for this mission. If exactly 100 days is needed, iterate until that value is obtained. Increasing a_t by 10% produces a $T = 96$ days and we have bracketed the desired value for a_t; a 5% decrease might be in order for the next try. Now for the lead angle, using $T = 103.3$ days,

$$\psi = 89.46 - 0.524 \times 100 = 37.06°$$

which is somewhat less than the 41.3° of the preceding example and the 44.3° of the Hohmann transfer.

Notice how the injection velocity for the transfer trajectory has been increasing: from 32,730 m/s for the Hohmann transfer of 259 days to 33,920 m/s for a trip of 150.2 days to 36,860 m/s for a 103.3-day transfer. To establish a reference for these velocities, let us calculate the injection velocity required to escape from the solar system with injection at the Earth's orbit.

Sec. 5-7 The Intercept Problem and Fast Transfers

$$V_i = \sqrt{\frac{2\mu_s}{a_E}} = 42{,}120 \text{ m/s}$$

Our last injection velocity of 36,860 m/s is 87.5% of the solar system escape velocity.

Before leaving this section, we need to talk about what might be described as *secondary interception and rendezvous* when both the target and the interceptor are in the same orbit, either heliocentric or planetocentric, and at a reasonably large distance from each other. This might happen intentionally when the proper launch window (lead angle) for a direct intercept was not available or a plane change was to be made at the target orbit. Or it might happen unintentionally due to injection or navigation or guidance and control errors.

Let us assume that our spacecraft did not intercept the SOI of Mars and was placed in the Mars orbit either behind or ahead of Mars. There are various schemes for performing a secondary interception. The minimum-energy rendezvous technique is to apply a braking ΔV so as to enter an interior heliocentric Hohmann elliptical transfer that has a period equal to the time for Mars to reach the aphelion (point A) of the parking ellipse as shown in Fig. 5-7-4a with the spacecraft behind Mars and in Fig. 5-7-4b with the spacecraft ahead of Mars. It is important to check that the spacecraft does not approach too closely to or impact the Sun, particularly when the spacecraft is ahead of Mars. In such a case it will be necessary to increase the period of the parking orbit by another heliocentric revolution by Mars.

Example 5-7-3

A spacecraft was placed in the Mars orbit 20° (43.2×10^6 nmi) behind Mars. Find the ΔV_T and time required to execute a rendezvous with an interior Hohmann transfer.

Solution The period of the parking ellipse P_t is

$$P_t = P_M - \frac{20}{360} P_M = 0.9444 P_M = 5.606 \times 10^7 \text{ s} = 648.8 \text{ days}$$

Using the expression for the period of an ellipse gives us

$$a_t = \left[\mu_s \left(\frac{P_t}{2\pi} \right)^2 \right]^{1/3} = 2.194 \times 10^{11} \text{ m}$$

$$E_t = \frac{-\mu_s}{2a_t} = -3.024 \times 10^8 \text{ m}^2/\text{s}^2$$

From the energy equation,

$$V_{ta} = 23{,}650 \text{ m/s}$$

$$\Delta V_1 = 23{,}650 - 24{,}140 = -490 \text{ m/s}$$

Since $|V_2| = |\Delta V_1|$,

$$\Delta V_T = 2\Delta V_1 = 980 \text{ m/s}$$

To execute this minimum-energy rendezvous requires that $\Delta V_T = 980$ m/s and a time of 648.8 days, a long time considering the fact that it took 259 days to reach the Mars orbit.

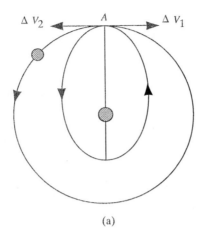

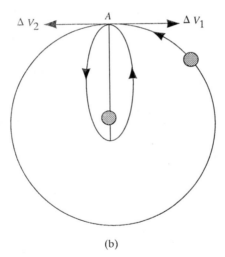

Figure 5-7-4. Secondary rendezvous of spacecraft with Mars: (a) behind Mars; (b) ahead of Mars.

As a matter of interest, let us find the other characteristics of the ellipse:

$$H = a_M V_{ta} = 5.392 \times 10^{15} \text{ m}^2/\text{s}$$

Knowing E and H yields

$$\epsilon_t = 0.0373$$

$$r_p = a_t(1 - \epsilon) = 2.112 \times 10^{11} \text{ m}$$

which is almost at the orbit of Mars and is a long way from the surface of the Sun.

If the spacecraft were 20° ahead of Mars, as in Fig. 5-7-4b, the time for Mars to reach point A would be $0.0556 P_M$, or 38.17 days. An interior parking ellipse with

Sec. 5-7   The Intercept Problem and Fast Transfers

that period would have a semimajor axis $a_t = 3.319 \times 10^{10}$ m. Since this is less than a_M, such an ellipse would be impossible. If the spacecraft goes into a parking orbit with a period of $1.0556 P_M$ while Mars makes a complete revolution plus in order to return to point A, the rendezvous time would be 725.2 days and the ellipse would cross the Mars orbit.

There are several ways to accomplish a faster rendezvous, albeit at the expense of larger ΔV's. The first of these would be an exterior non-Hohmann ellipse with interception of Mars at a predetermined point and time, as illustrated in Fig. 5-7-5. If we look closely at this trajectory, we see that it resembles that of a ballistic missile, normally launched from the surface of the Earth but in this case from the Mars orbit. You may wish to return to this trajectory after looking at ballistic missile trajectories in Chapter 9. This rendezvous trajectory requires two noncollinear firings.

A second possibility is a combination of two intersecting interior elliptical trajectories, as shown in Fig. 5-7-6; there will be three ΔV's. If the two interior ellipses are Hohmann ellipses, there will be two collinear firings and one non-collinear ΔV. A third technique might be a fast (high-energy) interior ellipse with two noncollinear firings, as shown in Fig. 5-7-7.

All of these intercept and rendezvous maneuvers assume impulsive thrusting, high thrust over a "small" period of time. With the thrusting force T given by the approximation

$$T \cong m \frac{\Delta V}{\Delta t}, \quad \text{then} \quad \Delta t \cong \frac{\Delta V}{T/m}$$

With a large thrust per unit mass T/m (commonly and imprecisely referred to as the thrust-to-weight ratio), Δt is small with respect to the other times of the problem and approaches "zero" in the limit for increasing T/m.

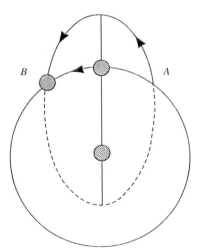

Figure 5-7-5. Rendezvous using an exterior ellipse.

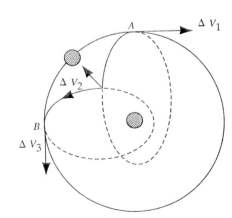

Figure 5-7-6. Rendezvous with two interior Hohmann transfers.

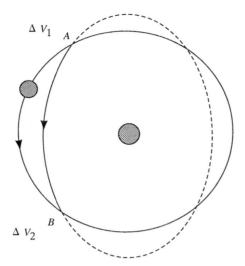

Figure 5-7-7. Rendezvous with interior ellipse and two noncollinear ΔV's.

As mentioned in passing at the beginning of this section, all of the heliocentric intercept and rendezvous techniques can be applied to planetocentric situations. For example, the rescue of an ill astronaut from a space station orbiting at an altitude of 414 km (225 nmi) is an example of a rendezvous where the target location is known and the injection velocity is limited by the booster being used. Another example is the interception of a GEO satellite for service and repair from a space shuttle at 322 km (175 nmi).

When the target and intercepting bodies (spacecraft) are close to each other, either in the same or closely adjacent orbits, we are apt to speak of *proximity operations and maneuvers* in which impulsive ΔV's and transfer ellipses are not necessary or appropriate. In such cases, relative motion equations are developed using the target orbit as the reference. Velocity corrections are made with low-thrust thrusters and are generally long-term (not impulsive) and may even be continuous. We do not treat such operations in this book.

The words *intercept* and *rendezvous* have been rather freely interchanged in this section without any obvious differentiation. To many people, however, *intercept* implies a matching of position only as in the hostile interception (and destruction) of a spy satellite, whereas *rendezvous* implies a matching of both position and velocity, as in a friendly joining up of two spacecraft. Another possible distinction might be to use intercept to describe impulsive maneuvers and reserve rendezvous for proximity operations. Whichever word is used and whatever the context, the operation in hand should be obvious.

5-8 CANONICAL UNITS

With the dimensional units that we have been using, numbers are not only large but are also imprecise because of round-off errors, time-varying changes in orbital parameters, and uncertainties as to the values of such fundamental quantities as the

Sec. 5-8   Canonical Units

distance of the Earth from the Sun and the universal gravitational constant. As a consequence, some authors favor and use canonical units, which form a system of normalized units for a particular central force field and use distance units (DU), time units (TU), and speed units (SU).

The advantages of such a canonical system are lack of dependence on the value of uncertain or time-varying quantities, smallness and greater precision of numbers, and the commonality of numbers for identical maneuvers in the various central force fields. One of the major disadvantages is tutorial, in that the physical significance of a canonical unit is not obvious; we can visualize 11,800 m/s, but can we visualize 1.414 DU/TU? Furthermore, we need dimensional units for ΔV when sizing boosters in Chapter 6; canonical units need to be transformed back to dimensional units. Final disadvantages are the necessity to identify the central force field being used as a reference and to convert to a common system when calculating velocity budgets for patched conic (interplanetary) missions.

Canonical units are based on a circular reference orbit for the central force field of interest. In each case the gravitational parameter and a characteristic distance, the radius of the reference orbit, are set equal to unity.

Looking at *heliocentric canonical units*, the reference orbit is the circular orbit of the Earth about the Sun and the distance unit is the distance of the Earth from the Sun; that is,

$$1 \text{ DU} = 1 \text{ AU} \cong 1.496 \times 10^{11} \text{ m}$$

With respect to the speed unit,

$$1 \text{ SU} = 1 \text{ DU/TU}$$

The speed unit can be defined directly as the circular velocity of the Earth in the reference orbit, so that

$$1 \text{ SU} = \frac{1 \text{ DU}}{\text{TU}} = \sqrt{\frac{\mu_S}{1 \text{ DU}}}$$

Solving for μ_S, we find that

$$\mu_S = 1 \text{ DU}^3/\text{TU}^2$$

If we define the time unit as the period of the reference orbit divided by 2π (1 TU = $P/2\pi$), we also obtain the result that

$$\mu_S = 1 \text{ DU}^3/\text{TU}^2$$

Obviously, these canonical units have dimensional values that must be used when combining canonical units from different central force fields. The values associated with the heliocentric canonical units are

$$1 \text{ DU}_S = 1 \text{ AU} = 1.496 \times 10^{11} \text{ m}$$

$$1 \text{ TU}_S = 5.023 \times 10^6 \text{ s} = 58.14 \text{ days}$$

$$1 \text{ SU}_S = 29,790 \text{ m/s}$$

$$\mu_S = 1 \text{ DU}^3/\text{TU}^2 = 1.327 \times 10^{20} \text{ m}^3/\text{s}^2$$

Unless the referenced central force field is obvious, it is wise to use subscripts, such as the S above, to identify it.

Using the heliocentric canonical units, the circular orbital velocity of the Earth is given by

$$V_{cs} = \sqrt{\frac{\mu_S}{a_E}} = \sqrt{\frac{1 \text{ DU}^3/\text{TU}^2}{1 \text{ DU}}} = 1 \text{ DU/TU} = 1 \text{ SU}_S = 29{,}790 \text{ m/s}$$

The velocity required to escape from the solar system is

$$V_{esc} = \sqrt{\frac{2\mu_S}{a_E}} = \sqrt{2} \text{ SU}_S = 42{,}130 \text{ m/s}$$

Now looking at *planetocentric canonical units*, specifically for the Earth,

$$1 \text{ DU}_E = r_E = 6.378 \times 10^6 \text{ m}$$

$$1 \text{ TU}_E = \frac{P}{2\pi} = 806.8 \text{ s} = 13.44 \text{ min}$$

$$1 \text{ SU}_E = 1 \text{ DU/TU} = 7905 \text{ m/s}$$

$$\mu_E = 1 \text{ DU}^3/\text{TU}^2 = 3.986 \times 10^{14} \text{ m}^3/\text{s}^2$$

The circular satellite velocity at the surface of the Earth (1 DU) is

$$V_{cs} = 1 \text{ SU}_E = 7905 \text{ m/s}$$

and at 300 nmi (1.0865 DU) is

$$V_{cs} = \sqrt{\frac{1 \text{ DU}^3/\text{TU}^2}{1.0865 \text{ DU}}} = 0.9594 \text{ SU} = 7584 \text{ m/s}$$

TABLE 5-8-1 PLANETOCENTRIC SPEED UNIT CONVERSION FACTOR[a]

Planet	Planetocentric speed unit / Heliocentric speed unit
Sun	1.0000
Mercury	0.1001
Venus	0.2459
Earth	0.2654
Mars	0.1194
Jupiter	1.4140
Saturn	0.8438
Uranus	0.5067
Neptune	0.5628
Pluto	0.0274

[a] 1 SU_S = 29,790 m/s.

Sec. 5-9   Closing Remarks   **137**

The escape velocity from the surface of the Earth is 1.414 SU, as was the case for escape from the solar system, but this time the escape velocity is 11,180 m/s as compared with 41,230 m/s for escape from the solar system.

Similar canonical units exist for the other planets. Since the distance and time units in each reference system are different, so are the speed units and conversion among systems is required. A speed conversion based on the heliocentric speed unit is listed in Table 5-8-1 for all the planets.

5-9 CLOSING REMARKS

Although this has been a long and busy chapter, the sections have been reasonably self-contained. Even though the examples have concentrated on transfers from inner to outer orbits (from Earth to Mars), we did look at the problem of returning from Mars on a Hohmann transfer and at a Hohmann transfer to Venus. Some of the problems at the end of this chapter deal with heliocentric transfers from outer to inner orbits (from Earth to Venus, for example) and with planetocentric intercepts and rendezvous in both directions.

In preliminary and feasibility analyses of all types of missions and maneuvers, determination of an approximate velocity budget and comparative ΔV values[2] is of paramount importance. As we shall see in Chapter 6, the velocity requirements of a mission affect and define the size and nature of the propulsion system and of the booster systems that are used to place the spacecraft in the trajectories called for by the mission.

Two notable omissions in this chapter and book are a full treatment of noncoplanar transfer trajectories and any treatment of the navigation, guidance, and control required to accomplish an assigned mission, to enter an SOI at the desired location with the desired V_∞ and to handle multiple planetary flybys (swingbys), for example. Unfortunately, these topics are beyond the scope of this book, as is any treatment of three-body problems, even the restricted three-body Earth–Moon problem. The presence of a third body, such as Jupiter, can affect a heliocentric transfer trajectory under the right set of conditions, as can the Sun affect a planetary passage if the alignment of the planet, spacecraft, and Sun is appropriate.

After we see how to get into space in Chapter 6, we shall return to the SOI of the Earth and look, primarily in conceptual and general terms, at the problem of entering an atmosphere.

If at this point you wish to learn about the other orbital elements and see how the track of a geocentric satellite looks from the surface of the Earth, you may want to skip to Chapter 8. Otherwise, let us press on and look at space boosters.

---

[2] C_3, the minimum launch energy (km²/s²), is frequently used in evaluating various swingby schemes.

PROBLEMS

You are encouraged to make sketches to aid you in solving the problems and to retain only four significant digits.

5-1. Verify the value given in Table 5-3-1 for r_{SOI} of the following planets with respect to the Sun:
  (a) Mercury
  (b) Neptune
  (c) Jupiter
  (d) Venus

5-2. A spacecraft is approaching Venus with $V_\infty = 2760$ m/s and $\phi_{SOI} = -65°$.
  (a) Find the impact parameter b (m and nmi).
  (b) Find the approach distance d and describe what will happen to the spacecraft. Find the impact velocity in the event of impact.
  (c) Find r_p and verify the results of part (b).
  (d) Find V_p.

5-3. Do Problem 5-2 for $V_\infty = 2760$ m/s and $\phi_{SOI} = -88.97°$.

5-4. A spacecraft is approaching Jupiter with $V_\infty = 10,000$ m/s and an approach distance $d = 500,000$ nmi.
  (a) Use the impact parameter to determine if the spacecraft will hit Jupiter or fly by; if impact, find V_{impact}.
  (b) Find r_p and verify part (a).
  (c) Find V_p and ϕ_{SOI}.

5-5. Do Problem 5-4 for $V_\infty = 10,000$ m/s and $d = 150,000$ nmi.

5-6. A spacecraft is to escape from the surface of the Earth with a $V_\infty = 10,280$ m/s and to be parallel to the Earth's orbital velocity at the time of escape and launch is to be at perigee.
  (a) Find the launch (injection) velocity with an Oberth escape.
  (b) Find the angle between the Earth's orbital velocity and the perigee position vector for $r = \infty$; for $r = r_{SOI}$.
  (c) Sketch the escape trajectory, showing its relationship to the Earth's orbital velocity.

5-7. At the time of launch of a spacecraft into an Oberth escape trajectory, the angle between the Earth's orbital velocity and the position vector of the launch site is 80°. V_∞ is to be parallel to the Earth's velocity and equal to 10,280 m/s.
  (a) Find the launch velocity and elevation angle, ϕ_L. (*Hint:* You may have to use iteration.)
  (b) Locate perigee for the escape trajectory.
  (c) Sketch the escape trajectory, showing perigee.

5-8. For the spacecraft of Problem 5-7, the launch conditions and requirements are the same with the exception that launch is at perigee.
  (a) Will V_∞ be parallel to the Earth's orbital velocity? If not, what are the values of the heliocentric elevation angle and velocity?
  (b) Describe the escape hyperbola.

5-9. With a direct Oberth escape from the surface of the Earth, find the velocity budget (ΔV_T) and transfer time for a Hohmann transfer to:
  (a) Mercury
  (b) Venus
  (c) Jupiter

(d) Saturn
(e) Uranus
(f) Neptune
(g) Pluto
(h) The surface of the Sun

5-10. With a direct escape (with V_∞ parallel to V_E) from the surface of the Earth and a Hohmann transfer to Saturn: Find the velocity budget and transfer time for:
  (a) an Oberth escape.
  (b) a parabolic escape.
  (c) Compare the ΔV_T requirements.

5-11. Do Problem 5-9 with escape from a 300-nmi parking orbit.

5-12. A spacecraft is on a Hohmann transfer from Earth to Mars with $d = 0$. As the spacecraft passes the end of the minor axis, it is given a collinear $\Delta V = 10$ m/s. Find the new approach distance and velocity.

5-13. You are doing the planning for a minimum-energy mission to Venus with a direct Oberth escape from the surface of the Earth, a phase II Hohmann transfer, and a phase III elliptical Venusian parking orbit with a periapsis height of 500 km and a period of 12 h.
  (a) Find the total velocity budget (m/s and km/s).
  (b) What should the approach distance at Venus be, and what is the planetocentric elevation angle at r_{SOI}?
  (c) What is the eccentricity of the Venusian orbit and the apoapsis height?
  (d) What is the transfer time between Earth and Venus?

5-14. A spacecraft enters the SOI of Saturn with a heliocentric velocity of 6000 m/s, a heliocentric elevation angle of 15°, and $d = 100{,}000$ nmi.
  (a) What is the value of V_∞?
  (b) Describe the incoming planetocentric hyperbola to include the location of periapsis. Will the spacecraft impact without a ΔV? If so, what are the impact velocity and elevation angle?
  (c) Find the ΔV required to enter a 300-nmi circular parking orbit. Find the period.
  (d) Find the ΔV required to enter an elliptical orbit with periapsis at 300 nmi and $\epsilon = 0.90$. Find the period. Compare the results with those of part (c).

5-15. Do Problem 5-14 with $d = 500{,}000$ nmi. Find the d required to place periapsis of the arrival hyperbola at an altitude of 300 nmi.

5-16. Plans are being made for an Earth–Saturn mission with a sunny-side (in front of) swing by Jupiter. Phase I will be an Oberth escape from the surface of the Earth, phase II will be a Hohmann transfer to Jupiter, and phase III will be a hyperbolic flyby of Jupiter with an approach distance of 450,000 nmi.
  (a) Find the heliocentric velocity and V_∞ as the spacecraft enters the SOI of Jupiter and the characteristics of the flyby trajectory to include the altitude of closest approach to Jupiter.
  (b) Find the turning angle δ, the heliocentric velocity, and the heliocentric elevation angle as the spacecraft leaves the SOI of Jupiter. Draw the vector diagram representing the turning of V.
  (c) Describe the modified heliocentric trajectory.
  (d) Will the modified trajectory intercept the orbit of Saturn? If so, what will V_∞ and the intersection angle (heliocentric elevation angle) be?
  (e) What is the velocity budget for phases I through III?

5-17. Plans are being made for a sunny-side flyby of Venus with an approach distance of 10,000 nmi and $V_i = 38,000$ m/s.
  (a) Find the change in the heliocentric velocity after the flyby with $\beta_i = 0°$.
  (b) Describe and sketch the modified trajectory after the flyby.
  (c) Do part (a) with $\beta_i = 10°$.

5-18. Do Problem 5-17 for a shady-side flyby of Venus and point out any differences in the modified trajectories.

5-19. A minimum-energy round trip to Venus is being planned with escape from the surface of the Earth and a 175-nmi circular parking orbit about Venus.
  (a) Find the total velocity budget to include the return to Earth with a hard landing.
  (b) Find the time to reach Venus to include the time for escape and the lead angle at the time of injection into the transfer trajectory.
  (c) Find the wait time at Venus and the return lead angle.
  (d) What is the total time required for the trip?

5-20. An orbital service unit (OSV) is dispatched from a space station in a 300-nmi (552-km) circular orbit to retrieve a satellite in a GEO at an altitude of 19,490 nmi (35,860 km). The velocity relative to the space station at dispatch is 3090 m/s and is tangent to the orbit of the space station.
  (a) Determine the elements of the transfer trajectory.
  (b) Find the velocity, elevation angle, and true anomaly when the OSV reaches the upper orbit.
  (c) What ΔV is required to inject the OSV into the upper orbit?
  (d) If the OSV intercepts the satellite, what was the lead angle at departure?
  (e) If, after entering the upper orbit, the OSV is 20° behind the satellite, what was the lead angle at departure and find a, ϵ, and P of an interior Hohmann ellipse that will allow the OSV to rendezvous with the satellite.

5-21. The spacecraft of Problem 5-16 leaves the SOI of Jupiter with a heliocentric velocity of 18,380 m/s and an elevation angle of $+7.0°$.
  (a) What is the time of flight to Saturn?
  (b) What lead angle is required for an interception of the SOI of Saturn?

5-22. Find the lead angle for a 50-day transfer from Earth to Mars.

CHAPTER 6

Vehicle and Booster Performance

Three Rocketdyne engines used in the Saturn boosters of the Apollo program: the H-1 on the left (890 kN of thrust); the F-1 in the center (6672 kN of thrust); and the J-2 on the right (890 kN). (Courtesy of NASA.)

6-1 INTRODUCTION

Now that we know how to determine the ΔV (energy) requirements of various missions (the velocity budget), it is time to consider method(s) for satisfying these requirements; that is, how to get a spacecraft into the trajectories and orbits of interest by increasing or decreasing its specific energy.

Through the centuries various schemes have been proposed for getting into space, principally to the Moon. In the seventeenth century, two Elizabethan bishops separately proposed reaching the Moon by being towed by birds and Cyrano de Bergerac in his *Voyage to the Moon* proposed using rockets lit in bursts, a crude form of staging. Then in the early nineteenth century Edgar Allen Poe proposed a balloon trip to the Moon in *The Unparalleled Adventures of One Hans Pfall*. With the assumption of an atmosphere extending to the Moon, Poe's hero ascended in his balloon, filled with an undiscovered-until-then gas whose density was about 37.4 times lighter than hydrogen, until the gravity of the Moon pulled him down to its surface.

Jules Verne, on the other hand, used a giant cannon to launch his manned spacecraft to the Moon in his two-book sequence: *From the Earth to the Moon* (1866) and *Round the Moon* (1870). Verne recognized the need for a launch window and for retro-rockets at the surface of the Moon and had his space travelers lie down to minimize the effects of the launch acceleration (which he grossly underestimated). In 1901, H. G. Wells in his *The First Men on the Moon* used a spacecraft covered with roller blinds coated with an antigravity material. By manipulating these blinds the crew could expose the spacecraft to only the gravitational attraction of interest, of the Moon for example.

Although all of these schemes were ingenious, creative, and impractical, they were exciting and can be given credit for inspiring an interest in space travel that eventually led to practical propulsion systems. A Russian engineer by the name of Konstantin Tsiolkowski (1857–1935) is considered to be the first to realize that the non-airbreathing rocket was the appropriate propulsion system for travel in space.

Sec. 6-1 Introduction

He is also credited with proposing the use of liquid propellants (the military rockets that had been used since at least the twelfth century all used solid propellants) and multistage boosters.

Tsiolkowski, however, was a theoretician and not an experimenter; it was Robert H. Goddard (1882–1945) who independently reached the same conclusions as Tsiolkowski and built and launched the first liquid-fueled rocket in 1926. Unfortunately, his work was largely unnoticed and unappreciated until after his death and until the start of the missile and space programs.

The propulsion system is the crux of the solution to the problem of getting into the desired trajectories in space. Although there are many possible propulsion schemes, current space systems rely primarily on chemical rockets, whose major characteristics and performance parameters are discussed in the next section. Since this is not a propulsion book, the coverage is necessarily limited.

In the section following the discussion of rocket engines, the performance equations are developed and the mass fraction concept introduced, both to be applied in the succeeding section to the sizing and configuration of a single-stage vehicle, which may be a booster or a spacecraft.

As propellant is consumed during engine operation, an increasing portion of the tankage or motor case obviously is emptied and becomes a dead weight that must be accelerated, along with the payload, to the required velocity. This dead weight handicaps and limits the performance of single-stage vehicles by increasing the total energy required. *Staging* is a mechanism used to reduce the amount of dead weight by discarding portions of the vehicle at discrete intervals. The advantages and disadvantages of staging are discussed and illustrated in the two sections subsequent to the one on single-stage vehicles.

Although a spacecraft spends most of its life in a space that is essentially free of external forces, it must initially be launched from the surface of the Earth and undergo a period of continuous thrusting during which the spacecraft and its booster pass through the atmosphere, where it is subjected to external aerodynamic forces (lift and drag) as well as the gravitational attraction of the Earth. This period of continuous thrusting is referred to as the *powered-boost trajectory*; the two-body equations of motion are obviously not valid. Furthermore, the ΔV required by the mission (the velocity budget) must be increased to compensate for the drag and gravity losses. Powered boost is covered in the next-to-last section and the chapter ends with some closing remarks that include a brief discussion on the use of lift in the boost phase.

SI units will be primary as before, with the mass in kg (kilograms) replacing the weight in kN (kilonewtons), as is the custom throughout most of the world, in describing weights of launch vehicles and components. Furthermore, Customary English units are shown in parentheses in many cases, as they are still used at times to describe rocket and booster performance. It is important to be consistent in the use of the various units. We shall express force (thrust) in kN, pressure in kPa (kilopascals), temperature in kelvin, energy in kJ (kilojoules), velocity in meters per second, mass in kg, and "weight" in kN.

6-2 ROCKET ENGINES

The major objective of a propulsion system is to produce thrust. A generalized functional diagram of a propulsion system is shown in Fig. 6-2-1, where the elements shown do not necessarily have to be separate; a component of a propulsion system may well accomplish more than one function. We do see from this figure that there is a need for an energy source (and possibly a conversion device to transform the energy into a usable form), a mechanism for transferring the energy to the propellant, which may or may not be separate from the energy subsystem and which is somehow fed into the thrust producer. Ancillary but important components are the control devices and the means for dissipating any waste energy.

Some of the possible energy sources are:

1. Chemical reactions (including combustion)
2. Nuclear reactions:
    a. Fission
    b. Fusion
    c. Radioactive decay (isotopes)
3. Solar energy
4. Gas under pressure
5. Batteries
6. Flywheels
7. Springs
8. Electromagnetic and gravity fields

Some thrust-producing devices are:

1. Hot-gas expanders (nozzles)
2. Propellers

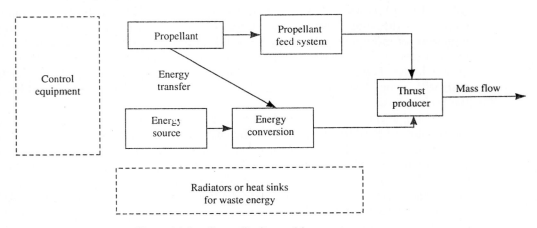

Figure 6-2-1. Generalized propulsion system.

Sec. 6-2 Rocket Engines

3. Particle accelerators:
   a. Electrostatic accelerators
   b. Magnetic accelerators
4. Direct impact from external forces (e.g., solar sailing and nuclear bombs)

Two major divisions of chemical propulsion systems are:

1. *Airbreathers*, in which the oxidizer (oxygen) comes from the atmosphere and the vehicle carries only the fuel
2. *Non-airbreathers*, in which the vehicle must carry the oxidizer as well as the fuel

Although the airbreathers are more efficient, space flight at present relies on non-airbreathing engines, primarily chemical rockets. We therefore restrict ourselves to chemical rockets and develop operational parameters and performance relationships that are specifically directed toward (chemical) rocket engines.

The first parameter we shall mention is the *total impulse* I_t, which has the units of N-s and is defined by the expression

$$I_t = \int T \, dt \tag{6-2-1}$$

where T is the thrust. If we consider an average thrust (or assume the thrust to be constant, as is frequently the case),

$$I_t = T_{\text{ave}} t_b$$

where t_b is the *burn time* (in seconds) of the engine (motor). We see from this expression that

$$t_b = \frac{I_t}{T_{\text{ave}}}$$

Consequently, impulsive thrusting, where t_b approaches zero in the limit, calls for high values of thrust.

The *specific impulse* I_{sp}, with units of seconds, is a very important parameter that is used more frequently than is the total impulse. One definition is that it is the total impulse per unit *weight* of propellant used and is given by the expressions

$$I_{\text{sp}} = \frac{I_t}{W_p} = \frac{I_t}{m_p g_0} \tag{6-2-2}$$

where the subscript p denotes the propellant and g_0 is the acceleration due to gravity at the Earth's surface.

The specific impulse is also defined as the thrust divided by the propellant *weight* flow rate $\dot{W}_p$, so that

$$I_{\text{sp}} = \frac{T}{\dot{W}_p} \tag{6-2-3}$$

With a constant or average thrust, the two definitions are equivalent, as can be seen by starting with Eq. (6-2-3) and ending with Eq. (6-2-2):

$$I_{sp} = \frac{T}{\dot{W}_p} = \frac{T}{\dot{m}_p g_0} = \frac{Tt_b}{m_p g_0} = \frac{I_t}{m_p g_0} = \frac{I_t}{W_p} \qquad (6\text{-}2\text{-}4)$$

At this point we need an expression for the thrust. Considering only the rocket motor and neglecting both the motion (velocity) of the motor itself and any external forces on the body, the idealized rocket motor can be represented by Fig. 6-2-2. In this figure, P_a is the ambient pressure, T_a is the ambient temperature, V_e is the exit velocity of the exhaust gases with respect to the rocket motor, P_e is the pressure in the exhaust gases at the exit, and P_c and T_c are the combustion chamber pressure and temperature, respectively. The thrust T is the resultant of the forces acting on the inner surface of the motor, and only the axial forces (normal to y) will contribute to the thrust.

Although it is customary to derive the thrust equation by use of the momentum equation, it is also possible, although less rigorously, to develop the thrust equation from Newton's law in its most general form:

$$\mathbf{F}_{\text{applied}} = \int \mathbf{a}\,dm \qquad (6\text{-}2\text{-}5)$$

where $\mathbf{a}$ is the vector acceleration of a particle of the propellant with respect to the rocket motor. With the assumptions of uniform, one-dimensional steady flow and with integration over the entire mass of particles, the resulting scalar equation can be written as

$$F_x = \int a_x\,dm \qquad (6\text{-}2\text{-}6)$$

Since $a_x = dV_x/dt$,

$$F_x = \int \frac{dV_x}{dt}\,dm = \int \dot{m}\,dV_x$$

where $\dot{m}$ (dm/dt) is the mass flow rate of the propellant. With the assumption of steady flow, $\dot{m}$ is constant and

$$F_x = \dot{m}\int_c^e dV_x = \dot{m}(V_e - V_c)$$

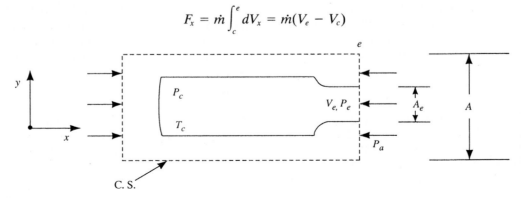

Figure 6-2-2. Idealized rocket motor.

Sec. 6-2 Rocket Engines

But the velocity in the combustion chamber V_c is so much less than the exhaust velocity V_e that, to a first approximation, it can be neglected. Accordingly,

$$F_x \cong \dot{m} V_e \qquad (6\text{-}2\text{-}7)$$

Summing up the axial forces both inside and outside the motor yields

$$F_x = T + P_a A - P_a(A - A_e) - P_e A_e$$

or

$$F_x = T - (P_e - P_a) A_e \qquad (6\text{-}2\text{-}8)$$

Substituting the expression of Eq. (6-2-7) for T in Eq. (6-2-8) and rearranging yields an expression for the thrust whereby

$$\boxed{T = \dot{m} V_e + 1000(P_e - P_a) A_e} \qquad (6\text{-}2\text{-}9)$$

where the 1000 in the second term of Eq. (6-2-9) is introduced to convert the kPa of P_e and P_a to Pa. In Eq. (6-2-9), the term $m V_e$ is known as the *momentum thrust*, and the term $1000(P_e - P_a) A_e$ is known as the *pressure thrust*.

Looking at the pressure thrust term, we see that:

1. If the exhaust gases are expanded to P_a, the pressure thrust goes to zero. In space, where $P_a = 0$, a P_e of zero imposes severe requirements on the length and exit area of the nozzle (i.e., on the expansion ratio of the nozzle).
2. If $P_e \neq P_a$ and is greater than P_a at sea level (at launch), the pressure thrust increases with altitude as the ambient pressure P_a decreases. Since, as will be shown, the momentum thrust is independent of altitude, the overall thrust will increase as the altitude increases.

With respect to the momentum thrust term, an expression for the exhaust velocity V_e can be obtained by applying the first law of thermodynamics to an idealized nozzle such as the convergent–divergent nozzle of Fig. 6-2-3. The assumptions associated with an ideal nozzle are an ideal gas that obeys the perfect gas law ($P = \rho R T$) and has constant specific heats, an isentropic (reversible and adiabatic)

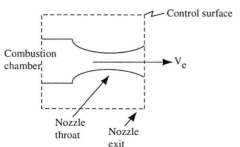

Figure 6-2-3. Convergent–divergent nozzle for supersonic exhaust velocity.

expansion so that there are no viscous losses or external heat transfer, frozen equilibrium (no dissociation), and one-dimensional flow.

The continuity (mass balance) equation yields the condition that the mass flow rate $\dot{m}$ is constant and can be expressed in terms of the exit conditions as

$$\dot{m} = \rho_e A_e V_e$$

where ρ_e is the density at the exit.

The first law (the energy equation) for this particular single flow can be written in terms of the specific enthalpy and specific kinetic energy as

$$h_c + \frac{V_c^2}{2} = h_e + \frac{V_e^2}{2} \qquad (6\text{-}2\text{-}10)$$

so that

$$\frac{V_e^2}{2} = h_c - h_e + \frac{V_c^2}{2} \cong h_c - h_e = C_p(T_c - T_e) \qquad (6\text{-}2\text{-}11)$$

where C_p is the constant-pressure specific heat and T_c and T_e are the absolute temperatures in the combustion chamber and at the exit, respectively. Since $V_c \ll V_e$, V_c^2 was assumed to be very small (approximately zero) and was dropped from the relationships. Solving Eq. (6-2-11) for V_e yields

$$V_e = \sqrt{2 C_p T_c \left(1 - \frac{T_e}{T_c}\right)} \qquad (6\text{-}2\text{-}12)$$

We now are going to expand Eq. (6-2-12), starting with an expression for R, the gas constant, in which

$$C_p - C_v = R = \frac{\overline{R}}{M} \qquad (6\text{-}2\text{-}13)$$

where C_v is the constant-volume specific heat, $\overline{R}$ the universal gas constant and M the molecular mass of the exhaust gases. But

$$C_p - C_v = C_p\left(1 - \frac{C_p}{C_v}\right) = C_p\left(1 - \frac{1}{k}\right) = C_p \frac{k-1}{k}$$

where k, the ratio of specific heats, is equal to C_p/C_v. With the relationship just developed for $C_p - C_v$, Eq. (6-2-13) can be rewritten to obtain an expression for C_p:

$$C_p = \frac{k}{k-1} \frac{\overline{R}}{M} \qquad (6\text{-}2\text{-}14)$$

We need one more relationship, that of the polytropic temperature ratio for an isentropic expansion, namely,

$$\frac{T_e}{T_c} = \left(\frac{P_e}{P_c}\right)^{(k-1)/k} \qquad (6\text{-}2\text{-}15)$$

If we substitute Eq. (6-2-15) along with Eq. (6-2-14) into Eq. (6-2-12), we finally obtain the expanded SI expression for the exhaust velocity in m/s whereby

$$V_e = \sqrt{2000 \frac{\overline{R}}{M} \frac{k}{k-1} T_c \left[1 - \left(\frac{P_e}{P_c}\right)^{(k-1)/k}\right]} \quad (6\text{-}2\text{-}16)$$

In Eq. (6-2-16), $\overline{R} = 8.314$ kJ/mol-K, T_c is the chamber temperature in kelvin, P_c/P_e is known as the *pressure ratio*, and the factor of 1000 is inserted to convert the joules of the specific kinetic energy expression into kilojoules.

Remember that the higher the exhaust velocity V_e, the larger the thrust for a given mass flow rate $\dot{m}$. From Eq. (6-2-16) we see that V_e is a function of the pressure ratio P_c/P_e, which is the inverse of the P_e/P_c term in Eq. (6-2-16). As the chamber pressure P_c increases, so does V_e. As the exit pressure P_e decreases, V_e increases. In both of these cases, the pressure ratio P_c/P_e increases, thus decreasing P_e/P_c and increasing V_e.

If $P_e > P_a$ (the ambient pressure), the nozzle is *underexpanded*; this is the usual and preferable condition. If $P_e < P_a$, the nozzle is *overexpanded*. The pressure ratio P_c/P_e is also a function of the *expansion ratio* ε, which is the ratio of the area of the exit of a convergent–divergent nozzle (see Fig. 6-2-4) to the throat area (i.e., $\varepsilon = A_e/A_t$). At the throat the flow Mach number is unity and is supersonic at the exit. When the exit pressure P_e is zero (when the nozzle is completely expanded into a vacuum), the pressure ratio is infinite and the exhaust velocity is at a theoretical maximum. It is still finite, however, and is given by

$$V_{e,\max} = \sqrt{2000 \frac{\overline{R}}{M} \frac{k}{k-1} T_c} \quad (6\text{-}2\text{-}17)$$

This mathematical and theoretical maximum cannot be reached physically because the exhaust gases cool so rapidly in such an expansion that the temperature will fall below the liquefaction or freezing temperature. Consequently, the exhaust no longer behaves as a gas, much less as an ideal gas.

We also see from Eq. (6-2-16) that

$$V_e \sim \sqrt{\frac{T_c}{M}} \quad (6\text{-}2\text{-}18)$$

so that the lower the molecular mass of the propellant(s), the larger the exhaust velocity. This can be an important consideration in the choice of propellants.

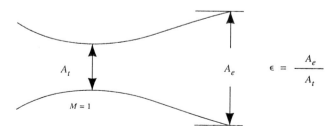

Figure 6-2-4. Expansion ratio of a convergent–divergent nozzle.

With respect to the mass flow rate, which is constant, it can be expressed in terms of the throat conditions as

$$\dot{m} = \rho_t A_t V_t$$

where V_t is the corresponding velocity for a unity Mach number (the sonic velocity). In practice, the mass flow rate at the exit is affected somewhat by the nozzle shape and internal flow conditions.

Defining an *effective exhaust velocity* c as the thrust divided by the mass flow rate and using the thrust equation of Eq. (6-2-9), c can be expressed as

$$c = \frac{T}{\dot{m}} = V_e + \frac{1000(P_e - P_a)A_e}{\dot{m}} \qquad (6\text{-}2\text{-}19)$$

As P_a decreases with increasing altitude, c will increase somewhat.

From the definition of c, a simplified thrust equation can be written as

$$\boxed{T = \dot{m}c = \frac{\dot{W}c}{g_0}} \qquad (6\text{-}2\text{-}20)$$

Since the pressure term in Eq. (6-2-19) is often small with respect to V_e, it is not uncommon to ignore it and to interchange V_e and c, approximating the thrust by the simple relationship

$$\boxed{T \cong \dot{m}V_e} \qquad (6\text{-}2\text{-}21)$$

Returning to the thrust equation of Eq. (6-2-20) and the definition of the specific impulse, we have

$$\boxed{I_{\text{sp}} = \frac{T}{\dot{W}_p} = \frac{c}{g_0}} \qquad (6\text{-}2\text{-}22)$$

Equation (6-2-22) shows the equivalence between the specific impulse and the effective exhaust velocity.

The specific impulse (or the effective exhaust velocity) is a measure of the efficiency of the propellant–nozzle combination. With the assumption of an ideal nozzle with an ideal expansion ratio, it is also a measure of the propellant efficiency. The higher the I_{sp}, the more thrust (and thus energy and ΔV) per kilogram of propellant there will be.

For a high I_{sp} we want the following:

1. High T_c and high P_c limited only by the thermal and structural characteristics of the combustion chamber and nozzle.
2. Low P_e but not less than P_a (the ambient pressure).
3. Low value of molecular mass M. With a chemical rocket, M is determined by

Sec. 6-2 Rocket Engines

the composition of the propellants. When the propellant is separate from the energy source, such as in a nuclear propulsion system, a light propellant, such as H_2 with $M = 2$, is an appropriate choice for the propellant.

Example 6-2-1

A propellant combination has a molecular mass of 23 kg/mol and a ratio of specific heats (k) equal to 1.24. The combustion chamber temperature and pressure are 3400 K and 6900 kPa with $P_e = 101.3$ kPa (1 atm).
(a) Find the value of C_p (kJ/kg-K); of C_v.
(b) Find the exhaust velocity V_e (m/s) at sea level, where $P_a = 1$ atm; at 9150 m, where $P_a = 37.88$ kPa.
(c) If $\dot{m} = 100$ kg/s and the exit diameter is 0.62 m, find the effective exhaust velocity c, the I_{sp}, and the thrust T at sea level; at 9150 m; in space, where $P_a = 0$.
(d) What is the theoretical maximum value of V_e and of I_{sp}?

Solution

(a)
$$R = \frac{\bar{R}}{M} = \frac{8.314}{23} = 0.3615 \text{ kJ/kg-K}$$

$$C_p = \frac{k}{k-1}R = 1.868 \text{ kJ/kg-K}$$

$$C_v = C_p - R = 1.506 \text{ kJ/kg-K}$$

(b)
$$\frac{P_e}{P_c} = 0.01468 \qquad \frac{P_c}{P_e} = 68.12$$

Substituting values into Eq. (6-2-16) results in

$$V_e = 2663 \text{ m/s (8737 ft/s)}$$

Since V_e is independent of the ambient pressure P_a, if P_e remains unchanged (the expansion ratio is constant), V_e does not change with altitude.

(c) With $V_e = 2663$ m/s and $A_e = \pi(0.31)^2 = 0.3019$ m^2, Eq. (6-2-19) can be written as

$$c = 2663 + \frac{1000(101.3 - P_a) \times 0.3019}{100}$$

At sea level with $P_a = 101.3$ kPa,

$$c = V_e = 2663 \text{ m/s}$$

$$I_{sp} = \frac{c}{g_0} = 271.5 \text{ s}$$

$$T = \dot{m}c = 266.3 \text{ kN (59,870 lb)}$$

At 9150 m with $P_a = 37.88$ kPa,

$$c = 2854 \text{ m/s}$$

$$I_{sp} = 290.9 \text{ s}$$

$$T = 285.4 \text{ kN (64,160 lb)}$$

All values are 7% higher than those at sea level.
In space (in a vacuum),

$$c = 2969 \text{ m/s}$$

$$I_{sp} = 302.6 \text{ s}$$

$$T = 296.9 \text{ kN } (66{,}750 \text{ lb})$$

All of these values are 11.5% higher than those at sea level.

(d) To find the theoretical maxima, use Eq. (6-2-17).

$$V_{e,\max} = \sqrt{2000 \times 0.3615 \left(\frac{1.24}{0.24}\right) \times 3400} = 3564 \text{ m/s}$$

and

$$I_{sp,\max} = \frac{3564}{9.81} = 363.3 \text{ s}$$

Some liquid bipropellant performance parameters are listed in Table 6-2-1. Other propulsion system parameters of varying degrees of importance and usefulness are listed and defined below.

The *impulse-to-weight ratio* does not have an individual symbol but is defined as I_t/W_i, where I_t is the total impulse and W_i is the total initial weight of the item in question, which might be a motor, a stage, or even a complete vehicle. We will not use this parameter.

The *propellant mass (weight) fraction* ζ_p, on the other hand, is an important parameter and is defined as the ratio of the mass (weight) of the propellant to the initial mass (weight) of the item in question, usually a stage or complete vehicle. ζ_p is always less than unity and is expressed as

$$\zeta_p = \frac{m_p}{m_i} = \frac{W_p}{W_i} < 1 \qquad (6\text{-}2\text{-}23)$$

The *thrust-to-weight ratio* is another useful parameter and can be expressed as

$$\frac{T}{W} = \frac{T}{mg_0} = \frac{\dot{m}c}{mg_0} = \frac{\dot{m}I_{sp}}{m} \qquad (6\text{-}2\text{-}24)$$

where the weight is normally the weight of the booster. With a constant $\dot{m}$ and a constant I_{sp}, the T/W ratio increases with time as propellant is consumed and the

TABLE 6-2-1 TYPICAL VALUES OF SOME LIQUID BIPROPELLANTS

Propellant[a]	r (mass)	M (kg/mol)	T_c (K)	k	I_{sp} (s)
LOX/RP-1	2.5	23	3300	1.24	310
LOX/LH$_2$	4	9	2700	1.26	400
F$_2$/H$_2$	7	10	3500	1.33	420
N$_2$O$_4$/UDMH (storable)	2	21	3000	1.24	290

[a] LOX, liquid oxygen; RP-1, kerosene; LH$_2$, liquid hydrogen; F$_2$, fluorine; N$_2$O$_4$, nitrogen tetroxide; UDMH, unsymmetrical dimethylhydrazine.

Sec. 6-2 Rocket Engines

weight decreases, reaching a maximum at t_{bo} (burnout). The farthermost expression in Eq. (6-2-24) can be used to relate the burn time t_b to I_{sp}. With the average $\dot{m} = m_p/t_b$, the initial thrust-to-weight ratio can be written as

$$\frac{T}{W_i} = \frac{m_p}{m_i}\frac{I_{sp}}{t_b}$$

or

$$\boxed{t_b = \frac{\zeta_p I_{sp}}{T/W_i}} \quad (6\text{-}2\text{-}25)$$

where $\zeta_p = m_p/m_i = W_p/W_i$, the *propellant mass fraction*, which is always less than unity. Therefore, if the initial T/W ratio is greater than unity, say at lift-off at launch, the burn time t_b must be less than the I_{sp} for the propellant being used.

Example 6-2-2

A launch vehicle has a lift-off T/W ratio of $1.5g_0$, a propellant mass fraction $\zeta_p = 0.9$, and an $I_{sp} = 260$ s.
(a) Find the time to burnout.
(b) Find the T/W ratio at burnout, assuming T to be constant.
Solution

(a) $$t_b = \frac{0.9 \times 260}{1.5} = 156 \text{ s}$$

(b) $$W_{bo} = W_i - \zeta_p W_i = W_i(1 - \zeta_p) = 0.1 W_i$$

$$\frac{T}{W_{bo}} = 10\frac{T}{W_i} = 15g_0$$

Equation (6-2-25) also confirms an earlier observation that a large T/W ratio is required if thrusting is to be "impulsive" (i.e., if t_b is to be "small").

Another propulsive parameter is the *thrust coefficient* C_F, which is defined, without proof or further discussion, as

$$C_F = \frac{T}{P_c A_t} \quad (6\text{-}2\text{-}26)$$

where T is the thrust, P_c the combustion chamber pressure, and A_t the throat area of the nozzle. Equation (6-2-26) gives us another thrust equation, namely,

$$T = C_F P_c A_t \quad (6\text{-}2\text{-}27)$$

Since the thrust increases with altitude, so does C_F.

To add to the confusion, there is also a *characteristic exhaust velocity* c^*, which is defined as the product of P_c and A_t divided by the mass flow rate $\dot{m}$. This definition and the relationships of c^* with I_{sp} and c are shown below.

$$c^* = \frac{P_c A_t}{\dot{m}} = \frac{I_{sp} g_0}{C_F} = \frac{c}{C_F} \quad (6\text{-}2\text{-}28)$$

We will not use or mention C_F or c^* again in this book, although you should be prepared to run into one or both in the literature.

Although the current practice is to keep the thrust constant for the majority of the propulsion applications (to minimize throttling losses), the thrust can be varied if need be by the following techniques:

1. Varying the propellant mass flow rate $\dot{m}_p$ of liquid propellant engines by the use of throttling valves. This is a "lossy" process with reduced efficiency.
2. Change the nozzle throat area A_t with a movable plug inserted into the nozzle. This is often used with throttling.
3. Changing the combustion chamber temperature T_c by varying the oxidizer-to-fuel mixture ratio of liquid propellant and hybrid engines, for example.

With chemical rockets the energy source is a chemical reaction (usually combustion) with the reaction products serving as the propellant. Combustion requires an oxidizer and a fuel. The various types of chemical rockets are:

1. Liquid propellant
   a. Bipropellant with separate oxidizer and fuel (the customary case)
   b. Monopropellant with the oxidizer and fuel combined
2. Solid propellant
3. Hybrid, a combination of liquid and solid propellants

As might be expected, there are advantages and disadvantages to each type of propellant. With liquid propellants the combustion chamber [with nozzle(s) attached] is separate from the propellant tanks and the pressure in the tanks is independent of the combustion chamber pressure. Furthermore, the thrust can be vectored by gimballing (moving) the combustion chamber and nozzle(s) as a unit, but not the tanks, which can be as large as necessary to give the desired burn time. Stopping a liquid engine (terminating thrust) is a simple matter of closing a valve and cutting off the flow of propellant to the combustion chamber. Unfortunately, many of the liquid propellants are cryogenic and impose storage problems. Propellant feed can be accomplished by using turbopumps driven by exhaust gases or a hot gas generator or by pressurizing the propellant tanks.

Solid propellants, on the other hand, form the combustion chamber itself so that the physical properties (e.g., strength) must be considered as well as control of the burning rate and the avoidance of hot spots at the walls of the motor. The longer the burning time, the larger the motor must be and the larger and heavier the pressure vessel (the motor case). Furthermore, thrust vectoring calls for swiveled nozzles (introducing seal problems) unless vanes or other devices that intrude into the exhaust are used. However, solid propellants are storable, whereas liquids are often cryogenic. Thrust termination can be difficult, and restart for most configurations is not possible. However, solids do not require a propellant feed system. Although the I_{sp} of solids is lower than that of liquids, the density (specific gravity) is higher and the density specific impulse is competitive with that of liquids.

Sec. 6-2 Rocket Engines

The hybrid rocket engine combines a liquid oxidizer (either storable or cryogenic) with a solid fuel (usually an inert, rubberlike material or even a plastic) and exploits the advantages of both liquid and solid propellants, plus a few advantages peculiar to the hybrid, while minimizing their disadvantages. The advantages of the hybrid include safety (it is inert prior to mixing of the propellants), a start–stop–restart capability (by use of an oxidizer cutoff valve), variable thrust (by controlling the oxidizer flow rate), and I_{sp} values (and densities) between those of the solids and the liquids.

As for usage, the initial large space boosters were liquids and then were augmented by strap-on solids. Upper stages for placing payloads into geocentric orbits or into heliocentric trajectories [often called inertial upper stage(s) (IUS)] were predominantly liquid in the early days but are now being supplemented and in some cases replaced by solid propellant IUSs. Orbital transfers and maneuvering are generally accomplished with solid propellant motors. The hybrid is attracting attention and has the potential of replacing the liquids and the solids being used for orbital transfers and maneuvering. They could conceivably be used in IUSs and eventually in boosters; there is a current proposal for using hybrids in a medium booster.

Now that we are discussing propellants, another parameter of interest for liquid propellants is r, the *oxidizer-to-fuel ratio* or *mixture ratio*, which is simply equal to the oxidizer flow rate $\dot{m}_{ox}$ divided by the fuel flow rate $\dot{m}_f$, or

$$r = \frac{\dot{m}_{ox}}{\dot{m}_f} \qquad (6\text{-}2\text{-}29)$$

Sometimes this ratio is expressed as a ratio of the volumetric flow rates. The oxidizer-to-fuel ratio affects the ratio of the chamber temperature to the molecular mass (i.e., T_c/M) by changing both T_c and M.

The propellant mass flow rate $\dot{m}_p$ is the sum of the component mass flow rates:

$$\dot{m}_p = \dot{m}_{ox} + \dot{m}_f \qquad (6\text{-}2\text{-}30)$$

Some typical sea-level values of I_{sp} for several of the more popular bipropellant combinations are listed in Table 6-2-1. The values given in this table are valid for the following conditions:

1. $P_c = 6895$ kPa and $P_e = 101.3$ kPa.
2. Ideal nozzles with optimum expansion ratios.
3. Adiabatic combustion and isentropic expansion with frozen equilibrium (as opposed to shifting equilibrium).

As the altitude at which a rocket engine is operating increases, so does the specific impulse (and thrust). In a vacuum, the I_{sp} is on the order of 10% higher than the sea-level value.

The following example is introduced to give an idea of the order of magnitude of the performance of high-energy liquid bipropellant combinations.

Example 6-2-3

Find the theoretical maximum value of V_e and I_{sp} for a fluorine–hydrogen combination with $M = 10$ kg/mol, $T_c = 3500$ K, and $k = 1.33$.

Solution Using Eq. (6-2-17), we obtain

$$V_{e,\max} = \sqrt{2000 \times \frac{8.314}{10}\left(\frac{1.33}{1.33-1}\right) \times 3500} = 4843 \text{ m/s}$$

and

$$I_{sp,\max} = \frac{V_e}{g_0} = 493.7 \text{ s}$$

Solid propellant combinations fall into three categories: double-base, composite, and composite modified double-base. With double-base propellants the oxidizer and fuel are combined in the same molecule and form a homogeneous grain. The composite propellants, on the other hand, are individual but mixed throughout a heterogeneous grain in a rubberlike binder that may also serve as the fuel. The composite modified double-base, as might be expected, is a combination of the other two.

The composite solid propellant is preferred for most uses. The oxidizer of a typical composite is usually a perchlorate, such as ammonium perchlorate, and the fuel a polymer, such as polybutadiene (a synthetic rubber), with aluminum added. Molecular masses are on the order of 25 to 30 kg/mol and sea-level I_{sp} values are on the order of 270 s.

Chemical rockets can be characterized as high-thrust devices of relatively low specific impulse. Solid-core nuclear rockets with separate propellants that are not involved in the production of the energy are also high-thrust devices, albeit heavy and still temperature limited, with the potential for higher I_{sp} values. Using H_2 with $M = 2$ kg/mol, the potential I_{sp} is on the order of 1000 s; with ammonia (NH_3) and $M = 8.5$ kg/mol, the I_{sp} would be on the order of 600 s. A successful gaseous core reactor, on the other hand, has the potential for I_{sp} values on the order of 10,000 s. Although nuclear propulsion systems are high-thrust systems, they may not be suitable for first (lift-off) stages because of the possibility of nuclear radiation in the exhaust gases.

Electric propulsion systems also have the potential for I_{sp} values on the order of 10,000 s, but the thrust output, although continuous, would be very low, on the order of $10^{-3} g_0$. Such propulsion systems are obviously unsuited for lift-off stages and do not meet the criterion of small burn times for impulsive thrusting. Consequently, our two-body (force-free) equations of motion are not valid with low-thrust propulsion systems; low-thrust trajectories are powered trajectories without applied gas dynamic (aerodynamic) forces.

We close this section with mention of *hypergolic ignition*, which is a spontaneous ignition when a fuel and oxidizer are mixed. Most hydrazine derivatives and F_2 are hypergolic.

6-3 THE PERFORMANCE EQUATIONS

Considering the one-dimensional translational motion of a vehicle, the scalar equation of motion of a rigid body is

$$F_{\text{external}} = ma = m\frac{dV}{dt} \qquad (6\text{-}3\text{-}1)$$

where m is the instantaneous mass of the vehicle to include the propellant mass m_p and a and V are the acceleration and velocity of the cm of the rigid body with respect to an inertial reference frame.

When rocket engines are fired in space, since the thrust T is an applied external force, the two-body Keplerian equations of motion are not valid during this period of propulsive burning. Since, as was derived in the preceding section,

$$T = \dot{m}_p c = \frac{dm_p}{dt} c \qquad (6\text{-}3\text{-}2)$$

and the only change in the mass of the vehicle is the consumption of propellant, then $dm_p/dt = -dm/dt$ and Eq. (6-3-1) can be written as

$$-c\frac{dm}{dt} = m\frac{dV}{dt} \qquad (6\text{-}3\text{-}3)$$

If time is explicitly removed and the variables m and V are separated, Eq. (6-3-3) becomes

$$dV = -c\frac{dm}{m} \qquad (6\text{-}3\text{-}4)$$

Equation (6-3-4) can now be integrated from i to f, where i denotes the conditions before firing and f the conditions at the completion of firing. With the assumption of a constant c, the results of the integration are the expressions

$$\boxed{\Delta V = c \ln \frac{m_i}{m_f} = c \ln \text{MR}} \qquad (6\text{-}3\text{-}5)$$

or

$$\boxed{\text{MR} = \exp\left(\frac{\Delta V}{c}\right)} \qquad (6\text{-}3\text{-}6)$$

In these equations it is apparent that MR, the *mass ratio*, is defined by

$$\text{MR} = \frac{m_i}{m_f} = \frac{W_i}{W_f} \qquad (\text{always} > 1) \qquad (6\text{-}3\text{-}7)$$

where m_i and W_i are the initial mass and weight (before firing) and m_f and W_f are the final mass and weight (after firing). Now

$$m_f = m_i - \Delta m_p$$

where Δm_p is the propellant consumed during the burn. If the *propellant mass fraction* ζ_p is redefined in terms of the propellant consumed, then

$$\zeta_p = \frac{\Delta m_p}{m_1} = \frac{\Delta W_p}{W_i} \qquad (\text{always} < 1) \qquad (6\text{-}3\text{-}8)$$

and

$$\text{MR} = \frac{1}{1 - \zeta_p} \qquad \text{and} \qquad \zeta_p = \frac{\text{MR} - 1}{\text{MR}} \qquad (6\text{-}3\text{-}9)$$

m_i, the *initial gross (total) mass* of a vehicle or of a particular stage of the vehicle can be expressed as the sum of its major components: for example,

$$m_i = m_s + m_p + m_{\text{PL}} \qquad (6\text{-}3\text{-}10)$$

where m_{PL} is the *payload mass*, m_p the total *propellant mass* (to include Δm_p plus residuals and reserves), and m_s, which will be referred to as the *overall structural mass*, actually includes everything else, such as engines, tanks, cases, guidance and control, and pumps. If we divide each term in Eq. (6-3-10) by the initial gross mass, the resulting mass fractions must always add up to unity, as shown by

$$1 = \frac{m_s}{m_i} + \frac{m_p}{m_i} + \frac{m_{\text{PL}}}{m_i} \qquad (6\text{-}3\text{-}11)$$

where m_s/m_i is the *overall structural mass fraction*, m_p/m_i the *total propellant mass fraction*, and m_{PL}/m_i the *payload mass fraction*, which is generally referred to as the *payload ratio*. It is convenient at times, particularly in performing preliminary analyses, to neglect any propellant residuals and reserves and to approximate m_p/m_i by ζ_p.

Rearranging Eq. (6-3-11) yields an expression for the payload mass fraction (payload ratio) in terms of the other two mass fractions, namely,

$$\boxed{\frac{m_{\text{PL}}}{m_i} = \left(1 - \frac{m_p}{m_i}\right) - \frac{m_s}{m_i}} \qquad (6\text{-}3\text{-}12)$$

With the approximation that the total propellant mass fraction is equivalent to the propellant mass fraction ($m_p/m_i \cong \zeta_p$), the payload ratio expression of Eq. (6-3-12) can be written as

$$\boxed{\frac{m_{\text{PL}}}{m_i} \cong (1 - \zeta_p) - \frac{m_s}{m_i} = \frac{1}{\text{MR}} - \frac{m_s}{m_i}} \qquad (6\text{-}3\text{-}13)$$

Sec. 6-3 The Performance Equations

The overall structural mass fraction can be written as

$$\frac{m_s}{m_i} = \frac{m_i - m_{PL} - m_p}{m_i} = 1 - \left(\frac{m_{PL}}{m_i} + \frac{m_p}{m_i}\right) \qquad (6\text{-}3\text{-}14)$$

At this point it is convenient to define the *step mass* as the sum of $m_s + m_p$ [i.e., the initial gross mass less the mass of the payload $(m_i - m_{PL})$]. With this definition it is possible to introduce a more refined structural factor known as the *effective structural factor*, sometimes referred to as the *step structural factor*, which is denoted by the symbol λ and defined as

$$\lambda = \frac{m_s}{m_p + m_s} = \frac{m_s}{m_i - m_{PL}} = \frac{m_s/m_i}{1 - m_{PL}/m_i} \qquad (6\text{-}3\text{-}15)$$

With the approximation that $\zeta_p \cong m_p/m_i$, Eq. (6-3-11) can be written as

$$1 \cong \frac{m_s}{m_i} + \zeta_p + \frac{m_{PL}}{m_i} \qquad (6\text{-}3\text{-}16)$$

By manipulating Eqs. (6-3-15) and (6-3-16) it is possible to develop several additional relationships that can be useful, specifically,

$$\frac{m_s}{m_i} = \lambda\left(1 - \frac{m_{PL}}{m_i}\right) = \frac{\lambda\zeta}{1 - \lambda} \qquad (6\text{-}3\text{-}17)$$

and

$$\boxed{\frac{m_{PL}}{m_i} = \left(1 - \frac{m_s}{m_i}\right) - \zeta = 1 - \frac{\zeta}{1 - \lambda} = \frac{1 - \lambda - \zeta}{1 - \lambda}} \qquad (6\text{-}3\text{-}18)$$

In Eqs. (6-3-17) and (6-3-18), ζ is ζ_p and can be expressed as

$$\zeta = \frac{MR - 1}{MR} = (1 - \lambda)\left(1 - \frac{m_{PL}}{m_i}\right) \qquad (6\text{-}3\text{-}19)$$

where the mass ratio MR is

$$MR = \frac{1}{1 - \zeta} = \frac{1}{\lambda + (1 - \lambda)(m_{PL}/m_i)} \qquad (6\text{-}3\text{-}20)$$

A major design objective for spacecraft and boosters is the highest practicable payload ratio, m_{PL}/m_i. Examination of Eq. (6-3-18) shows that a high payload ratio calls for low values of both ζ, the propellant mass fraction, and λ, the effective structural factor, as might be expected.

It should be noted that all of the mass ratios and mass fractions may be replaced by equivalent weight ratios and weight fractions where the weight is equal to the product of the mass and the acceleration of gravity at the surface of the planet of interest. In other words,

$$W = mg_0$$

where g_0 is the acceleration due to gravity of the attracting body at the surface of the attracting body.

Another vehicle parameter of interest is the *thrust-to-weight ratio T/W*, which is a function of the instantaneous mass (weight) and the instantaneous thrust. Although the thrust-to-weight ratio is dimensionless, it has the appearance of an acceleration and can be expressed in terms of g_0 values, the acceleration due to gravity. This can be seen from

$$\frac{T}{W} = \frac{T}{mg_0} = \frac{ma}{mg_0} = \frac{a}{g_0} \qquad (6\text{-}3\text{-}21)$$

where a is the acceleration of the vehicle due to the thrust. In the absence of any other forces, such as drag, the thrust-to-weight ratio can represent the acceleration of a vehicle. In such a case, such as at lift-off or in space, a thrust-to-weight ratio of 1.25 indicates an acceleration equal to $1.25g_0$. In the SOI of the Earth with $g_0 = 9.81$ m/s², this acceleration would be 12.26 m/s², whereas within the SOI of Mars, where $g_0 = 3.63$ m/s², the acceleration would be 4.537 m/s².

Considering an essentially constant thrust (usually the case with liquid propellants) or an average thrust (typical of solid propellants), the thrust-to-weight ratio will be at a minimum at lift-off (ignition) and will increase to reach a maximum at burnout (shutdown). If there is to be a lift-off from a planetary surface, the thrust-to-weight ratio obviously needs to be greater than unity; it is on the order of 1.2 to $1.5g_0$ for liquids and somewhat higher for solids.

We close this section with the reminder that the burn time t_b is inversely proportional to the initial thrust-to-weight ratio, $(T/W)_i$:

$$t_b \cong \frac{\zeta I_{sp}}{(T/W)_i} \qquad (6\text{-}3\text{-}21)$$

6-4 SINGLE-STAGE PERFORMANCE

Let us discuss the characteristics and performance of single-stage vehicles by example. There are essentially two categories of vehicles: those with small or moderate ΔV values that are used for orbital transfers and adjustments and those with large ΔV values used for launching from a planetary surface. We shall refer to the former group as *spacecraft* and to the latter group as *boosters*. Within our two-body assumptions, spacecraft are subjected only to thrust as a force, whereas boosters are also subjected to aerodynamic and gravitational forces.

Our first example is that of a spacecraft that needs a limited ΔV capability for maneuvering and orbit trim and whose payload is the mission element. The gross weight (mass) is the payload plus the propellant plus everything else, the rocket engine(s), propellant tanks or cases, structure, and so on.

Example 6-4-1

A spacecraft with a payload of 450 kg (4414 N or 992.4 lb) requires a ΔV capability of 450 m/s. The propulsion system selected has a vacuum I_{sp} value of 400 s.

Sec. 6-4 Single-Stage Performance

(a) Select a value for the effective structural factor λ that gives a reasonable payload ratio. What are the component weights and the gross weight of the spacecraft?
(b) Find the approximate burn times if all the propellant is burned at one time with $(T/W)_i = 0.5, 1.0,$ and $1.5g_0$, respectively.

Solution

(a)
$$c = g_0 I_{sp} = 9.81 \times 400 = 3924 \text{ m/s}$$

$$\text{MR} = \exp\left(\frac{\Delta V}{c}\right) = \exp\left(\frac{450}{3924}\right) = 1.122$$

$$\frac{m_p}{m_i} \cong \zeta = \frac{\text{MR} - 1}{\text{MR}} = 0.108$$

A ζ of 0.108 means that a little more than 10% of the mass (weight) of the spacecraft will be propellant; this 10.8% does not include the mass of the motors, cases, or nozzles—only that of the propellant.

The expressions for the payload ratio and overall structural mass fraction can be written as

$$\frac{m_{PL}}{m_i} = 1 - \frac{1}{1 - \lambda} = 1 - \frac{0.108}{1 - \lambda}$$

$$\frac{m_s}{m_i} = \lambda\left(1 - \frac{m_{PL}}{m_i}\right) = \frac{\lambda \zeta}{1 - \lambda} = \frac{0.108\lambda}{1 - \lambda}$$

At this point we need a value for λ. Since we do not have one, we shall use iteration to find a value that appears to give reasonable results. Picking a low value of λ, say 0.05, gives a payload ratio of 0.886 for an initial weight of 507.7 kg (4981 N). However, the overall structural mass fraction is only 0.00568 and the weight of the structure is 2.88 kg (28.2 N), a ridiculously low and patently impossible figure.

By increasing λ incrementally and examining the results, we arrive at a value of 0.7 for λ. The corresponding values for the payload ratio and initial mass are 0.64 and 703 kg (6896 N). The overall structural mass fraction is 0.252 and the mass is 177 kg; the propellant mass is 75.9 kg. These seem to be reasonable and realistic figures and would probably be the ones we would use until more data were available.

To show how sensitive the sizing of this, and of any other, spacecraft is to the choice of λ, let us increase λ to 0.8. Now the payload ratio drops to 0.46, the initial mass increases to 978 kg (9588 N), the structural mass fraction and mass increase to 0.432 and 423 kg, and the propellant mass is 105.6 kg.

(b) With $\zeta = 0.108$ and $I_{sp} = 400$ s,

$$t_b \cong \frac{43.2}{(T/W)_i}$$

If $(T/W)_i = 0.5$, $t_b \cong 86.4$ s. If $\lambda = 0.7$, $T_i = 3448$ N. If $(T/W)i = 1.0$, $t_b \cong 43.2$ s and $T_i = 6896$ N; and if $(T/W)_i = 1.5$, $t_b \cong 28.8$ s and $T_i = 10{,}340$ N. These firing times are on the order of 1.5% or less of the 5064-s (84.4-min) orbital period of an Earth-surface satellite.

In the example above, the ΔV requirements are modest, considerably less than that required to enter a low Earth orbit by way of a Hohmann transfer from the Earth. In such a case there is no apparent maximum value for λ per se. As λ

increases, so does the gross weight of the spacecraft, but it is still possible to accomplish the mission, namely to increase the velocity of the spacecraft by the specified ΔV. In the case of boosters, however, where the required ΔV is large, we shall find that there is an upper limit on the value of λ; in fact, λ must be surprisingly (perhaps unachievably) small.

In the booster example to follow, the spacecraft of Example 6-4-1 (with $\lambda = 0.7$) will be the payload of the booster; the ΔV requirement will be increased to a value consonant with a boost from the surface of the Earth to a low Earth orbit (LEO).

Example 6-4-2

A spacecraft with a gross weight of 703 kg (6896 N) is to be the payload of a single-stage booster capable of $\Delta V = 7930$ m/s. Since the booster will be launched from sea level, the average I_{sp} will be lower than the vacuum value, so that $I_{sp} = 350$ s.
(a) Find the minimum gross lift-off weight (GLOW). Start with an initial value of 0.7 for λ and determine an acceptable value for λ.
(b) If the initial $T/W = 1.5g_0$, find the lift-off thrust, the approximate burn time t_b, and the T/W value at burnout with constant thrust.
Solution (a) First we find the booster mass ratio MR and propellant mass fraction ζ_p:

$$c = g_0 I_{sp} = 9.81 \times 350 = 3433 \text{ m/s}$$

$$\text{MR} = \exp\left(\frac{7930}{3433}\right) = 10.07$$

$$\frac{m_p}{m_i} \cong \zeta = \frac{10.07 - 1}{10.07} = 0.90$$

Notice that for this booster with the much higher ΔV requirement, 90% of the booster weight will be propellant as opposed to the 10% requirement for the spacecraft. This increase in ζ will severely reduce the acceptable values of λ. For example, with the payload ratio expressed as

$$\frac{m_{PL}}{m_i} = 1 - \frac{\zeta}{1 - \lambda} = 1 - \frac{0.9}{1 - \lambda}$$

it is apparent that λ must be less than 0.1; a larger value, such as 0.7, will result in a negative payload ratio, an obvious impossibility. The trade-off between λ and the *gross lift-off weight* (GLOW), and the sensitivity of the latter to changes in the former are portrayed in Fig. 6-3-1.

With $\lambda = 0.095$, the payload ratio is 0.0055 and the initial mass is 127,200 kg (1.248×10^6 N or 280,600 lb), all to put 703 kg (6896 N or 1550 lb) into a low Earth orbit. The structural mass ratio is 0.095, the structural mass is 12,090 kg, and the propellant mass is 114,500 kg.

Reducing the value of λ to 0.05, which is not very likely, the GLOW is reduced by an order of magnitude, to 13,360 kg (130,000 N or 29,460 lb). The structural mass fraction becomes 0.04737, the structural mass becomes 633 kg, and the propellant mass is 12,024 kg. The allowance for rocket engines, tankage, nozzles, and other structure is unreasonably low and undoubtedly unattainable.

Since the larger value of 0.095 is more realistic for λ than is the value of 0.05, we shall use it to size our booster and to determine the required thrust.

Sec. 6-4 Single-Stage Performance

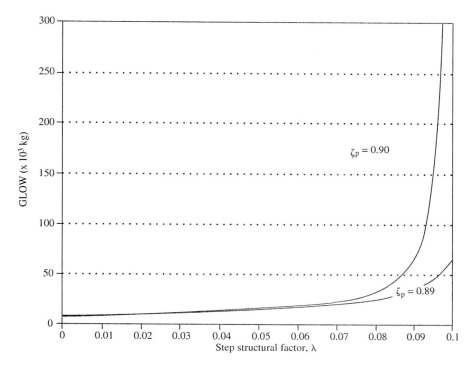

Figure 6-3-1. Gross lift-off weight (GLOW) as a function of λ for two values of ζ_p.

(b) Since at launch, $W_i = 1.248 \times 10^6$ N, $T/W = 1.5g_0$ calls for

$$T_{SL} = 1.5W_i = 1.872 \times 10^6 \text{ N } (420{,}900 \text{ lb})$$

The approximate burn time is

$$t_b \cong \frac{\zeta I_{sp}}{(T/W)_i} = \frac{0.9 \times 350}{1.5} = 210 \text{ s}$$

This is a typical burn time for current large boosters. At burnout, $W_{bo} = W_i(1 - \zeta) = 124{,}800$ N (28,060 lb),

$$\left(\frac{T}{W}\right)_{bo} \cong \frac{1.872 \times 10^6}{124{,}800} = \frac{(T/W)_i}{1 - \zeta} = 15g_0$$

The approximate sign was used above because the thrust increases somewhat with altitude, so that the $15g_0$ represents a lower value.

As should be apparent by now, λ, the effective structural factor, is a very important parameter. As a reminder, let us repeat two expressions, one for λ and the other for the payload ratio:

$$\lambda = \frac{m_s}{m_s + m_p} = \frac{m_s}{m_i - m_{PL}} \tag{6-4-1}$$

$$\frac{m_{PL}}{m_i} = 1 - \frac{\zeta}{1 - \lambda} \tag{6-4-2}$$

We have seen from these two examples that the payload ratio is very sensitive to small changes in λ, whose acceptable magnitude is a function of m_i, the gross weight of the vehicle. λ is always less than unity and is on the order of 0.1 for large boosters and can be on the order of 0.6 to 0.8 for upper stages and spacecraft.

Let us return to our basic equations and see if we can establish limiting relationships for ΔV and m_{PL}/m_i in terms of λ and other parameters of interest. Since the mass ratio MR is

$$MR = \frac{1}{1 - \zeta} = \exp\left(\frac{\Delta V}{c}\right) \tag{6-4-3}$$

the nondimensional required velocity increment, $\Delta V/c$, can be found from

$$\frac{\Delta V}{c} = \ln MR = \ln \frac{1}{1 - \zeta} = \ln \frac{1}{\lambda - (1 - \lambda)(m_{PL}/m_i)} \tag{6-4-4}$$

The second right-hand expression in Eq. (6-4-4) was developed by solving Eq. (6-4-2) for ζ and substituting the result for ζ in $\ln[1/(1 - \zeta)]$.

To find the *limiting velocity* $(\Delta V/c)_{max}$ for a given value of λ, we can let the payload (ratio) go to zero so that

$$\boxed{\left(\frac{\Delta V}{c}\right)_{max} = \ln \frac{1}{\lambda}} \tag{6-4-5}$$

This equation shows that the maximum ΔV value that can be developed by a vehicle with no payload, only structure, is a logarithmic function of the structural factor. As λ becomes smaller, approaching zero in the limit, the velocity increases, becoming infinite in the limit. As λ, the structural weight, increases, the velocity decreases, approaching zero in the limit as λ becomes infinite. Figure 6-4-1 shows the relationship between the maximum value of the nondimensional velocity increment and λ. If $\lambda = 0.1$, $(\Delta V/c)_{max} = 2.3$; then with $I_{sp} = 350$ s the maximum possible increase in the velocity of the vehicle (with no payload) would be 7897 m/s. If a larger increase were needed, λ would have to be reduced or I_{sp} would have to be increased, or both, and consideration would have to be given to the reducing effect of a finite payload.

Example 6-4-3

Find the maximum possible increase in ΔV for:
(a) The spacecraft of Example 6-4-1 with $\lambda = 0.7$.
(b) The booster of Example 6-4-2 with $\lambda = 0.095$.
Solution

(a) $$\left(\frac{\Delta V}{c}\right)_{max} = \ln \frac{1}{0.7} = 0.3567$$

Since $I_{sp} = 400$ s and $c = 3924$ m/s,

$$\Delta V_{max} = 0.3567 \times 3924 = 1400 \text{ m/s}$$

Sec. 6-4   Single-Stage Performance

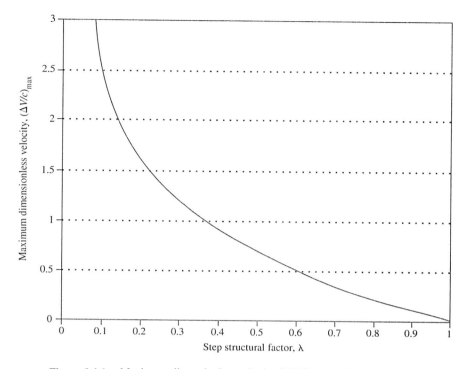

Figure 6-4-1. Maximum dimensionless velocity $(\Delta V/c)_{max}$ as a function of λ.

The actual (required) ΔV was 450 m/s, indicating that a larger value of λ would have been acceptable.

(b)
$$\left(\frac{\Delta V}{c}\right)_{max} = \ln \frac{1}{0.095} = 2.354$$

Since $I_{sp} = 350$ s and $c = 3433$ m/s,

$$\Delta V_{max} = 2.354 \times 3433 = 8081 \text{ m/s}$$

The actual (required) ΔV was 7930 m/s, indicating that λ was approaching its maximum possible value for this mission.

Equation (6-4-5) can be rewritten to show what the maximum possible value of λ can be for a required $\Delta V/c$,

$$\boxed{\lambda_{max} = \exp\left[-\left(\frac{\Delta V}{c}\right)_{req}\right]} \tag{6-4-6}$$

The maximum possible value of λ (with no payload) as a function of the required dimensionless velocity $(\Delta V/c)_{req}$ is sketched in Fig. 6-4-2.

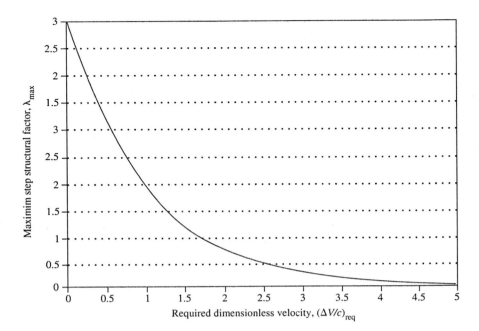

Figure 6-4-2. Maximum step structural fraction, λ_{max}, as a function of the required dimensionless velocity, $(\Delta V/c)_{req}$.

Example 6-4-4

Find the maximum possible value of λ for:
(a) The spacecraft of Example 6-4-1.
(b) The booster of Example 6-4-2.

Solution

(a)
$$\left(\frac{\Delta V}{c}\right)_{req} = \frac{450}{3924} = 0.1147$$

and

$$\lambda_{max} = \exp(-0.1147) = 0.892$$

The value of λ selected was 0.7.

(b)
$$\left(\frac{\Delta V}{c}\right)_{req} = \frac{7930}{3433} = 2.31$$

so that

$$\lambda_{max} = \exp(-2.31) = 0.0993$$

The value of λ selected was 0.0905.

Equation (6-4-4) can be rewritten to obtain the following expression for the payload ratio in terms of $(\Delta V/c)_{req}$ and λ:

Sec. 6-4 Single-Stage Performance

$$\boxed{\frac{m_{PL}}{m_i} = \frac{\exp[-(\Delta V/c)_{req}] - \lambda}{1 - \lambda}} \qquad (6\text{-}4\text{-}7)$$

The payload ratio as a function of $(\Delta V/c)_{req}$ for several values of λ is shown in Fig. 6-4-3. Notice that increasing the required velocity, which is a function of the mission, with λ held constant results in a rapid decrease in the payload ratio; a decrease in the payload ratio calls for an increase in the initial mass m_i or a decrease in the payload m_{PL}. It is also interesting to note from Eq. (6-4-7) that the limit of the payload ratio as λ goes to zero is identical to the expression for λ_{max} as the payload goes to zero; see Eq. (6-4-6).

Example 6-4-5

You are doing a feasibility study of a booster with a payload of 10,000 kg (98,100 N or 22,050 lb) and a required ΔV value of 9000 m/s.
(a) With a propellant whose $I_{sp} = 350$ s and with $\lambda = 0.1$, find the payload ratio and gross lift-off weight (GLOW) (i.e., m_i).
(b) Do part (a) with $\lambda = 0.05$.
(c) Do part (a) with $I_{sp} = 400$ s and $\lambda = 0.05$.
(d) Do part (a) with $I_{sp} = 400$ s and $\lambda = 0.1$, a more realistic value than 0.05.

Solution

(a) $\qquad c = 350 \times 9.81 = 3434$ m/s

$$\left(\frac{\Delta V}{c}\right)_{req} = \frac{9000}{3434} = 2.62$$

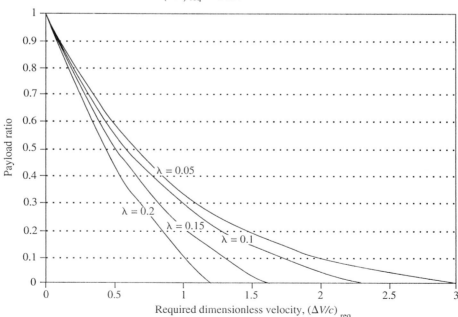

Figure 6-4-3. Payload ratio as a function of the required dimensionless velocity, $(\Delta V/c)_{req}$, for several values of λ.

Substituting into Eq. (6-4-7) yields

$$\frac{m_{\text{PL}}}{m_i} = \frac{\exp(-2.62) - 0.1}{1 - 0.1} = -0.032$$

Since a negative payload ratio is a physical impossibility, the mission cannot be accomplished with a booster with these characteristics.

(b) The required nondimensional velocity remains the same and Eq. (6-4-7) becomes

$$\frac{m_{\text{PL}}}{m_i} = \frac{\exp(-2.62) - 0.05}{1 - 0.05} = 0.024$$

Now with a positive payload ratio, the booster is possible albeit with an m_i (GLOW) of 416,700 kg (4,088 × 10^6 N or 919,000 lb).

(c) Increasing the I_{sp} to 400 s increases c to 3924 m/s and reduces the required $\Delta V/c$ value to 2.294. Now the payload ratio is

$$\frac{m_{\text{PL}}}{m_i} = \frac{\exp(-2.294) - 0.05}{1 - 0.05} = 0.0535$$

The increase in the payload ratio with the higher-performance propellant reduces the GLOW to 186,900 kg (1.83 × 10^6 N or 412,300 lb), a significant reduction of 55%.

(d) With the required $\Delta V/c$ value unchanged at 2.294 and λ increased to 0.1, the payload ratio becomes

$$\frac{m_{\text{PL}}}{m_i} = \frac{\exp(-2.294) - 0.1}{1 - 0.1} = 9.58 \times 10^{-4}$$

This very low payload ratio translates into an $m_i = 1.044 \times 10^7$ kg (1.024×10^8 N or 23.02×10^6 lb). Although finite, this value is so large as to be impossible, certainly impracticable, yet a value of 0.1 for λ is a realistic value.

This might be a good time to remind ourselves that:

1. λ is a "dry" structural factor in that it assumes that all of the propellant has been consumed at the end of engine burn. Therefore, λ does not consider the weight increase from propellant reserves and residuals.
2. ΔV is an ideal velocity increment in that it does not consider any additional ΔV required to overcome such losses as the drag and gravity losses.
3. It is not a simple matter to obtain low values of λ. As mentioned in Example 6-4-5, 0.1 is a realistic low value for λ for large boosters.

It probably is not possible with current technology to put a payload into a geocentric orbit using a single-stage booster with only rocket engines. Historically and currently, there is interest in a single-stage-to-orbit (STO) vehicle using air-breathing engines in combination with rocket engines.

Single-stage performance is limited by the fact that *all* of the structural weight has to be accelerated to the burnout velocity when all we want to do is to accelerate the payload. As propellant is used, the empty tankage becomes dead weight and

Sec. 6-5 Restricted Staging

might as well be blocks of lead ballast. If this dead weight could be jettisoned as it appears, the effective thrust-to-weight ratio would increase.

Although it is not yet possible to jettison tankage continuously, it is possible to discard discrete amounts of tankage and other hardware at specific points on the trajectory. This jettisoning is called *staging* and each stage contributes its individual increment of velocity ΔV_i to the overall ΔV_T of the mission. With appropriate staging it is now possible to place payloads into geocentric orbits and into escape trajectories for interplanetary transfers and trips. Staging is the subject of the next section.

6-5 RESTRICTED STAGING

Before discussing the details of staging, it is necessary to distinguish between a stage and a step. If a *stage* is defined as all of the vehicle that is being accelerated by the rocket engine(s), then a *step* is the stage less the payload for that stage. Even though the steps, not the stages, are jettisoned after their engines cease operating, the process of jettisoning is known as *staging*.

Referring to the two-stage booster shown in Fig. 6-5-1, Stage 1 is the entire vehicle since it is the entire vehicle that is accelerated by the engine(s) of stage 1. Step 1 is stage 1 (the entire vehicle) less stage 2, which is the payload of stage 1. Stage 2, the payload of stage 1, is accelerated by the engines of stage 2 and comprises step 2 plus the stage payload PL, where the stage payload is the payload of the booster. The booster payload is the piece of hardware that is being accelerated to the final burnout velocity; it may be a satellite, a reentry vehicle with warhead(s), or a spacecraft heading for space and possibly another planet.

To add to the possible confusion, the term *stage* is generally used to refer to a step of a multistage vehicle when describing the physical characteristics of a step. For example, the overall characteristics of the complete three-stage Minuteman missile (which technically is the first stage) are referred to as the missile character-

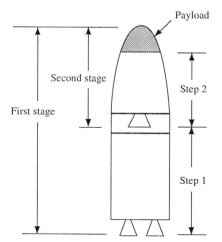

Figure 6-5-1. Series (tandem) booster.

istics, the so-called stage 1 characteristics are actually the step 1 characteristics, and the stage 2 and stage 3 characteristics are actually the step 2 and step 3 characteristics. In performing the analyses and feasibility designs in this and the following sections, it is necessary to maintain the distinction between stages and steps.

The booster in Fig. 6-5-1 is an example of series or tandem staging. Among other possible configurations are partial staging, parallel staging, and piggyback staging. With *partial staging*, shown in Fig. 6-5-2, only some of the engines are dropped. The Atlas missile dropped two of its three main engines and was referred to as a one-and-a-half-stage missile.

With *parallel staging*, sketched in Fig. 6-5-3, staging is accomplished by dropping attached or strap-on engines, which are usually (but do not have to be) solid-propellant rockets. To add further confusion, the strap-on engines are often referred to as strap-on boosters. The *Titan III* and *Titan IV* are examples of parallel staging in combination with tandem staging; there are upper stages in series with the first step.

In *piggyback staging*, shown in Fig. 6-5-4, the second stage is attached to the first step. The engines of both steps operate from launch, with those of the first step burning out first, at which time the first step is jettisoned, separated from the second stage. The Space Transportation System (STS), the Space Shuttle, represents a combination of parallel and piggyback staging.

Let us consider a vehicle with n stages. Applying Eq. (6-3-5) to an individual stage, denoted by the subscript j,

$$\Delta V_j = c_j \ln \mathrm{MR}_j \tag{6-5-1}$$

Since the total velocity increment for the vehicle as a whole (ΔV_T) is the sum of the ΔV values of the individual stages,

Figure 6-5-2. Partial staging.

Sec. 6-5 Restricted Staging

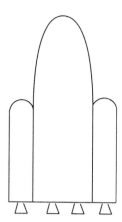

Figure 6-5-3. Example of parallel staging with strap-on boosters.

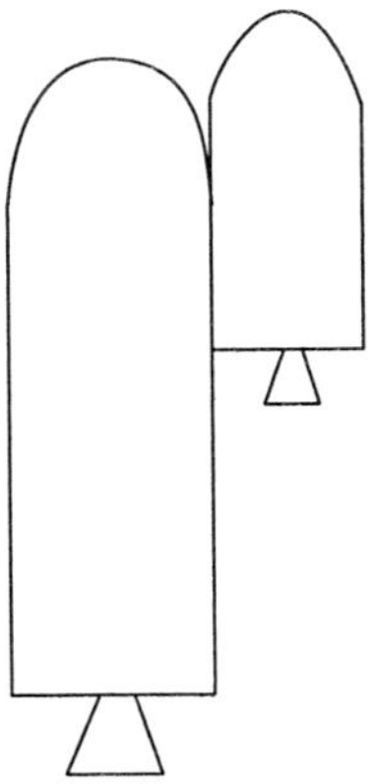

Figure 6-5-4. Example of piggyback staging.

$$\Delta V_T = \sum_{j=1}^{n} \Delta V_j = \sum_{j=1}^{n} c_j \ln \mathrm{MR}_j \qquad (6\text{-}5\text{-}2)$$

where the subscript j denotes the jth stage.

We first examine the simplest (unrealistic) case, in which each stage has identical engines with identical characteristic exhaust velocities c. (We realize that c increases with altitude within the atmosphere, so that c changes somewhat with altitude.) Since

$$c_j = c = g_0 I_{\mathrm{sp}}$$

$$\Delta V_T = c \sum_{j=1}^{n} \ln \mathrm{MR}_j = c(\ln \mathrm{MR}_1 + \ln \mathrm{MR}_2 + \cdots + \ln \mathrm{MR}_n) \qquad (6\text{-}5\text{-}3)$$

Equation (6-5-3) can be rewritten as

$$\frac{\Delta V_T}{c} = \ln(\mathrm{MR}_1 \times \mathrm{MR}_2 \times \cdots \times \mathrm{MR}_n) \qquad (6\text{-}5\text{-}4)$$

MR_1, the mass ratio of the first stage, is equal to $(m_i/m_f)_1 = m_{i1}/m_{f1}$ and MR_2, the mass ratio of the second stage, is equal to $(m_i/m_f)_2 = m_{i2}/m_{f2}$. The initial mass of the second stage, m_{i2}, is equal to the final mass of the first stage (its payload) less the mass of any interstage structure, structure joining the two stages. The relationship of subsequent stages to the preceding stage is similar.

The following assumptions were made to obtain Eq. (6-5-4):

1. Identical characteristic velocities c for each stage.
2. Each upper stage fires immediately upon shutdown of the engine(s) of the preceding stage.

Let us further simplify our vehicle by letting all of the individual stage mass ratios be identical (i.e., $MR_1 = MR_2 = \cdots MR_n \equiv MR_j$). Now Eq. (6-5-4) can be written as

$$\frac{\Delta V_T}{c} = \ln(MR_j)^n = n \ln MR_j \qquad (6\text{-}5\text{-}5)$$

where MR_j is the mass ratio of an individual stage.

Equation (6-5-5) can be solved to find the mass ratio of the individual stage in terms of the other parameters, namely,

$$MR_j = \exp\left(\frac{\Delta V_T}{nc}\right) = \exp\left(\frac{\Delta V_T}{n g_0 I_{sp}}\right) \qquad (6\text{-}5\text{-}6)$$

In all of these expressions,

$$MR_j = \frac{m_i}{m_f} = \frac{1}{1 - \zeta_j} > 1$$

where ζ_j, the propellant mass fraction, which has been assumed to be equal to $(m_p/m_i)j$, can be written as $(MR_j - 1)/MR_j$. Since the stage mass ratios are identical, the propellant mass fractions of the stages will also be identical. (Remember our assumptions of identical c and MR values.)

Using the relationships of Eq. (6-5-6), ζ_j can be written as

$$\zeta_j = 1 - \exp\left(\frac{-\Delta V_T}{nc}\right) = 1 - \exp\left(\frac{-\Delta V_T}{n g_0 I_{sp}}\right) \qquad (6\text{-}5\text{-}7)$$

Let us return to the single-stage booster of Example 6-4-5, which had a requirement to accelerate a payload of 10,000 kg through a ΔV_T value of 9000 m/s. With $I_{sp} = 400$ s and $\lambda = 0.100$, the first-stage mass m_i, the GLOW (gross lift-off weight), was 1.044×10^7 kg (1.024×10^8 N or 23 million pounds). This is not a realistic booster even though the values for I_{sp} and λ are optimistic. In Example 6-5-1 we shall see if staging can reduce the GLOW to an acceptable value, using more realistic values for I_{sp} and λ.

Example 6-5-1

You are examining the feasibility of increasing the velocity of a 10,000-kg payload by 9000 m/s. For your first analysis you will use identical engines with $I_{sp} = 380$ s if more than one stage is required and identical values of λ but not less than 0.1.

(a) Find the minimum number of stages to yield what appears to be a reasonable value for the GLOW.

(b) For the configuration selected, list the characteristics and mass fractions of the stages and steps and find the overall mass fractions of the complete booster.

(c) What are the ΔV contributions of the individual stages to ΔV_T?

Solution

(a) $\qquad c = g_0 I_{sp} = 9.81 \times 380 = 3728$ m/s

Sec. 6-5 Restricted Staging

and
$$\frac{\Delta V_T}{c} = \frac{9000}{3728} = 2.414$$

Let us start with the minimum number of stages, $n = 1$, even though we do not think that a single-stage configuration will be acceptable; it will, however, establish a baseline. With $n = 1$,

$$\text{MR} = \exp\left(\frac{\Delta V}{c}\right) = \exp(2.414) = 11.18$$

$$\zeta = \frac{\text{MR} - 1}{\text{MR}} = \frac{11.18 - 1}{11.18} = 0.9105$$

and
$$\frac{m_{\text{PL}}}{m_i} = 1 - \frac{\zeta}{1 - \lambda} = 1 - \frac{0.9105}{1 - \lambda}$$

With the minimum allowable value of 0.1 for λ, the payload ratio is -0.0117, which is obviously impossible. Consequently, this mission cannot be accomplished with a single-stage configuration using an I_{sp} of 380 s.

With $n = 2$, *a two-stage configuration*, and identical engines for both stages, the MR values of both stages will be identical to MR_j, where MR_j is

$$\text{MR}_j = \exp\left(\frac{\Delta V}{nc}\right) = \exp\left(\frac{9000}{2 \times 3728}\right) = 3.344$$

$$\zeta_j = \frac{\text{MR}_j - 1}{\text{MR}_j} = \frac{3.344 - 1}{3.344} = 0.70$$

In sizing this or any other multistage space vehicle, the procedure is to *start with the last (the uppermost) stage*, which in this example is the second stage, *and find the payload ratio and thus the gross weight of that stage*. Then move to the next-lower stage, where the payload of that stage is the gross weight of the upper stage.

It is important to note that the payload ratio of the second stage $(m_{\text{PL}}/m_i)_2$ is equal to m_{PL}/m_2, where m_{PL} is the booster payload (10,000 kg) and m_2 is the total mass of the second stage. This second-stage payload ratio can be found with $\lambda = 0.1$ to be

$$\frac{m_{\text{PL}}}{m_2} = 1 - \frac{\zeta_2}{1 - \lambda_2} = 1 - \frac{0.70}{1 - 0.1} = 0.2222$$

Since the booster payload is 10,000 kg, the mass of the upper stage will be

$$m_2 = \frac{m_{\text{PL}}}{m_2/m_{\text{PL}}} = \frac{10,000}{0.222} = 45,000 \text{ kg} \ (99,220 \text{ lb})$$

As an aside, a λ of 0.1 for an upper stage is probably much too low and this value of m_2 is lower than it should be.

The payload for the next stage, the first stage, is the entire second stage, so that

$$m_{\text{PL}1} = m_2 = 45,000 \text{ kg}$$

and the first-stage payload ratio will be

$$\frac{m_{\text{PL}}}{m_i} = \frac{m_2}{m_1} = 1 - \frac{0.70}{1 - 0.1} = 0.2222$$

Note that the payload ratios for the first and second stages are identical; this is to be expected inasmuch as ζ and λ have identical values for each stage.

Now the mass of the first stage, which is the overall mass of the booster (the GLOW), is

$$m_1 = m_i = \frac{45,000}{0.2222} = 202,500 \text{ kg } (1.987 \times 10^6 \text{ N or } 446,600 \text{ lb})$$

This appears to be a reasonable mass and the analysis will stop here with the two-stage vehicle. This decision will be justified later in this section.

The overall (booster) payload ratio is

$$\frac{m_{PL}}{m_1} = \frac{10,000}{202,500} = 0.04938$$

The overall payload ratio is also equal to the product of the individual stage payload ratios, as can be seen from

$$\frac{m_{PL}}{m_1} = \left(\frac{m_{PL}}{m_i}\right)_1 \times \left(\frac{m_{PL}}{m_i}\right)_2 = \frac{m_2}{m_1} \times \frac{m_{PL}}{m_2} = 0.2222 \times 0.2222 = 0.04938$$

This relationship between the overall payload ratio and the individual stage payload ratios can be generalized and extended to a vehicle of n stages in the following form:

$$\boxed{\frac{m_{PL}}{m_1} = \prod_{j=1}^{n} \left(\frac{m_{PL}}{m_i}\right)_j} \qquad (6\text{-}5\text{-}8)$$

Equation (6-5-8) *is valid for any configuration and any combination of stage* I_{sp} *and* λ *values*, even though it was developed and demonstrated for this particular configuration.

(b) The weights (masses) of the two stages and steps of this example are summarized below:

STAGE CHARACTERISTICS AND WEIGHTS

STAGE	λ	$(m_{PL}/m_i)_i$	m_{PL}	m_i	m_p	m_s	STEP WEIGHT
Payload							10,000
Second	0.1	0.222	10,000	45,000	31,500	3,500	35,000
First	0.1	0.222	45,000	202,500	141,800	15,750	157,500
					173,300	19,250	202,500

The overall (booster) mass fractions for this vehicle are

$$\frac{m_p}{m_1} = \frac{173,300}{202,500} = 0.8558$$

$$\frac{m_s}{m_1} = \frac{19,250}{202,500} = 0.0951$$

$$\frac{m_{PL}}{m_1} = \frac{10,000}{202,500} = 0.0494$$

These three mass fractions add up to unity, as they should.

Sec. 6-5 Restricted Staging

(c) *In general, the ΔV contribution of any jth stage is*

$$\boxed{\Delta V_j = c_j \ln \mathrm{MR}_j} \quad (6\text{-}5\text{-}9)$$

Since both stages have identical c and MR values, their ΔV values are also identical and are

$$\Delta V_1 = \Delta V_2 = 3728 \ln(1.344) = 4500 \text{ m/s} = \frac{\Delta V_T}{2}$$

With the completion of the requirements of Example 6-5-1, we shall return to our discussion of staging and its effects. *In general, the payload ratio for any stage can be expressed as*

$$\boxed{\left(\frac{m_{\mathrm{PL}}}{m_i}\right)_j = 1 - \frac{\zeta_j}{1 - \lambda_j} = \frac{\exp[-(\Delta V_T/nc)] - \lambda_j}{1 - \lambda_j}} \quad (6\text{-}5\text{-}10)$$

The last expression in Eq. (6-5-10) was developed by substituting the expression for ζ_j in Eq. (6-5-7) in the middle expression.

In the simplified analysis of Example 6-5-1 the assumption of identical characteristic velocities in each stage c_j means that the individual mass ratios MR, and thus the individual propellant mass fractions ζ_j, will also be identical. Then the further assumption of identical structural factors λ_j means that the stage payload ratios will also be identical, as can be seen from Eq. (6-5-10). If the stage payload ratios are identical, the overall (booster) payload expression of Eq. (6-5-8) becomes

$$\frac{m_{\mathrm{PL}}}{m_1} = \left(\frac{m_{\mathrm{PL}}}{m_i}\right)^n = \left\{\frac{\exp[-(\Delta V_T/nc)] - \lambda}{1 - \lambda}\right\}^n \quad (6\text{-}5\text{-}11)$$

Although *Eq. (6-5-11) is valid only when the n stages are identical*, it can be used to give us insight into the advantages and limitations of staging.

Example 6-5-2

The booster of Example 6-4-1 has a required $\Delta V_T = 9000$ m/s, a payload of 10,000 kg, and identical stages with $I_{\mathrm{sp}} = 380$ s ($c = 3728$ m/s) and $\lambda = 0.1$. Find the overall payload ratio and the booster mass m_1 (GLOW) and discuss the effects on m_1 of adding additional stages.
Solution

$$\frac{\Delta V_T}{nc} = \frac{9000}{3728n} = \frac{2.414}{n}$$

Substituting into Eq. (6-5-11) yields the following expression for the overall payload ratio as a function of n.

$$\frac{m_{\mathrm{PL}}}{m_1} = \frac{[\exp(-2.414/n) - 0.1]^n}{0.9}$$

The results of substituting increasing values of n are:

n	m_{PL}/m_1	m_1 (kg)	Δm_1 (kg)	PERCENT CHANGE
1	−0.0117	Impossible		
2	0.0489	204,500		
3	0.0574	174,100	−30,400	−14.8
4	0.0608	164,500	−9,600	−5.5
5	0.0626	159,700	−4,800	−2.9
6	0.0637	157,000	−2,700	−1.6
7	0.0644	155,300	−1,700	−1.1

We see that the payoff decreases rapidly as the number of stages increases. Furthermore, this analysis does not consider the extra weight and complexity associated with adding engines and interstage structure to each additional stage and the additional firings at altitude. For this mission, there appears to be no advantage to having more than three stages, and in all probability two stages would probably be adequate.

As n increases without limit, the overall payload ratio approaches an asymptotic limit. As n goes to infinity, the limit of the payload ratio, from Eq. (6-5-11), becomes

$$\lim_{n \to \infty} \left(\frac{m_{PL}}{m_1}\right) = \left(\frac{m_{PL}}{m_1}\right)_{max} = \exp\left[\frac{-(\Delta V_T/c)}{1 - \lambda}\right] \quad (6\text{-}5\text{-}12)$$

Equation (6-5-12) gives the maximum payload ratio (the minimum weight) of a booster with an infinite number of stages as a function of the required $\Delta V_T/c$, which is a function of the mission and type of engine used in all stages, and of λ, the effective structural factor for all stages.

Example 6-5-3

For the booster of Example 6-5-2 with $\Delta V_T/c = 2.414$ and a payload of 10,000 kg, find the maximum payload ratio and minimum m_1 (GLOW) with an infinite number of stages.
Solution

$$\left(\frac{m_{PL}}{m_i}\right)_{max} = \exp\left(\frac{-2.414}{1 - \lambda}\right)$$

λ	$(m_{PL}/m_1)_{max}$	m_{1min} (kg)
0.00	0.0894	111,900
0.05	0.0788	126,900
0.10	0.0684	146,200
0.20	0.0489	204,500
0.30	0.0318	314,500
0.40	0.0179	558,600
0.50	0.0080	1,250,000

Notice that an infinite number of stages with $\lambda = 0.1$ results in a m_i value of 146,200 kg, whereas with three stages (see Example 6-5-2), m_1 is 174,100 kg.

Sec. 6-5   Restricted Staging   177

Figure 6-5-5 is a plot of the maximum payload ratio with an infinite number of stages [Eq. (6-5-12)] versus λ as a function of two values of $\Delta V_T/c$, 2 and 3. As the velocity requirement increases, the payload ratio decreases, as might be expected inasmuch the propellant requirement increases.

If we return to Eq. (6-5-11) and *let the payload* m_{PL} *go to zero* in the limit, we can find an expression that relates $\Delta V/c$, n, and λ:

$$\frac{\Delta V}{c} = n \ln \frac{1}{\lambda} \qquad (6\text{-}5\text{-}13)$$

Equation (6-5-13) can be used in three different ways, as shown in Fig. 6-5-6. One is to find the maximum value of $\Delta V/c$ with a specified n and λ, a second is to find the minimum number of stages for a specified $\Delta V/c$ and λ, and the last is to find a minimum value of λ for a specified $\Delta V/c$ and a specified number of stages.

Example 6-5-4

Find the maximum possible value of $\Delta V/c$ for a two-stage booster with no payload and stage λ values = 0.1. If the stage engines have an I_{sp} of 350 s, what is the maximum ΔV possible?

Solution

$$\left(\frac{\Delta V}{c}\right)_{max} = 2 \times \ln \frac{1}{0.1} = 4.605$$

$$c = 9.81 \times 350 = 3434 \text{ m/s}$$

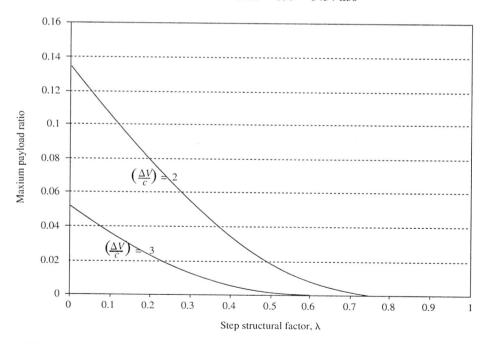

Figure 6-5-5. Maximum payload ratio as a function of λ for two values of dimensionless velocity.

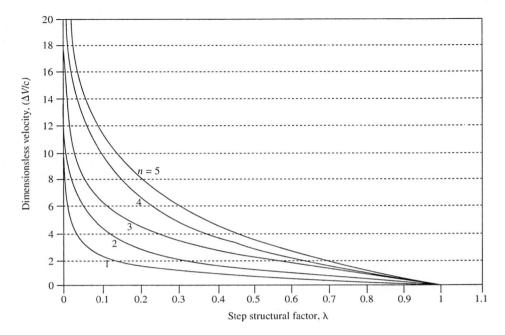

Figure 6-5-6. Dimensionless velocity as a function of λ for various numbers of stages.

and
$$\Delta V_{max} = 15{,}810 \text{ m/s}$$

Example 6-5-5

The booster of Example 6-5-2 has a required $\Delta V_T/c = 2.414$ and a step λ of 0.1. Find the minimum number of stages with the payload set equal to zero.
Solution
$$n = \frac{2.414}{\ln 10} = 1.048$$

Since $n > 1$ and there are no fractional stages with tandem boosters, the minimum number of stages is two. There may be advantages to using three stages, particularly when a finite payload is introduced.

Example 6-5-6

The booster of Example 6-5-2 has $\Delta V/c = 2.414$. With a two-stage configuration and no payload, find the maximum allowable value for λ.
Solution
$$\lambda_{max} = \exp\left(\frac{-\Delta V_T}{nc}\right) = \exp\left(\frac{-2.414}{2}\right) = 0.2991$$

With a payload, the structural factor λ must be less than this value of 0.2991.

Sec. 6-6   Generalized Staging

In this section we have examined staging and its impact from various viewpoints but with unrealistic restrictions, such as identical engines and structural factors in each stage. In the next section these restrictions are removed and a more general approach to staging is considered.

6-6 GENERALIZED STAGING

In normal practice the engines in each stage of a booster and in a spacecraft (which may itself be multistaged) are not identical, for reasons related to the compatibility of various propellants and engines with the size and operating regime of the different stages and spacecraft. Nor are the lambdas (the effective structural weight factor) the same because the stage weights vary, often considerably.

Let us start with the equation that states that the total ΔV value is the sum of the ΔV values of the individual stages:

$$\Delta V_T = \sum_{j=1}^{n} \Delta V_j \qquad (6\text{-}6\text{-}1)$$

where ΔV_j, the velocity contribution of the jth stage, is

$$\Delta V_j = c_j \ln \text{MR}_j = g_0 I_{\text{sp},j} \ln \text{MR}_j \qquad (6\text{-}6\text{-}2)$$

With the substitution of Eq. (6-6-2), Eq. (6-6-1) becomes

$$\Delta V_T = \sum_{j=1}^{n} g_0 I_{\text{sp},j} \ln \text{MR}_j \qquad (6\text{-}6\text{-}3)$$

Since Eq. (6-6-2) indicates that ΔV_j is directly proportional to $I_{\text{sp},j}$, we shall arbitrarily let

$$\Delta V_j = K_j I_{\text{sp},j} \qquad (6\text{-}6\text{-}4)$$

From Eq. (6-6-1) we can write

$$\Delta V_T = \sum_{j=1}^{n} K_j I_{\text{sp},j} \qquad (6\text{-}6\text{-}5)$$

where

$$K_j = g_0 \ln \text{MR}_j = g_0 \ln \frac{1}{1 - \zeta_j} \qquad (6\text{-}6\text{-}6)$$

At this point we shall *assume the stage mass ratios*, MR_j, *to be equal* (the ζ_j's will also be equal). This is a reasonable approximation that is offered without justification but will be partially justified by example. With this approximation, $K_j = K$ and

$$\Delta V_T \cong K \left(\sum_{j=1}^{n} I_{\text{sp},j} \right) \qquad (6\text{-}6\text{-}7)$$

From Eq. (6-6-7),

$$K \cong \frac{\Delta V_T}{\sum_{j=1}^{n} I_{sp,j}} \cong \frac{\Delta V_T}{n I_{sp,ave}} \qquad (6\text{-}6\text{-}8)$$

where

$$I_{sp,ave} = \frac{\sum_{j=1}^{n} I_{sp,j}}{n} \qquad (6\text{-}6\text{-}9)$$

Consequently, the velocity contribution of the jth stage can be written as

$$\Delta V_j = K I_{sp,j} = \frac{K c_j}{g_0} \qquad (6\text{-}6\text{-}10)$$

and the (identical) mass ratios for each stage can be found from

$$\mathrm{MR}_j = \exp\left(\frac{\Delta V_j}{g_0 I_{sp,ave}}\right) \qquad (6\text{-}6\text{-}11)$$

Examination of Eq. (6-6-10) shows that with the assumption of identical MR values for each stage, the $(\Delta V_j/c_j)$ for each stage will also be identical. However, the ΔV_j's themselves will *not* be identical unless the c_j's (the stage I_{sp}'s) are identical.

We shall rework our two-stage booster of Example 6-5-2 with different and more realistic I_{sp}'s and λ's for the two stages.

Example 6-6-1

A two-stage booster to place a 10,000-kg spacecraft into a low Earth orbit is under study; the ΔV required is 9000 m/s. The first stage will use LOX/RP-1 engines with $I_{sp} = 330$ s, and the second stage will use LOX/H$_2$ engines with $I_{sp} = 420$ s.
(a) Find the ΔV contribution, mass ratio, and propellant mass fraction for each stage.
(b) Find m_1 (the GLOW) and the stage, step, and component weights.
Solution

(a) $$I_{sp,ave} = \frac{330 + 420}{2} = 375 \text{ s}$$

and

$$K = \frac{9000}{2 \times 375} = 12$$

$$\Delta V_1 = 12 \times 330 = 3960 \text{ m/s}$$

$$\Delta V_2 = 12 \times 420 = 5040 \text{ m/s}$$

Sec. 6-6 Generalized Staging

As a check,

$$\Delta V_1 + \Delta V_2 = 3960 + 5040 = 9000 \text{ m/s} = \Delta V_T$$

The second-stage MR and ζ are

$$\text{MR}_2 = \exp\left(\frac{5040}{9.81 \times 420}\right) = 3.398$$

and

$$\zeta_2 = \frac{3.398 - 1}{3.398} = 0.7057$$

The first-stage MR and ζ are

$$\text{MR}_1 = \exp\left(\frac{3960}{9.81 \times 330}\right) = 3.398 = \text{MR}_2$$

and

$$\zeta_1 = \frac{3.398 - 1}{3.398} = 0.7057 = \zeta_2$$

(b) Starting with the uppermost (second) stage, the payload ratio can be found from

$$\frac{m_{\text{PL}}}{m_2} = 1 - \frac{0.7057}{1 - \lambda_2}$$

Thinking that this stage will be a "smaller" stage than the first stage and that λ_2 will be larger than λ_1, let us somewhat arbitrarily set $\lambda_2 = 0.25$. (Notice that λ_2 must be < 0.2943 if the payload ratio is to be positive.) Then

$$\frac{m_{\text{PL}}}{m_2} = 0.059 \quad \text{and} \quad m_2 = \frac{10{,}000}{0.059} = 196{,}460 \text{ kg}$$

This is not a small stage at all, so a smaller λ is in order. With $\lambda_2 = 0.15$, the second-stage payload ratio is 0.170 and

$$m_2 = 58{,}820 \text{ kg } (577{,}000 \text{ N} \quad \text{or} \quad 129{,}700 \text{ lb})$$

This is still not a small stage, but the booster payload is not small either. The second-stage step mass is $m_2 - m_{\text{PL}} = 48{,}820$ kg. As for the first stage, which will be larger than the second, a smaller λ_1 is appropriate. With $\lambda_1 = 0.1$ and the first-stage payload $= m_2$,

$$\frac{m_2}{m_1} = 1 - \frac{0.7057}{1 - 0.9} = 0.216 \quad \text{and} \quad m_1 = 272{,}300 \text{ kg}$$

The first-stage step weight is $m_1 - m_2 = 213{,}500$ kg.

This GLOW of 272,300 kg is 33% higher than the GLOW of Example 6-5-2, which had an effective average I_{sp} that was 5 s higher and a second stage λ that was 0.1 rather than 0.15. This difference is indicative of the sensitivity of the GLOW to I_{sp} and λ.

The need to ensure that the payload ratio is not negative means that the term $(1 - \lambda)$ must be less than the propellant mass fraction ζ. This restriction can lead to the assumption of values for λ that may be unrealistically low. Let us redo Example 6-6-1 with three stages rather than two.

Example 6-6-2

Rework Example 6-6-1 using three stages, keeping $m_{PL} = 10{,}000$ kg, $I_{sp1} = 330$ s, $I_{sp2} = 420$ s, and, also using LOX/H$_2$, $I_{sp3} = 420$ s. Find the GLOW and stage weights.

Solution

$$I_{sp,\,ave} = \frac{330 + 420 + 420}{3} = 390 \text{ s}$$

and

$$K = \frac{9000}{3 \times 390} = 7.692$$

By adding the third stage we have increased $I_{sp,\,ave}$ and decreased K. The stage velocity increments are

$$\Delta V_1 = 7.692 \times 330 = 2538 \text{ m/s}$$
$$\Delta V_2 = 7.692 \times 420 = 3231 \text{/s}$$
$$\Delta V_3 = 7.692 \times 420 = 3231 \text{ m/s}$$
$$\Delta V_T = \Sigma \Delta V_j = 9000 \text{ m/s}$$

The stage velocities have been reduced, which in turn will reduce the mass ratios and propellant mass fractions. Note that $(\Delta V/c)$ is 0.784 for each stage even though the ΔV's are not necessarily the same.

Starting with the third stage, the mass ratio is

$$MR_3 = \exp\left(\frac{3231}{9.81 \times 420}\right) = 2.191$$

and the propellant mass fraction is

$$\zeta_3 = \frac{2.191}{2.191 - 1} = 0.5436$$

The values of MR and ζ are less than those of the second-stage version. Since the third stage will be the smallest, λ should be the largest. Let us arbitrarily set $\lambda_3 = 0.2$ and find the payload ratio.

$$\frac{m_{PL}}{m_3} = 1 - \frac{0.5436}{1 - 0.2} = 0.3205$$

and

$$m_3 = 31{,}200 \text{ kg}$$

For the second stage, m_3 is the payload and the MR and ζ are the same as for the third stage. Since this stage will be larger than the third stage, let us use a $\lambda_2 = 0.15$. With these values the payload ratio will be

$$\frac{m_3}{m_2} = 1 - \frac{0.5436}{1 - 0.15} = 0.3605$$

Sec. 6-6 Generalized Staging

$$m_2 = \frac{31{,}200}{0.3605} = 86{,}550 \text{ kg}$$

The first-stage payload is the entire second stage ($m_2 = 86{,}500$ kg) and ζ is 0.5436. Since this is the largest stage, we will use the smallest value of 0.1 for λ_1 to find the payload ratio.

$$\frac{m_2}{m_1} = 1 - \frac{0.5436}{1 - 0.9} = 0.396$$

$$\text{GLOW} = m_1 = \frac{86{,}500}{0.396} = 218{,}400 \text{ kg}$$

Thus the GLOW for the three-stage configuration is 218,400 kg versus the GLOW of 272,300 kg for the two-stage booster. The three step weights are 21,200 kg for the third stage, 55,350 kg for the second stage, and 131,850 kg for the first stage (versus 213,500 kg for the first-stage step weight of Example 6-6-1).

Let us now remove the assumption that the stage MRs be identical and return to the two-stage booster of Example 6-6-1 and see what happens to the configuration with arbitrarily chosen and unequal mass ratios.

Example 6-6-3

The two-stage booster of Example 6-6-1 had a payload of 10,000 kg, a second-stage $I_{sp} = 420$ s with $\lambda_2 = 0.15$, and a first-stage $I_{sp} = 330$ s with a $\lambda_1 = 0.10$. With ΔV_T still equal to 9000 m/s, ΔV_1 is to be $2\Delta V_2$ so that $\Delta V_1 = 6000$ m/s and $\Delta V_2 = 3000$ m/s. Find the stage MRs and GLOW.
Solution
Looking at the second stage first,

$$\text{MR}_2 = \exp\left(\frac{3000}{9.81 \times 420}\right) = 2.071$$

and

$$\zeta_2 = \frac{2.071 - 1}{2.071} = 0.517$$

With $\lambda_2 = 0.15$, the payload ratio is

$$\frac{m_{PL}}{m_2} = 1 - \frac{0.517}{1 - 0.15} = 0.392$$

and

$$m_2 = 25{,}110 \text{ kg}$$

As for the first stage,

$$\frac{m_2}{m_1} = \exp\left(\frac{6000}{9.81 \times 330}\right) = 6.38$$

and

$$\zeta_2 = \frac{6.38 - 1}{6.38} = 0.843$$

With $\lambda_1 = 0.1$, the payload ratio is

$$\frac{m_2}{m_1} = 1 - \frac{0.843}{1 - 0.1} = 0.063$$

so that the

$$\text{GLOW} = m_2 = 398{,}600 \text{ kg}$$

This GLOW is considerably larger than that obtained with equal mass ratios. Although equal mass ratios approach optimal staging ratios, the actual determination of the optimal staging ratios and the choice of appropriate engines is not a simple task. Furthermore, other factors, such as the stability and control of the remaining stages when a stage is jettisoned, must be considered.

With respect to the choice of engines, the higher I_{sp} propellants are not as a normal practice used in the first stage and in larger second stages for one or more of the following reasons:

1. Both propellants are cryogenic (e.g., oxygen, hydrogen, and fluorine).
2. They may be caustic, such as fluorine.
3. The high-energy fuels, such as hydrogen and fluorine, have low densities and require tanks with greater volumes and larger cross sections, thus increasing the structural weight and drag in the atmosphere.

There is another engine parameter called the *density specific impulse* that is sometimes used to include the effects of low density upon the volume and size of the propellant tankage. The density specific impulse is simply the product of the specific impulse of a propellant combination and its density or, as is more often the case, its average specific gravity. Values for one solid and several liquid propellants are:

PROPELLENTS	SPECIFIC GRAVITY	I_{sp}	DENSITY I_{sp}
LOX/RP-1	1.0	310	310
LOX/H$_2$	0.26	400	104
F$_2$/H$_2$	0.4	420	168
Solids	1.7	270	459

This table gives an indication as to why liquid oxygen and hydrocarbons and solid (strap-on) motors are most commonly used in the first stage(s) of a booster, where the cross-sectional area and the tankage mass are to be minimized.

Boosters are launched vertically for structural integrity reasons (side forces are to be avoided) and to reduce the time spent in the atmosphere so as to minimize the drag losses and the gravity losses. We have not considered these losses in sizing our boosters but rather have restricted ourselves to satisfying the ΔV_{ideal} requirement(s) developed by application of the force-free two-body equations. The drag and gravity losses, which are described and discussed in the next section, appear as additional ΔV's, so that the total ΔV that must be imparted to the vehicle (ΔV_{req}) becomes

$$\Delta V_{req} = \Delta V_{ideal} + \Delta V_{drag} + \Delta V_{gravity} \tag{6-6-12}$$

The gravity loss, which is proportional to the burn time, is on the order of 2000 to 3000 m/s, and the drag loss will be on the order of 300 to 1000 m/s.

In sizing a booster, the "ideal" velocity budget must be apportioned in accordance with the mission scheme or plan and there may be more than one acceptable configuration and more than one acceptable booster. For example, that portion of an Earth–Mars mission that deals with the task of placing a spacecraft on a transfer orbit to Mars can be accomplished in two ways: an Oberth escape from the surface of the Earth into the heliocentric transfer orbit or a Hohmann transfer to a low Earth orbit followed by an Oberth escape into the heliocentric transfer orbit. In the case of escape from the surface of the Earth, there is only one booster with the task of placing the spacecraft in the heliocentric orbit. In the second case, we can talk of two boosters: the first booster, whose ΔV includes the gravity and drag losses, places the second booster (and the spacecraft) into the low Earth orbit. Then the second booster (which might be multistage) generates sufficient ΔV to place the spacecraft into the desired heliocentric trajectory. Since the second booster will be operating outside the Earth's atmosphere, drag losses will not be significant; however, gravity losses should still be considered.

6-7 POWERED BOOST

Before examining the vertical launch trajectory of a booster, let us look at the gravity and drag losses, the ΔV's that should be added to the ideal ΔV to obtain the required ΔV needed to size the booster. Consider a vehicle in an atmosphere and in a significant gravity field: for example, near the surface of the Earth during a boost phase. If the velocity vector of an axisymmetric booster is kept aligned with the vehicle axis, there will be no lift or side forces. The forces acting on the vehicle are shown in Fig. 6-7-1. The equation of motion in the direction of the velocity is

$$T - D - mg \cos\theta = m \frac{dV}{dt} \tag{6-7-1}$$

where T is the thrust, D the drag, and θ is equal to $(90° - \phi)$, where ϕ is the elevation angle. The thrust T, however, is the product of the propellant mass flow rate $\dot{m}_p$ and the effective exhaust velocity c, so that

$$T = \dot{m}_p c = -c \frac{dm}{dt} \tag{6-7-2}$$

where dm/dt is the rate of change of the vehicle mass, the negative of the propellant mass flow rate.

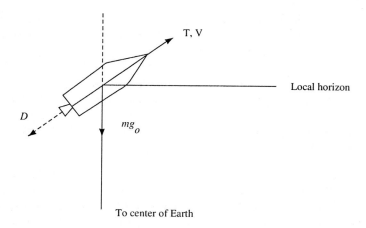

Figure 6-7-1. Booster in the Earth's atmosphere.

Substituting Eq. (6-7-2) into Eq. (6-7-1), rearranging, and separating variables results in

$$dV = -c\,\frac{dm}{m} - \frac{D}{m}\,dt - g\cos\theta\,dt \tag{6-7-3}$$

Integrating from ignition of the engines to burnout yields

$$\Delta V = c\ln\text{MR} - \int \frac{D}{m}\,dt - \int (g\cos\theta)\,dt \tag{6-7-4}$$

We recognize $c\ln\text{MR}$ as ΔV_{ideal}, the ΔV of our velocity budgets. The other two integrals are the *drag loss* and *gravity loss*, which have the dimensions of velocity, can be represented as ΔV's, and are defined as

$$\Delta V_{\text{drag}} = -\int \frac{D}{m}\,dt \tag{6-7-5}$$

and

$$\Delta V_{\text{grav}} = -\int (g\cos\theta)\,dt \tag{6-7-6}$$

where D, m, and θ are functions of time and g is a function of r, which in turn is also a function of time.

Let us first evaluate the order of magnitude of the gravity loss by developing an approximate closed-form expression. In Eq. (6-7-6), g can be found from the expression

$$g = g_0\left(\frac{r_E}{r_E + h}\right)^2 \tag{6-7-7}$$

Since final burnout in the boost phase (the end of powered flight) occurs relatively close to the surface of the Earth, the change in g is small.

Sec. 6-7   Powered Boost

Example 6-7-1

The final engine burnout in the boost phase of an Earth-launched boost is at 125 nmi. Find the value of g at burnout along with the average value of g during.
Solution   With $h = 125$ nmi $= 230$ km $= 230{,}000$ m, $r_{bo} = 6.608 \times 10^6$ m and

$$g_{bo} = 9.81 \left(\frac{6.378 \times 10^6}{6.608 \times 10^6} \right)^2 = 9.139 \text{ m/s}^2$$

and

$$g_{\text{ave}} = 9.474 \text{ m/s}^2 \ (96.6\% \text{ of } g_0)$$

In addition, since θ is kept small, particularly in the atmosphere, in order to minimize any lateral forces on the lightweight booster structure, using an average value of $\cos\theta$ seems to be in order. Consequently, Eq. (6-7-6) can be approximated by

$$\Delta V_{\text{grav}} \cong -(g \cos\theta)_{\text{ave}} \int_0^{bo} dt = -(g \cos\theta)_{\text{ave}} t_b \tag{6-7-8}$$

This equation shows that the gravity loss is directly proportional to the total burn time, which can be approximated by

$$t_b \cong \frac{\zeta_{\text{ave}} I_{\text{sp, ave}}}{(T/W)_i} \tag{6-7-9}$$

To minimize the gravity loss, we want to minimize the burn time, that is, to remain under the influence of gravity for as short a time as practicable. This calls for a large initial thrust-to-weight ratio.

The final approximation is to assume, with the variation in g small and with θ small, that $(g \cos\theta)_{\text{ave}} \cong g_0$. Therefore, the expression for the gravity loss reduces to the absurdly simple approximation that

$$\boxed{\Delta V_{\text{grav}} \cong -g_0 t_b} \tag{6-7-10}$$

Example 6-7-2

The two-stage booster of Example 6-6-1 has an average $I_{\text{sp}} = 375$ s and an average propellant mass fraction $\zeta = 0.2057$. If the initial $T/W = 1.5 g_0$, find the order-of-magnitude value of the gravity loss, ΔV_{grav}.
Solution

$$t_b \cong \frac{375 \times 0.7057}{1.5} = 176.4 \text{ s}$$

and

$$\Delta V_{\text{grav}} \cong 9.81 \times 176.4 = 1730 \text{ m/s}$$

Although this is only an order-of-magnitude value, it does indicate that the gravity loss cannot be ignored. Since the ideal ΔV, which was established by the mission, is 9000 m/s, this gravity loss increases the ΔV that must be generated by the booster by almost 20%, resulting in a significant increase in the GLOW of the booster.

Unfortunately, it is not possible to obtain an equivalent closed-form approximation for the drag loss; to obtain numbers for the drag loss requires integration and a numerical solution. However, an approximate expression can be developed that will give insight into the factors affecting the magnitude of the drag loss.

The instantaneous drag is given by the expression

$$D = \tfrac{1}{2}\rho_{SL}\sigma V^2 A_f C_D \qquad (6\text{-}7\text{-}11)$$

where ρ_{SL} is the sea-level atmospheric density, σ the density ratio for any altitude ($\sigma = \rho/\rho_{SL}$), A_f the frontal cross-sectional area, and C_D the overall drag coefficient that includes zero-lift drag, wave drag, and base drag. C_D is a function of Mach number, elevation angle, shape, and so on. Substituting Eq. (6-7-11) into the definition of the drag loss yields

$$\Delta V_{drag} = -\int \frac{\rho_{SL}\sigma V^2 A_f C_D}{2m}\, dt \qquad (6\text{-}7\text{-}12)$$

With expressions for σ (ρ) and C_D, numerical solutions of Eq. (6-7-2) can be obtained with a computer. Although there is no closed-form solution, with some simplifications and rearrangement it is possible to get a feel for the impact of the booster characteristics and trajectory on the magnitude of the drag loss. Since only the first stage is normally in the *sensible atmosphere* (where the density is significant), the drag analysis from this point on will be limited to the first stage. First we define the *cross-sectional density* Γ as the initial mass divided by the frontal cross-sectional area (m_i/A_f). Note that the units of Γ are kg/m². Furthermore, the instantaneous mass can be expressed as

$$m = m_i - \dot{m}_p t = m_i\!\left(1 - \frac{\dot{m}_p t}{m_i}\right) \qquad (6\text{-}7\text{-}13)$$

The expression for the drag loss (for the first stage) can now be written as

$$\Delta V_{drag} = -\frac{\rho_{SL}}{2\Gamma}\int \frac{C_D \sigma V^2}{1 - (\dot{m}_p/m_i)\,t}\, dt \qquad (6\text{-}7\text{-}14)$$

In this equation, the density ratio σ can be approximated for the Earth's atmosphere by the exponential relationship[1] that

$$\sigma \cong \exp\!\left(\frac{-h}{7254}\right) \qquad (6\text{-}7\text{-}15)$$

where h is the altitude in meters. At 73.6 km (40 nmi), approaching the edge of the sensible atmosphere, σ has dropped from unity at sea level to 3.95×10^{-5}. Since the altitude h is obviously a function of time, so is σ.

---

[1] See Eq. (7-3-7) for a discussion of this approximation.

Sec. 6-7 Powered Boost

Rather than attempt to integrate Eq. (6-7-14), we restrict our discussion to a qualitative look at the drag loss. From Eq. (6-7-14) we see that ΔV_{drag} is directly proportional to σ (the atmospheric density), C_D (the drag coefficient), the square of the airspeed (the magnitude of the velocity), and inversely proportional to Γ (m/A_f):

$$\Delta V_{\text{drag}} \sim \frac{\sigma V^2 C_D}{\Gamma} \tag{6-7-16}$$

Looking at Eq. (6-7-16), we see that the heavier and the more slender the vehicle is (the larger the value of Γ) the smaller is the drag loss. Furthermore, as the vehicle becomes more slender, C_D decreases somewhat, further reducing the drag loss. We also see that it is desirable to minimize σV^2 (which represents the dynamic pressure) as well as the time in the atmosphere. Minimizing the dynamic pressure calls for a trade-off between altitude (σ) and airspeed, to keep the velocity low in the lower (denser) altitudes without remaining too long. Low airspeed at low altitude implies a low acceleration, which in turn implies a low initial thrust-to-weight ratio $(T/W)_i$. But a low initial T/W increases the burn time and the gravity loss. Thus there is a conflict between ΔV_{grav} and ΔV_{drag}. A low ΔV_{grav} calls for a rapid acceleration to the burnout velocity to keep the time the vehicle is exposed to gravity to a minimum (a low t_b), whereas a low ΔV_{drag} calls for keeping V low until σ has significantly decreased with altitude. Since ΔV_{grav} is considerably larger than ΔV_{drag}, its reduction takes priority and the emphasis in powered-boost trajectories is on the higher $(T/W)_i$'s with the lower t_b's. Whereas gravity losses are on the order of 1500 to 2500 m/s, the accompanying drag losses are on the order of 300 to 1000 m/s. Booster designs and staging favor lowering the ΔV_{grav} to take advantage of the greater payoff. It should be noted that with manned launches the burn time and T/W ratio may be influenced by the need to keep the acceleration within acceptable limits for the occupants of the spacecraft.

The Hohmann transfers and Oberth escapes from the surface of the Earth of the previous chapters show the trajectories starting with V perpendicular to the radius of the Earth ($\phi = 0°$), thus implying a horizontal launch. However, boost vehicles are actually launched vertically and then undergo a gradual transition during powered boost to a ϕ (at burnout) of $0°$ at the start of Hohmann transfers or Oberth escapes. Other than the obvious (and perhaps absurd) reason that with a truly horizontal launch the booster would hit any projecting structures or topography, there are sound reasons for vertical launches. Since, as we have seen, the major design objective is to minimize the mass of the vehicle, the structure is the lightest weight possible and has no tolerance for side forces. Vertical launch enables the boost vehicle to take advantage of the greater strength and rigidity of a column in compression. In addition, with a vertical launch the booster will spend less time in the denser portions of the atmosphere, reducing the drag losses, and will have lower gravity losses.

Although the booster is launched vertically, it is necessary for it to achieve the proper elevation angle at burnout for the desired Keplerian trajectory. If the desired trajectory is a Hohmann transfer or an Oberth escape from perigee, the velocity vector at burnout needs to be perpendicular to the position vector from the center of the Earth ($\phi_{bo} = 0°$).

A typical powered-boost sequence, as sketched in Fig. 6-7-2, might comprise four phases, all designed to keep any side forces as close to zero as possible:

1. A vertical launch and liftoff.
2. Shortly after liftoff (10 to 15 s), when the velocity is low, a small (3 to 5°) programmed turn downrange at a very low pitch rate.
3. Then a gravity turn (with zero angle of attack and zero lift) until the booster is out of the sensible atmosphere.
4. A constant-attitude (increasing angle of attack) turn until burnout is attained with the appropriate conditions (V, r, and ϕ) for the trajectory desired.

The *gravity turn* is a clever maneuver in which thrust vectoring is used to keep the velocity vector aligned with the body axes (to preclude the generation of any side forces) and the vertical component of gravity is used to rotate the booster slowly about its center of gravity. Referring to Fig. 6-7-1, the equations of motion along and normal to the body axis can be written as

$$T - D - mg \cos\theta = m \frac{dV}{dt} \qquad (6\text{-}7\text{-}17)$$

and

$$mg \sin\theta = mV \frac{d\theta}{dt} \qquad (6\text{-}7\text{-}18)$$

where

$$\theta = 90° - \phi \qquad (6\text{-}7\text{-}19)$$

Rearranging Eq. (6-7-18) results in a nonlinear equation for the turning rate of the body axis (of the vehicle), namely,

$$\dot\theta = \frac{d\theta}{dt} = \frac{g \sin\theta}{V} \qquad (6\text{-}7\text{-}20)$$

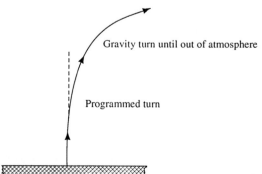

Figure 6-7-2. Vertical launch.

Although Eq. (6-7-20) is nonlinear, we can get a qualitative and somewhat quantitative feel for the magnitude of $\dot{\theta}$ by substituting numbers for a typical boost trajectory. For example, if early in the trajectory, shortly after the programmed turn, $\theta = 5°$ and $V = 110$ m/s, the turning rate of the gravity turn would be approximately 7.8×10^{-3} rad/s or 0.44 deg/s. As θ increases, so would $\dot{\theta}$ except that V is also increasing; the net effect is to reduce the turning rate. For example, if θ were to increase to 50° and V to 2800 m/s, the instantaneous turning rate would be 2.7×10^{-3} rad/s or 0.15 deg/s. These numbers show that the gravity turn is a gentle turn with the objective of changing the direction of the velocity vector without exerting side forces on the booster structure while it is in the sensible atmosphere. Since the turning force, gravity, acts at the cm, there is no torque on the booster.

Since the elevation angle $\phi = 90° - \theta$, as θ is increasing, ϕ obviously is decreasing on its way to reaching its desired value at burnout, which, along with the burnout velocity and position, is a function of the desired Keplerian trajectory or orbit. The powered-boost trajectory, therefore, is a boundary value problem in which the initial downrange turn is determined and programmed, as is the gravity turn. The missile is unguided (in an open-loop mode to preclude any side forces engendered by guidance commands) until it leaves the sensible atmosphere; during this period an autopilot activates the thrust vector control to keep the velocity vector and the missile axis aligned. When it is safe to generate an angle of attack, the guidance system takes over and either continues the gravity turn until the desired value of ϕ is achieved or initiates a faster (or possibly slower) turn, usually a constant attitude turn, to accomplish a satisfactory burnout.

Although burnout normally occurs at the start of a transfer trajectory, it is possible to intercept a low Earth circular orbit directly by extending the powered trajectory so that burnout occurs at the desired orbital altitude with $\phi = 0°$, as shown in Fig. 6-7-3. It is also possible to intercept such a low-Earth circular orbit by using a suborbital (ballistic) transfer ellipse that is entered at burnout with $\phi_{bo} \neq 0$ and that has its apogee at the orbital altitude, where $\phi = 0°$. Notice that in either of these cases a collinear (injection) ΔV is required to maintain the spacecraft in the desired orbit, a ΔV that is provided either by an engine in the payload or by a separate injection stage.

Rather than use either of the trajectories described in the preceding paragraph, it is more common to enter a Hohmann transfer (to a parking orbit) at burnout so

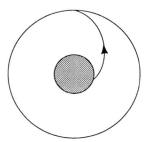

Figure 6-7-3. Powered boost into a low Earth orbit.

that $\phi_{bo} = 0°$ (i.e., the burnout velocity is perpendicular to the Earth's radius and the burnout point is the perigee of the Hohmann transfer). In the event of a direct hyperbolic escape (an Oberth maneuver) from the surface of the Earth, ϕ_{bo} will also be equal to zero and burnout will be a perigee of the escape hyperbola.

6-8 CLOSING REMARKS

As mentioned in Section 6-5, the probability of building a single-stage-to-orbit vehicle with rocket engines only is low. To do so would require relatively large increases in the I_{sp} value of the engines and/or a decrease in the already small values of λ, neither of which is likely in the near future. Furthermore, decreasing λ further exacerbates the problems with any side forces, namely the use of lift during the boost phase. Even if it were possible to use lift during the boost,[2] the fuel consumption rate(s) of rocket engines are so high that it would not be possible to provide the extra propellant needed for an extended cruise mode.

Vertically launched rocket booster vehicles can be categorized as *accelerators* inasmuch as the objective is to reach the desired burnout velocity as rapidly as possible so as to minimize: the total propellant consumption; the time in the atmosphere; and the time that the booster is subjected to gravity. With respect to the latter, keeping the boost trajectory predominantly vertical increases the magnitude of the component of gravity, which is why the gravity loss is so large. With respect to the propellant consumption rate, the I_{sp} is inversely proportional to the specific fuel consumption (the fuel consumption rate per unit of thrust). Obviously, the lower the specific fuel consumption (the higher the I_{sp}), the more fuel efficient is the propulsion system and the propellant requirement is less. It is interesting to note that chemical rockets have specific fuel consumptions on the order of 10 times higher than those of gas turbine engines and on the order of three times that of ramjets.

The successful use of lift during the boost phase requires a stronger (and heavier) structure and the addition of air-breathing engines, changing the nature of the boost vehicle from that of an accelerator to that of a *cruiser*. Whether launched vertically or horizontally (taking off from a conventional runway) a cruise booster would level off at some point in the ascent and accelerate (more slowly than an accelerator) with the elevation angle $\phi = 0°$ ($\theta = 90°$). During this period ΔV_{grav}, which is proportional to $\cos \theta$, will be equal to zero, so there is no gravity-loss penalty associated with the time spent in horizontal acceleration.

There obviously are problems associated with cruise boosters; otherwise, there would be one in operation today rather than in the study and exploratory development phases. The potential payoff, however, is large enough with respect to reducing the costs and difficulties of getting into and out of space to warrant pursuing the goal of an aerospace plane, a one-stage cruise vehicle that could take off and land

---

[2] The wings on the STS (the Space Shuttle) are to provide lift during entry and not during boost.

horizontally and that would have an acceptable payload capability. The current program is the National Aerospace Plane (NASP) program, which involves the X-30 aircraft and the demonstration of a hypersonic capability, the first step in the development of a single-stage-to-orbit booster.

The problems associated with the cruise booster are primarily propulsive and structural. Air-breathing engines (with their higher propulsive efficiencies) that can operate over a large Mach number range are needed to accelerate the vehicle to the highest possible hypersonic Mach number to reduce the magnitude of the ΔV to be provided by the rocket engines outside the atmosphere. Structural (and weight) problems arise from adding a wing with sufficient area to generate the lift to balance the weight of the booster during the period of level ($1g$) flight and from maintaining structural integrity in the presence of lateral forces along the trajectory, especially during the pull-up needed to leave the atmosphere. Although the gravity loss will be reduced, the presence of the wing will increase the drag loss.

Although this has been another long chapter, it should be obvious that the treatment of propulsion and staging has been selective and that we are not yet "rocket scientists." However, we have added to our understanding of how the ΔV's (energy changes) required for the establishment and modification of Keplerian trajectories and orbits can be generated.

We now know how to get into space and how to maneuver in space. In the next chapter, we look, again briefly and selectively, at the problem of returning from space and entering the Earth's atmosphere, what is often referred to as the *entry problem*. Since the problems associated with entry are common to entering any planetary atmosphere, the next chapter is entitled "Atmospheric Entry."

PROBLEMS

6-1. A monopropellant has a molecular weight $M = 32$ kg/mol and a ratio of specific heats $k = 1.3$. It is used in a rocket motor with $T_c = 900$ K.
  (a) Find the theoretical maximum V_e and I_{sp}.
  (b) With a pressure ratio, P_c/P_e, = 68, find the value of V_e and I_{sp}.
  (c) Do part (b) with $P_c/P_e = 300$.
  (d) Do part (b) with $P_c/P_e = 25$.
  (e) Comment on the effects of varying the pressure ratio.

6-2. Do Problem 6-1 for another monopropellant with $M = 22$, $k = 1.25$, and $T_c = 1275$ K.

6-3. Using the data from Table 6-2-1, find V_e and I_{sp} for LOX/LH$_2$ with:
  (a) $P_e = 0$ (i.e., find $V_{e,max}$ and $I_{sp,max}$).
  (b) With $P_c = 6900$ kPa and $P_e = 101.3$ kPa. Compare with I_{sp} from the table.
  (c) With $P_c = 6900$ kPa and $P_e = 38$ kPa.
  (d) With $P_c = 35{,}000$ kPa and $P_e = 101.3$ kPa.
  (e) Discuss any conclusions with regard to pressures.

6-4. Do Problem 6-3 for N$_2$H$_4$/UDMH.

6-5. Table 6-2-1 shows the I_{sp} of LOX/LH$_2$ to be 37% higher than that of N$_2$O$_4$/UDMH. Why is this so?

6-6. A booster has a lift-off $T/W = 1.3g_0$. The first-stage uses LOX/H$_2$ with a sea-level $I_{sp} = 400$ s and the propellant mass flow rate is 340 kg/s.
  (a) What is the gross lift-off weight (GLOW)?
  (b) What is the approximate first-stage burn time if $\zeta_p = 0.85$? If $\zeta_p = 0.80$?
  (c) How much first-stage propellant is used?

6-7. A single-stage ballistic missile has a GLOW = 27,120 kg (59,800 lb) and a sea-level thrust of 337,000 N. If $\zeta_p = 0.8$ and $I_{sp} = 247$ s:
  (a) Find T/W at lift-off.
  (b) Find $\dot{m}$ (kg/s).
  (c) Find the approximate t_b.

6-8. One limitation to the performance of a solid-core nuclear rocket is the allowable temperature of the walls. One advantage is the use of a separate propellant. Using H$_2$ with $M = 2.02$ and $k = 1.409$, find the maximum exhaust velocity and I_{sp} for:
  (a) $T_w = 3500$ K
  (b) $T_w = 5000$ K

6-9. Do Problem 6-8 using an inert gas (for safety reasons) such as helium: $M = 4.003$ and $k = 1.658$.

6-10. Do Problem 6-8 with nitrogen as the propellant: $M = 28.02$ and $k = 1.408$. Is this acceptable performance?

6-11. A gaseous-core nuclear rocket has the potential of eliminating the wall temperature barrier. With H$_2$ as a propellant ($M = 2.02$ and $k = 1.409$), find $V_{e,\max}$ and $I_{sp,\max}$ with the following propellant temperatures:
  (a) $T_p = 7500$ K
  (b) $T_p = 10{,}000$ K
  (c) $T_p = 15{,}000$ K

6-12. A ballistic missile has GLOW = 27,210 kg, $m = 137.7$ kg/s, and $I_{sp} = 247$ s.
  (a) Find $(T/W)_{LO}$.
  (b) Find t_b.
  (c) If $\zeta_p = 0.8$, find T/W at burnout. (Assume T constant.)
  (d) Find the total impulse I_T.

6-13. Do Problem 6-12 with GLOW = 90,000 kg (198,450 lb), $\zeta_p = 0.85$, $m = 383$ kg/s, and $I_{sp} = 350$ s.

6-14. An 11,340-kg orbiter needs a maneuvering and orbital trim capability of 550 m/s using LOX/H$_2$ with an $I_{sp,vac} = 420$ s.
  (a) Find the minimum propellant mass.
  (b) If $T_{ave} = 22{,}240$ N (2267 kg), find the average mass flow rate and total burn time.
  (c) Find the initial and final T/W ratios.
  (d) If the orbiter payload (crew, instruments, and cargo) weighs 2268 kg (22,240 N), find the overall structural weight (mass) fraction, the step weight fraction, and the structural weight.

6-15. The orbiter of Problem 6-14 is the payload of a two-stage booster with a required $\Delta V_T = 10{,}675$ m/s. The first stage has $I_{sp} = 330$ s and a $\lambda = 0.1$. The second-stage $I_{sp} = 420$ s and $\lambda = 0.2$. The stage MRs are to be identical.
  (a) What are the ΔV requirements for each stage?
  (b) What are the stage MRs? The stage payload ratios?
  (c) What is the gross weight of the second stage? Its step weight?
  (d) What is the GLOW of the booster?

(e) What is the total mass (weight) and mass fraction of the propellant? Of the structure?

(f) What is the overall payload ratio?

6-16. Your organization is shopping for a booster capable of accelerating a 453.5-kg payload to an ideal velocity of 5795 m/s (no gravity or drag losses and a nonrotating Earth). Two companies have submitted proposals. Are they acceptable, and which is better?

Company A: Single stage with V_e = 3050 m/s, all-up mass of 6803 kg, and empty (structural) mass of 907 kg.

Company B: Two stages with V_e = 3050 m/s for both stages: first-stage gross mass of 6803 kg and an empty mass of 720 kg; second-stage gross mass of 1757 kg and an empty mass of 186.4 kg.

6-17. A multistage booster is being designed to land an 1814-kg spacecraft on Mars, requiring a ΔV = 15,250 m/s. The first stage will use LOX/RP-1 with I_{sp} = 310 s. The upper stages will use LOX/H$_2$ with I_{sp} = 410 s. The schedule for λ (the step structural mass fraction) is

STAGE	λ	STAGE	λ
1	0.1	4	0.3
2	0.2	5	0.35
3	0.25	6	0.4

(a) Assuming identical MRs and identical approximate values for λ and c (0.2 and 3400 m/s), determine the minimum number of stages.

(b) Using the number of stages from part (a), see if that n-stage booster will meet the mission requirements. Do you recommend this configuration? Why?

6-18. For the booster of Problem 6-17, use four stages with identical MRs.

(a) Find ΔV and $\Delta V/c$ for each stage along as well as the MR and ζ_p.

(b) Find the stage masses (weights) and GLOW.

(c) Find the stage payload ratios and the overall payload ratio.

(d) If $(T/W)_{LO}$ = 1.3g_0, find the approximate burn time and gravity loss.

6-19. Do Problem 6-18 with five stages.

6-20. An early single-stage ballistic missile with ΔV = 5000 m/s and a 915-kg payload used LOX/RP-1 with I_{sp} = 247 s. Find the GLOW and ζ_p with:

(a) λ = 0.1
(b) λ = 0.05
(c) λ = 0.15

6-21. The ballistic missile of Problem 6-20 is fitted with an upgraded engine with I_{sp} = 300 s. Rework Problem 6-21.

CHAPTER 7

Atmospheric Entry

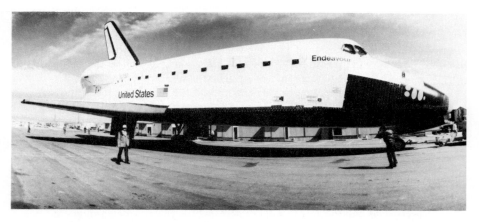

The official roll-out of the Endeavor, a Space Shuttle orbiter. (Courtesy of NASA.)

7-1 INTRODUCTION

A spacecraft entering the SOI of a planet (or moon) without an atmosphere requires a propulsive ΔV to enter a parking orbit or to make a "soft" landing on the surface. It is not necessary, however, to worry about heating; furthermore, the touchdown deceleration can be controlled by the choice and programming of the thrust-to-weight ratio of the propulsion system.

The presence of an atmosphere presents advantages and disadvantages. The major advantage is that the aerodynamic (gas dynamic) forces can be used to replace or supplement the propulsive ΔV's to brake and maneuver the vehicle, thus modifying the trajectory. These forces, however, introduce the dual problems of dissipating the generated heat and handling the load factors (g_0 values) associated with the aerodynamic deceleration and with any maneuvering. The resultant heat transfer and the deceleration must be controlled so that the spacecraft and its contents (to include living occupants) will survive.

Although it was a common practice in the early days of missiles and space flight to refer to a satellite, spacecraft, or warhead entering an atmosphere as a reentry vehicle (R/V), such a vehicle might be entering a planetary atmosphere for the first time, thus making the term *reentry vehicle* inappropriate. Therefore, let us not distinguish between entry and reentry and replace *reentry vehicle* (R/V) by *entry vehicle* (E/V) and *reentry* by *entry*.

It was also a common practice in those days to refer to the *sensible atmosphere*, which was defined somewhat arbitrarily as extending from the surface of the Earth to approximately 75 km (41 nmi or 250,000 ft). For the velocities and configurations of that period, it seemed reasonable to a first approximation to consider the upper edge of the sensible atmosphere as representing the start of atmospheric entry. Even though not as valid for the higher entry velocities currently encountered, the use of the sensible atmosphere to approximate the start of entry can be convenient.

An entry vehicle entering an atmosphere on a Keplerian orbit possesses a large amount of energy, predominantly kinetic energy (KE), that must be transformed to thermal energy and safely dissipated. Consider the kinetic energy at 76 km (41 nmi) of an E/V entering the Earth's atmosphere. If the E/V is descending from a low Earth orbit at 322 km (175 nmi), its velocity as it approaches the atmosphere will be on the order of 7945 m/s, and the specific kinetic energy will be on the order of 32,000 kJ/kg, energy that somehow must be dissipated. To put the KE dissipation problem in perspective, water vaporizes at approximately 2325 kJ/kg and carbon at approximately 60,460 kJ/kg.

If the E/V is returning on a geocentric hyperbola from outside the SOI of the Earth, the minimum velocity and specific energy at atmospheric entry will be on the order of 11,000 m/s and 60,000 kJ/kg, respectively. Obviously, if all of this kinetic energy were converted to thermal energy within the entry vehicle, the E/V and its contents would vaporize. Yet meteors survive, indicating that all of the thermal energy is not and does not need to be transferred to (and dissipated or absorbed by) the entry vehicle itself.

The two basic categories of atmospheric entry are:

1. The ballistic entry without any lift, with a lift-to-drag ratio (L/D) equal to zero.
2. Lifting entry with $0 < L/D \lesssim 3.5$.

Lifting entry can be further broken down as to the amount and direction of the generated lift and the entry conditions.

In the next section, the entry problem, its parameters, and general solution are discussed, followed by a section devoted to an examination of the deceleration problems associated with a ballistic entry trajectory. In the section following we show how a small lift-to-drag ratio dramatically reduces the deceleration loading of an entry vehicle and discuss possible uses of lift to modify entry trajectories and to increase the landing footprint.

In the penultimate section we look at the heating problem and the trade-offs between lift and heat dissipation and comment on the use of aerodynamic drag for trajectory modification to include aeroassisted transfers and aerocapture without remaining within the planetary atmosphere. As usual, in the last section we close with a few summary remarks.

Also, as usual, it should be pointed out that this chapter does not begin to represent a complete treatment of atmospheric entry but rather is a presentation and recognition of the general problems and solutions associated with entering a planetary atmosphere, primarily with the deceleration and heating problems. A full treatment of the entry problem and of entry vehicles involves many engineering disciplines: flight mechanics, atmospheric physics, trajectory analysis, heat transfer, real gases and aerothermochemistry, vehicle dynamics and stability, ablation and hot gas radiation, and hypersonic aerodynamics, as well as computational fluid dynamics and numerical analysis.

7-2 THE ENTRY PROBLEM

While outside the atmosphere and within the SOI of a planet, an entry vehicle can be assumed to a first approximation to be in a Keplerian force-free trajectory, either elliptical or hyperbolic, and to be subject only to gravitational attraction and cen-

Sec. 7-2 The Entry Problem

trifugal force. However, as the vehicle enters the atmosphere, its trajectory is modified by the aerodynamic (gas dynamic) forces: the drag, the lift, and possibly a side force.

The drag force opposes the vehicle's motion, affecting its velocity and thus modifying its trajectory. The lift and centrifugal force act in a direction normal to the motion, and the gravitational force is always directed toward the central force field (the center of the planet). The gravitational force g curves the trajectory toward the planetary surface, with the curvature increasing as the altitude decreases, and g becomes larger. This curved path is a gravity turn, similar to that used in the vertical launch and the boost trajectory. The forces and their relation to the trajectory are sketched in Fig. 7-2-1.

The first consideration for an entry trajectory is obviously that the E/V enter the atmosphere—that the incoming trajectory intercept the atmosphere. If periapsis of the incoming Keplerian entry trajectory is outside the sensible atmosphere, the vehicle will obviously miss the atmosphere completely and will continue on its Keplerian trajectory, which may be elliptical or hyperbolic (it is unlikely that it will be parabolic). If elliptical, the E/V will remain within the planetary SOI. If hyperbolic (or parabolic), the vehicle will leave the SOI with its energy and subsequent heliocentric trajectory modified by the flyby (see Section 5-6).

Although periapsis of the incoming trajectory must be within the atmosphere if there is to be entry, r_p can be less than the radius of the planet, indicating a hard surface impact in the absence of any modifying forces from a propulsive ΔV, from the deployment of aerodynamic decelerators such as parachutes and vanes or from the use of lifting surfaces such as wings or lifting bodies.

With periapsis greater than the planetary radius and with penetration of the atmosphere achieved, the E/V will either remain within the atmosphere or leave the

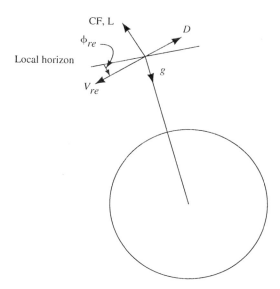

Figure 7-2-1. Geometry of an atmospheric entry trajectory.

atmosphere, in the latter case either to remain outside the atmosphere or to return for another atmospheric pass. In either case the trajectories will be modified by the atmospheric forces: by the drag in any event and by any lift that might be generated. Trajectories that remain within the atmosphere on the first pass lead to surface impact or landing, whereas those that leave the atmosphere, either intentionally or inadvertently, may be a one-time pass or lead to additional passes (further atmospheric entries).

The use of the atmosphere to modify a trajectory is referred to as *aeroassist* and the special case where atmospheric drag (*aerobraking*) is used to place an incoming vehicle into the proper position to enter a parking orbit outside the atmosphere is referred to as *aerocapture*. (*Aerocapture* is often reserved for the transformation of trajectories that are initially hyperbolic.) It should be pointed out that the use of aeroassist with trajectories that leave the atmosphere does not necessarily eliminate the need for propulsive ΔV's but rather has the potential of reducing the propulsive requirements, conceivably dramatically.

One aeroassist trajectory is the *grazing trajectory*, in which an E/V leaves the atmosphere with a reduced velocity, a reduced energy, and a smaller semimajor axis because of the atmospheric drag. The magnitude of the ΔV reduction is a function of the shape and drag coefficient of the E/V, of the velocity at entry, of the altitude of periapsis, and of the characteristics of the planetary atmosphere. Although it is still possible for a high-energy hyperbolic entry to be hyperbolic when it leaves the atmosphere so that it leaves the planetary SOI, it is more likely that the modified trajectory will be elliptical with apoapsis within the SOI, as shown in Fig. 7-2-2a.

If the E/V is allowed to reenter the atmosphere after a grazing pass, each successive pass will reduce the velocity and the height of apoapsis as well as reducing the eccentricity, making each orbital pass more circular, as shown in Fig. 7-2-2b. When apoapsis reaches the altitude for the desired circular orbit, a propulsive ΔV (from an AKM, an apogee kick motor) can be used to enter the circular orbit. To maintain a constant periapsis altitude and not to descend with each successive entry, a small trimming ΔV can be introduced at each apoapsis.

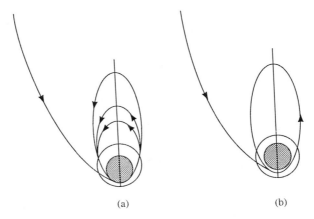

Figure 7-2-2. Grazing trajectories: (a) single pass; (b) multiple passes.

Sec. 7-2　The Entry Problem   201

Since the calculation of the ΔV reduction due to drag is beyond the scope of this book, an arbitrary ΔV reduction will be used in the following example to illustrate the fundamentals of a series of grazing trajectories and to show the gradual lowering of the periapsis without a trim ΔV at apoapsis.

Example 7-2-1

A returning spacecraft enters the SOI of Earth with a V_∞ of 2000 m/s on a hyperbolic trajectory with a perigee altitude of 55.2 km (30 nmi) that is within the planetary atmosphere. We shall arbitrarily assume that the velocity at perigee is reduced by 500 m/s on each pass through the atmosphere.
(a) Find the energy E and the eccentricity ϵ of the original entry hyperbola as well as the velocity at perigee.
(b) Find the characteristics of the modified trajectory after the first pass to include a, r_a and h_a, V_a, and the r_p of a second pass.
(c) What ΔV is required at apogee of the first pass if r_p is to remain at 55.2 km (30 nmi)?
(d) What will the apoapsis altitude and eccentricity be after the second pass?
(e) Without worrying about the change in perigee, what will the apoapsis altitude and eccentricity be after the third and fourth passes?

Solution

(a)
$$r_p = r_E + h_p = 6.433 \times 10^6 \text{ m}$$
$$E_0 = \frac{(2000)^2}{2} = 2 \times 10^6 \text{ m}^2/\text{s}^2 > 0$$
$$V_{p0} = \sqrt{2\left(E_0 + \frac{\mu_E}{r_p}\right)} = 11{,}310 \text{ m/s}$$
$$H_0 = V_{p0} r_p = 7.276 \times 10^{10} \text{ m}^2/\text{s}$$
$$\epsilon_0 = \sqrt{1 + \frac{2EH^2}{\mu_E^2}} = 1.064 > 0 \quad \text{(a hyperbola)}$$

(b) With ΔV at perigee = -500 m/s:
$$V_{p1} = 11{,}310 - 500 = 10{,}810 \text{ m/s}$$
$$H_1 = (10{,}810) \times (6.433 \times 10^6) = 6.954 \times 10 \text{ m}^2/\text{s}$$
$$E_1 = \frac{(10{,}810)^2}{2} - \frac{\mu_E}{r_p} = -3.534 \times 10^6 \text{ m}^2/\text{s}^2$$
$$a_1 = 5.640 \times 10^7 \text{ m}$$
$$\epsilon_1 = 0.886 < 0 \quad \text{(an ellipse)}$$
$$r_{a1} = a_1(1 + \epsilon_1) = 1.064 \times 10^8 \text{ m}$$
$$h_{a1} = 99{,}990 \text{ km} = 54{,}340 \text{ nmi}$$
$$V_{a1} = \frac{H_1}{r_{a1}} = 653.6 \text{ m/s}$$
$$r_{p1} = a_1(1 - \epsilon_1) = 6.430 \times 10^6 \text{ m}$$
$$h_{p1} = 52 \text{ km} = 28.2 \text{ nmi}$$

We see that the eccentricity of the entry trajectory has been reduced to the point that the trajectory leaving the atmosphere is now elliptical with an apogee, although high, well within the SOI of the Earth. Although these calculations indicate that the next perigee, that of the second pass, would be 3.2 km (1.8 nmi) lower than the original altitude, with the assumptions of an impulsive and tangential ΔV at perigee, theoretically there should be no change in the perigee altitude. In reality, however, these assumptions are not valid since the aerodynamic drag is not instantaneously applied at perigee; there will be a slight decrease in the perigee altitude.

(c) Using the perigee altitude of 52 km (28.2 nmi) of (b), the apogee velocity required to keep the altitude of the perigee of the second pass at 30 nmi (55.2 km) can be found from the characteristics of an ellipse with the apogee found above and the perigee at the original r_p.

$$2a_2 = r_{a1} + r_p = 1.064 \times 10^8 + 6.433 \times 10^6 = 1.128 \times 10^8 \text{ m}$$

$$E_2 = \frac{-\mu_E}{2a_2} = -3.533 \times 10^6 \text{ m}^2/\text{s}^2$$

$$V_{p2} = \sqrt{2\left(E_2 + \frac{\mu_E}{r_p}\right)} = 10{,}810 \text{ m/s}$$

$$H_2 = V_{p2} r_{p2} = 6.954 \times 10^{10} \text{ m}^2/\text{s}$$

and

$$V_{a2} = \frac{H_2}{r_{a1}} = 653.6 \text{ m/s}$$

Although we see that with four significant figures, V_a does not appear to require augmentation, some small ΔV would be required to increase the energy so as to keep r_p from gradually descending on subsequent passes. (Although inconsistent with the assumptions of an impulsive and tangential ΔV at perigee because of the atmospheric drag, a propulsive ΔV at apogee is required in practice.)

(d) With perigee maintained at 30 nmi (55.2 km), the new perigee velocity on the second pass will be

$$V_{p3} = 10{,}810 - 500 = 10{,}310 \text{ m/s}$$

$$E_3 = \frac{(10{,}310)^2}{2} - \frac{\mu_E}{r_p} = -8.814 \times 10^6 \text{ m}^2/\text{s}^2$$

$$a_3 = \frac{-\mu_E}{2E_3} = 2.261 \times 10^7 \text{ m}$$

$$H_3 = r_p V_{p3} = 6.632 \times 10^{10} \text{ m}^2/\text{s}$$

$$\epsilon_3 = 0.7155 \quad \text{(less elliptical)}$$

$$r_{a3} = a(1 + \epsilon_3) = 3.879 \times 10^7 \text{ m}$$

and

$$h_{a3} = 32{,}410 \text{ km} = 17{,}610 \text{ nmi}$$

The perigee for this trajectory on the next pass into the atmosphere would be on the order of

$$r_p = a(1 - \epsilon_3) = 6.432 \times 10^6 \text{ m}$$

Sec. 7-2 The Entry Problem

so that

$$h_p = 54.54 \text{ km} = 29.64 \text{ nmi}$$

which is slightly below the original 30 nmi. Consequently, as before a small $+\Delta V$ should be applied at the apogee of this new ellipse in order to trim the orbit.

(e) The perigee velocity at the next pass will be

$$V_p = 10{,}310 - 500 = 9810 \text{ m/s}$$
$$E = -1.384 \times 10^7 \text{ m}^2/\text{s}^2$$
$$a = 1.440 \times 10^7 \text{ m}$$
$$H = 6.311 \times 10^{10} \text{ m}^2/\text{s}$$
$$\epsilon = 0.5531 \quad \text{(continuing to circularize)}$$
$$r_a = 2.236 \times 10^7 \text{ m}$$
$$h_a = 15{,}990 \text{ km} = 8689 \text{ nmi} \quad \text{(continuing to lower)}$$

On the next pass, the perigee velocity will be on the order of

$$V_p = 9810 - 500 = 9310 \text{ m/s}$$
$$E = -1.862 \text{ m}^2/\text{s}^2$$
$$a = 1.070 \times 10^7 \text{ m}$$
$$H = 6.311 \times 10^{10} \text{ m}^2/\text{s}$$
$$\epsilon = 0.2404 \quad \text{(approaching a circle)}$$
$$r_a = 1.327 \times 10^7 \text{ m}$$
$$h_a = 6894 \text{ km} = 3747 \text{ nmi}$$

Although it is not realistic to assume that the drag ΔV is constant on each pass and occurs exactly at perigee, this example does show how it might be possible to circularize an incoming trajectory and reduce its altitude without a large expenditure of propulsive energy. There are disadvantages, however, to such multipass grazing trajectories. The guidance and control requirements are stringent, as is the correct determination and application of any trim ΔV's. In addition, the number of passes and time required to accomplish the desired trajectory are high, as is the time spent in the Van Allen radiation belts surrounding the Earth. As a consequence, such entry trajectories are not currently being given serious consideration for entry into the Earth's atmosphere but are being used and considered for entry into other planetary atmospheres, such as that of Venus.

There is, however, interest in the use of a single grazing trajectory to place an E/V into position to enter a circular parking orbit. Such a trajectory might be referred to as either an aerocapture maneuver, if it is the transition from a hyperbolic to an elliptical orbit, or an aeroassisted transfer, if it is the transition from one elliptical orbit to another, less eccentric even circular, orbit. This technique is being studied and tested with the intention of applying it to the return of an orbital transfer vehicle (OTV) from a GEO to a LEO such as the 175-nmi circular orbit of a space shuttle or the 300-nmi orbit of a space station. It might also be desirable to use the aerocapture maneuver to enter a parking orbit around Mars or any other planet or body of interest. In the following example, we look at a possible geocentric aerocapture.

Example 7-2-2

An orbital transfer vehicle (OTV) is returning a geosynchronous communications satellite to a space station in a 300-nmi (552-km) circular orbit.
(a) Find the ΔV required from the atmospheric drag to place the apogee of the modified orbit at 300 nmi with one pass and with perigee at 30 nmi (55.2 km). The entry trajectory is to be a Hohmann ellipse.
(b) With the OTV at 300 nmi, what propulsive ΔV is needed to inject the OTV into the circular orbit?
(c) If only propulsion is used, without atmospheric penetration and with a Hohmann transfer to 300 nmi, what circular injection ΔV is required?

Solution (a) With the geosynchronous orbit at an altitude of 19,490 nmi (35,680 km), the major axis of the Hohmann transfer ellipse, with perigee at 30 nmi (55.2 km), will be

$$2a_t = 4.224 \times 10^7 + 6.433 \times 10^6 = 4.867 \times 10^7 \text{ m}$$

$$E_t = \frac{-\mu_E}{2a_t} = -8.189 \times 10^6 \text{ m}^2/\text{s}^2$$

$$V_{ta} = \sqrt{2\left(E_t + \frac{\mu_E}{r_a}\right)} = 1579 \text{ m/s}$$

$$H_t = 1579 \times 4.224 \times 10^7 = 6.670 \times 10^{10} \text{ m}^2/\text{s}$$

$$\epsilon_t = 0.7358$$

$$V_{tp} = 10{,}370 \text{ m/s}$$

For apogee of the modified orbit to be at 300 nmi (552 km) and perigee at 30 nmi (955.2 km), the characteristics of the ellipse must be as follows:

$$2a = 6.433 \times 10^6 + 6.930 \times 10^6 = 13.36 \times 10^6 \text{ m}$$

$$E = \frac{-\mu}{2a} = -2.984 \times 10^7 \text{ m}^2/\text{s}^2$$

$$V_p = \sqrt{2\left(E + \frac{\mu_E}{6.433 \times 10^6}\right)} = 8015 \text{ m/s}$$

$$H = r_p V_p = 5.156 \times 10^{10} \text{ m}^2/\text{s}$$

$$\epsilon = 0.0368 \quad \text{(almost circular)}$$

$$V_a = \frac{H}{r_a} = 7440 \text{ m/s}$$

The ΔV to be generated by the atmospheric drag is

$$\Delta V_{\text{drag}} = 8015 - 10{,}370 = -2355 \text{ m/s}$$

Sec. 7-2 The Entry Problem

(b) At 300 nmi (552 km),

$$V_{cs} = 7584 \text{ m/s}$$

so that the propulsive injection ΔV (*apogee kick*) is

$$\Delta V = 7584 - 7440 = +144 \text{ m/s}$$

(c) To find the propulsive ΔV required to enter the circular orbit directly at perigee, the relevant Hohmann transfer characteristics, from GEO to 300 nmi, are

$$2a = 4.224 \times 10^7 + 6.930 \times 10^6 = 4.917 \times 10^7 \text{ m}$$

$$E = \frac{-\mu}{2a} = -8.107 \times 10^6 \text{ m}^2/\text{s}^2$$

Using the vis-viva integral gives us

$$V_p = 9940 \text{ m/s}$$

Therefore, the propulsive ΔV required to enter the circular orbit is

$$\Delta V_{\text{propulsive}} = 7584 - 9940 = -2357 \text{ m/s}$$

It should be noted that the drag ΔV's of Examples 7-2-2(a) and 7-2-1 can be replaced, partially or in entirety, by propulsive ΔV's.

Although aeroassisted entry trajectories decrease the velocity of an entry vehicle by entering the planetary atmosphere, the end result would normally be an orbit outside the atmosphere. Such an orbit would be the final objective of the maneuver(s), such as the return of a payload from a GEO to a LEO, the planetary capture of an interplanetary explorer, or the reduction of the energy of a returning spacecraft prior to landing, rather than remaining within the atmosphere and executing a landing on the planetary surface. Consequently, aeroassisted vehicles are not truly entry vehicles whose objective is a planetary surface landing but rather are spacecraft possessing only those E/V characteristics necessary to ensure survival (from heat and deceleration) during atmospheric penetration.

Since an aeroassisted maneuver does not have a planetary landing as an immediate objective, it must avoid being trapped in the atmosphere without the ability to return to a Keplerian orbit. Therefore, although the goal is velocity (and energy) reduction, the velocity must not be allowed to drop below the orbital velocity at any point on the atmospheric trajectory. Aeroassist and aerocapture are still concepts and are not techniques currently in use. They are, however, being studied thoroughly and experimental test vehicles and test programs are under way to evaluate the potential payoffs for various trajectories and vehicle configurations as well as the trade-offs between the associated weight penalties and the savings in propulsion. We discuss such trade-offs briefly in Section 7-5.

Let us now turn our attention to the problem of remaining and surviving within the atmosphere, with a safe landing as the objective of the maneuver. This is the essence of the classical entry problem and determines the design and trajectory of the E/V.

Atmospheric entry is generally considered to begin when the gas dynamic drag becomes on the order of 1% of the gravitational force. Since this point is not fixed and is a function of several variables, such as the velocity, vehicle configuration, and planetary characteristics, the entry analysis often starts at the edge of the sensible atmosphere.

By the time an entry vehicle reaches the edge of this sensible atmosphere, the gas dynamic drag will have increased to a value so much greater than that of the gravitational force that the approximations of the two-body equations are definitely no longer valid. Therefore, the actual trajectory will deviate from the incoming Keplerian trajectory as sketched in Fig. 7-2-3, which also defines the significant entry variables at the edge of the sensible atmosphere. As mentioned earlier, the edge of the sensible atmosphere of the Earth will be assumed to be on the order of 75 km (41 nmi or 250,000 ft). The significant entry conditions are the entry velocity, V_{re}, and the entry angle, ϕ_{re}, the elevation (flight path) angle at entry. (Although γ is the conventional symbol for the flight path angle, ϕ was used in the early days of missile warhead reentry and will be retained here so as to be consistent with the orbital symbology.)

At times the entry velocity may be expressed in nondimensional form as a function of the orbital velocity of a planetary surface circular satellite, V_{cso}, which is

$$V_{cso} = \sqrt{\frac{\mu_p}{r_p}} = \sqrt{g_0 r_p} \qquad (7\text{-}2\text{-}1)$$

With the nondimensionalized entry velocity denoted by V_{re}^*, where V_{re}^* is V_{re}/V_{cso}, the nature of an incoming trajectory can be determined from the value of V_{re}^*. For example, if V_{re}^* is slightly larger than unity, the entry is from a low planetocentric orbit (a value slightly less than unity implies a suborbital trajectory such as that of a ballistic missile). If V_{re}^* is on the order of 1.3, entry is from a high planetocentric orbit and if V_{re}^* is greater than 1.4, the entry trajectory originated outside the planetary SOI. (The entry velocity may also be nondimensionalized with respect to the circular orbital velocity at the entry altitude.)

In a successful entry the maximum deceleration $(-n_{max})$ and the heat trans-

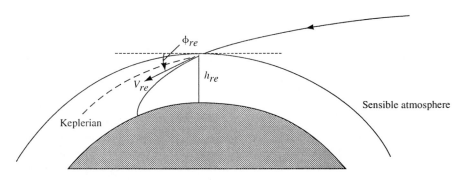

Figure 7-2-3. Entry and Keplerian trajectories.

Sec. 7-2 The Entry Problem

ferred to the E/V must be tolerated by the structure, the equipment, and any living occupants. In addition, at the end of the "entry phase" the vehicle must be in the proper position with an appropriate velocity for a predicted splash-down or impact with a ballistic entry or for a landing at a desired location with a lifting entry.

The E/V must remain within an *entry corridor* defined by an undershoot boundary and an overshoot boundary. The entry corridor for a ballistic entry is shown in Fig. 7-2-4a. Hitting the entry corridor tasks the guidance and control system if there is to be a successful entry on the first pass with a minimum expenditure of propulsive energy.

For a trajectory above the overshoot boundary, the drag is not large enough to keep the vehicle within the atmosphere; the E/V will leave the atmosphere on a new trajectory that could be either elliptical or hyperbolic. If elliptical, another pass

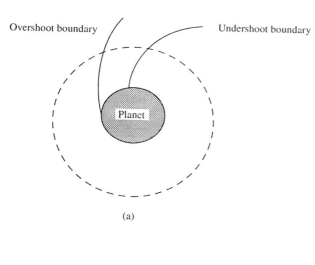

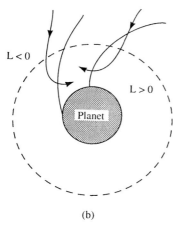

Figure 7-2-4. Entry corridor: (a) without lift; (b) with lift.

at entering the entry corridor might well require additional ΔV's as well as additional time and guidance and control activity. If hyperbolic, the vehicle would leave the SOI. The overshoot boundary can be extended by deploying large, lightweight, high-drag devices (such as a flared afterbody or an inflatable shield) that would increase the drag while leaving the lift unaffected. It is unlikely that such devices would survive the heating at the higher densities of the lower altitudes.

For a trajectory below the undershoot boundary, the deceleration or heating, or both, will exceed allowable limits. Both boundaries, and thus the size of the corridor, are determined by the entry velocity and entry angle in combination with the characteristics of the entry vehicle itself.

The judicious use of even small amounts of lift can be used to widen the entry corridor, as indicated in Fig. 7-2-4b. If the E/V strays into the undershoot area, positive lift can be used to return to the corridor. Negative lift can be used to return to the corridor from the overshoot area.

If a spacecraft is to make a safe planetary landing after entering the atmosphere, it must remain within this entry corridor. If, on the other hand, the spacecraft is executing an aeroassist or aerocapture maneuver, it must avoid the entry corridor or have the ability to leave the corridor (and the atmosphere) at the appropriate time.

Let us refer to Fig. 7-2-1a and write the Newtonian equations of motion for a point mass in an atmosphere, using wind axes (along and normal to the velocity vector), with the origin of inertial space located at the center of the planet. The two-dimensional dynamic equations can be written as

$$-D + mg \sin \phi = m \frac{dV}{dt} + m \frac{V^2}{r} \sin \phi \qquad (7\text{-}2\text{-}2a)$$

$$-L + mg \cos \phi = mV \frac{d\phi}{dt} + m \frac{V^2}{r} \cos \phi \qquad (7\text{-}2\text{-}2b)$$

where mV^2/r is the centrifugal acceleration and mg is the gravitational force.

The two kinematic equations are

$$\frac{dr}{dt} = \frac{dh}{dt} = V \sin \phi \qquad (7\text{-}2\text{-}3)$$

$$\frac{dS}{dt} = r \frac{d\theta}{dt} = V \cos \phi \qquad (7\text{-}2\text{-}4)$$

where S is the distance over the planetary surface and $d\theta/dt$ is the angular velocity of the radius vector **r** with respect to inertial space.

Although there are no closed-form solutions for this system of nonlinear differential equations as written, the use of appropriate assumptions and approximations can provide insight as to the characteristics of selected entry trajectories vis-à-vis the physical characteristics of the E/V.

At this point we shall look at the decelerations that the E/V might encounter during the entry process, first with no lift (ballistic entry) and then with varying degrees of lift.

7-3 BALLISTIC ENTRY

The basic assumption for a ballistic (direct) entry is the absence of lift ($L = L/D = 0$). Additional assumptions and approximations include the neglect of the gravitational (and centrifugal) force during the early initial high-velocity phase of the entry trajectory where the drag force is so much larger than the gravitational force and the E/V is decelerating. Since it is the gravitational force that curves the trajectory as the E/V slows down and approaches the planetary surface, with a flat-Earth model the trajectory during the deceleration phase can be approximated by a straight line with a constant elevation angle ($\phi = \phi_{re}$).

Implicit in the assumption that the lift is zero are the assumptions that the E/V is axisymmetric and that the angle of attack is zero (and remains equal to zero). Since an object in a trajectory maintains its initial attitude with respect to the inertial reference unless subjected to external forces, another implicit assumption is that the attitude of the E/V has somehow been appropriately adjusted prior to entry.

An expression for the drag is that

$$D = \frac{\rho_0 \sigma C_D A V^2}{2} \qquad (7\text{-}3\text{-}1)$$

where ρ_0 is the surface (sea level) atmospheric density, σ the atmospheric density ratio (ρ/ρ_0), and C_D and A the drag coefficient and cross-sectional area of the E/V, respectively.

Applying the assumptions and Eq. (7-3-1), Eq. (7-2-2a) can be written in the form

$$\frac{dV}{dt} = -\frac{C_D A}{m} \frac{\rho_0 \sigma}{2} V^2 \qquad (7\text{-}3\text{-}2)$$

Separating the variables and replacing m by W_0/g_0 (subscript 0 still denotes surface values) leads to

$$\frac{dV}{V^2} = -\frac{C_D A}{W_0} \frac{\rho_0 g_0}{2} \sigma \, dt \qquad (7\text{-}3\text{-}3)$$

Rewriting Eq. (7-2-3) (with a constant $\phi = \phi_{re}$) as

$$dt = \frac{dh}{V \sin \phi_{re}} \qquad (7\text{-}3\text{-}4)$$

and defining a *ballistic coefficient* (BC), with the dimensions of a pressure (pascal), as

$$BC = \frac{W_0}{C_D A} = \frac{mg_0}{C_D A} \qquad (7\text{-}3\text{-}5)$$

Eq. (7-3-3) can be written as

$$\frac{dV}{V} = -\frac{\rho_0 g_0}{2(BC) \sin \phi_{re}} \sigma \, dh \qquad (7\text{-}3\text{-}6)$$

To find an expression for the variation of σ, the atmospheric density ratio, with altitude requires some knowledge of the characteristics of the planetary atmosphere of interest. The Earth's atmosphere, for example, is modeled as a series of concentric layers; the exact number of layers defined is dependent on how the model is to be used and can range from three to seven. With respect to the atmospheric density, one useful model comprises four layers: the troposphere, the stratosphere, the ionosphere, and the exosphere. The troposphere and stratosphere are the two layers closest to the surface of the Earth and are of primary interest with respect to entry (and launch ascent); the remaining two layers are of interest with respect to orbital decay and lifetimes.

If the temperature profile in each layer is modeled by a constant temperature or by a constant temperature gradient (lapse rate) and hydrostatic equilibrium is assumed, the combination yields an exponential relationship that is a function of the density ρ, the acceleration due to gravity g, and the temperature T at a reference altitude. For the lower layers, from sea level to approximately 120 km (65 nmi), if g is assumed to be constant and an appropriate average value of T is used, the variation of σ with h can be approximated by the exponential expression

$$\sigma = \frac{\rho}{\rho_0} \cong e^{-\beta h} \quad (7\text{-}3\text{-}7)$$

where h is the altitude above the planetary surface and $1/\beta$ is the *scale height*. The validity of this exponential approximation decreases at the higher altitudes where the continuum tapers into free-molecule flow. Although it is possible to improve the accuracy of the approximation by using different values of β in certain regions, we shall use only one value of β for all altitudes. Furthermore, even though there is not complete agreement as to the best value of β, its value for the Earth will be taken to be $1/7524 \text{ m}^{-1}$ (0.1378 km^{-1}); typical values of β for Mars and Venus are given in Table 7-3-1.

Making use of Eq. (7-3-7), Eq. (7-3-6) can now be written in a form suitable for integration, namely,

$$\frac{dV}{V} = -\frac{\rho_0 g_0}{2(BC) \sin \phi_{re}} e^{-\beta h} dh \quad (7\text{-}3\text{-}8)$$

Before integrating, we need to look at the ballistic coefficient (BC) as defined in Eq. (7-3-5). In the absence of ablation or other mass and area changes, W/A can

TABLE 7-3-1 RELEVANT PLANETARY ATMOSPHERIC DATA

Planet	Radius (km)	Gravity (m/s²)	Sea-level density (kg/m³)	Beta (km⁻¹)
Venus	6052	8.85	16.02	0.1606
Earth	6378	9.81	1.226	0.1378
Mars	3393	3.73	0.0993	0.0361

Sec. 7-3 Ballistic Entry

be considered to be remain constant over the range of interest, and C_D, the drag coefficient, is essentially constant in the hypersonic velocity regime. Consequently, BC will be assumed to be constant.

Integrating from the entry altitude (h_{re}) to any other altitude on the straight-line trajectory yields the expression

$$\ln \frac{V}{V_{re}} = + \left[\frac{\rho_0 g_0}{2(BC)\beta \sin \phi_{re}} \right] e^{-\beta h} \Big|_{h_{re}}^{h} \qquad (7\text{-}3\text{-}9)$$

For the sake of simplification, define the bracketed term in Eq. (7-3-9) as B and evaluate the right-hand side to obtain

$$\ln \frac{V}{V_{re}} = B(e^{-\beta h} - e^{-\beta h_{re}}) \qquad (7\text{-}3\text{-}10)$$

which can be written as

$$\frac{V}{V_{re}} = \exp[B(e^{-\beta h} - e^{-\beta h_{re}})] \qquad (7\text{-}3\text{-}11)$$

Now for one last simplification. With $h < h_{re}$, as the E/V descends, the term $\exp(-\beta h)$, which is σ at the altitude of interest, rapidly becomes much larger than $\exp(-\beta h_{re})$, which is σ_{re}, so that the latter term can be neglected. Now the final approximation for the velocity along the entry trajectory can be written as

$$\boxed{V \cong V_{re} e^{Be^{-\beta h}}} \qquad (7\text{-}3\text{-}12)$$

In addition to the assumption made above with respect to the density ratios (σ_{re} and σ), we have made the following assumptions and approximations in deriving Eq. (7-3-12):

1. Neglected the combination of the gravitational and centrifugal terms ($g - V^2/r$) (i.e., assumed a straight-line trajectory).
2. Used an exponential model for the atmospheric density.
3. Assumed $C_D/\sin \phi_{re}$ to be constant.

Let us return and look at the variable B, which is defined as

$$B = \frac{\rho_0 g_0}{2(BC)\beta \sin \phi_{re}} \qquad (7\text{-}3\text{-}13)$$

B is dimensionless and is negative, since the entry angle ϕ_{re} is negative and all other terms are positive. With a specified planet so that β, ρ_0, and g_0 are specified, the magnitude of B is determined by the BC and by the entry angle (ϕ_{re}), being inversely proportional to both. Increasing the BC or the ϕ_{re} reduces the value of B and thus affects the entry velocity profile.

In comparing E/Vs with the same W/A ratio, the BC is a measure of the streamlining (the slenderness) of the body. For a blunt-body E/V, as sketched in Fig.

7-3-1a, $C_{D\infty}$ is on the order of unity ($C_{D\infty} \cong 1$), where the subscript ∞ denotes the hypersonic region. The qualitative variation of C_D with Mach number is shown in Fig. 7-3-1b. To a first approximation, $C_{D\infty}$ is a function of δ, the *half-cone angle*, as defined in Fig. 7-3-2, with

$$C_{D\infty} \cong 2 \sin^2 \delta \qquad (7\text{-}3\text{-}14)$$

For example, if $\delta = 20°$, $C_{D\infty} \cong 0.234$; if $30°$, $C_{D\infty} \cong 0.50$; and if $45°$ (a blunt body), $C_{D\infty} \cong 1.0$. As the E/V becomes more slender, the half-cone angle decreases as does $C_{D\infty}$, which in turn increases the BC (for a given W/A) and B becomes smaller.

Let us return to Eq. (7-3-12) and look at some velocity profiles for E/V's entering the *Earth's atmosphere*, where $\rho_0 = 1.226$ kg/m³ and $\beta = 1/7254 = 1.378 \times 10^{-4}$ m⁻¹.

Example 7-3-1

An E/V entering the Earth's atmosphere has a mass of 50 kg (490.5 N or 110 lb) and has a diameter of 3.534 m and a half-cone angle of $45°$ (a blunt body). $V_{re} = 8000$ m/s and ϕ_{re}, the parameter of interest, is unspecified at this time.

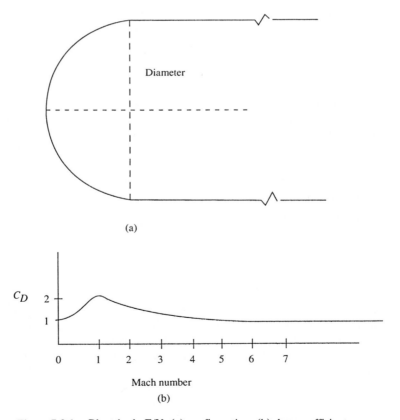

Figure 7-3-1. Blunt-body E/V; (a) configuration; (b) drag coefficient.

Sec. 7-3 Ballistic Entry

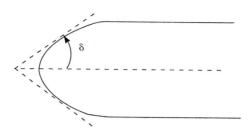

Figure 7-3-2. E/V half-cone angle.

(a) Find the value of the ballistic coefficient (BC).
(b) If $\phi_{re} = -22°$, find the velocity (and V/V_{re}) at an altitude of 50 km.
(c) Increase ϕ_{re} to $-45°$ (steepen the entry) and determine the effect on the velocity at 50 km.

Solution (a) Note that with $V_{cso} = 7905$ m/s, $V_{re}^* = V_{re}/V_{cso} = 1.012$, indicating entry from a low Earth orbit.

$$C_{D\infty} = 2\sin^2(45°) = 1.0$$

The cross-sectional area is

$$A = \pi r^2 = 9.81 \text{ m}^2$$

and

$$\frac{W}{A} = 50 \text{ N/m}^2 = 50 \text{ Pa}$$

Therefore, the ballistic coefficient is

$$BC = \frac{W/A}{C_{D\infty}} = \frac{50}{1} = 50 \text{ N/m}^2 = 50 \text{ Pa}$$

(b) With $\phi_{re} = -22°$ and BC = 50 Pa (1.14 lb/ft²),

$$B = \frac{1.226 \times 9.81 \times 7254}{2 \times 50 \times \sin(-22°)} = -2328$$

and at 50,000 m,

$$V = 8000e^{-2328e^{-(50,000/7254)}} = 752.9 \text{ m/s}$$

so that

$$\frac{V}{V_{re}} = 0.094$$

We see that the most of the deceleration seems to have taken place before the E/V reached 50 km.

(c) With ϕ_{re} increased to $-45°$ but with the BC remaining at 50 Pa, B decreases to -1233 and V at 50 km is 2288 m/s ($V/V_{re} = 0.286$). We see that with the steeper entry angle the E/V is traveling faster when it reaches 50 km and is still decelerating.

Example 7-3-2

The BC of the E/V of Example 7-3-2 has been increased to 94.4 Pa (1.97 lb/ft²), but V_{re} is still 8000 m/s.

(a) Find the velocity at 50 km with $\phi_{re} = -22°$ and compare the answer with that of Example 3-7-1(c).
(b) Find the velocity at 50 km with $\phi_{re} = -45°$ and compare with that of Example 7-3-1(c).

Solution

(a) $$B = \frac{1.226 \times 9.81 \times 7254}{2 \times 94.4 \times \sin(-22°)} = -1234$$

Substituting into Eq. (7-3-12) gives a $V/V_{re} = 0.2860$ and $V = 2288$ m/s. These values are identical with those found in the previous example for an E/V with BC = 50 Pa and $\phi_{re} = -45°$. Increasing the BC has the same effect on the velocity as increasing the entry angle.

(b) With $\phi_{re} = -45°$, B decreases, becoming -653.7, and V/V_{re} and V at 50 km increase to 0.515 and 4120 m/s.

In these two examples, we see that increasing either the entry angle or the ballistic coefficient decreases B, thus increasing the value of V at a specified altitude. The implication is that the E/V with the higher BC or ϕ_{re} does not slow down as quickly and continues its deceleration to lower altitudes. Because B is in essence a similarity parameter, changes in BC and ϕ_{re} result in equivalent changes in B.

Since B, which is not a physical variable, is inversely proportional to the product of BC and $\sin \phi_{re}$, it is not unusual to see the negative of this product defined as the *ballistic factor*[1] and used in lieu of the BC. The negative of the variable B itself could be used as a *planetary ballistic factor* inasmuch as it includes ρ_0 and g_0 as well as the E/V characteristics and entry values. We shall, however, use only the ballistic coefficient BC along with the entry angle ϕ_{re}, keeping them separate, since the BC represents the E/V characteristics and ϕ_{re} is a characteristic of the entry trajectory.

It is interesting to plot the nondimensionalized velocity (V/V_{re}) as a function of the altitude for various values of BC for a specified value of the entry angle, as is done in Fig. 7-3-3 for $\phi_{re} = -22°$. Notice that as the BC ($W/C_D A$) is increased, the E/V penetrates deeper into the atmosphere before it starts to slow down (decelerate) and that with BC = 50,000 Pa, the E/V is still decelerating as it hits the surface. *Increasing the entry angle has the same effect as increasing the BC.*

Let us now examine the deceleration ($-dV/dt$) along an entry trajectory. Returning to Eq. (7-3-12) and differentiating with respect to time yields

$$\frac{dV}{dt} = V_{re} B(e^{Be^{-\beta h}})(-\beta e^{-\beta h})\frac{dh}{dt} \qquad (7\text{-}3\text{-}15)$$

But

$$\frac{dh}{dt} = V \sin \phi_{re} = V_{re} \sin \phi_{re} e^{Be^{-\beta h}}$$

$$(7\text{-}3\text{-}16)$$

---

[1] To add to the confusion, at times the ballistic coefficient, as defined in this book, may be called the ballistic factor.

Sec. 7-3   Ballistic Entry   215

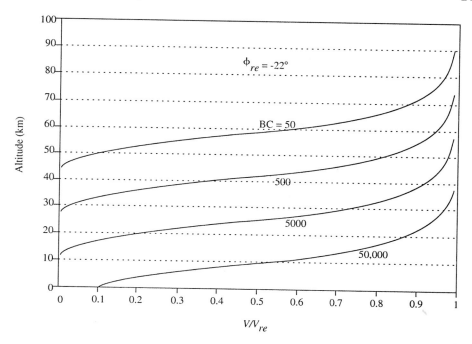

Figure 7-3-3. Altitude and velocity along a reentry trajectory for several values of the BC with $\phi_{re} = -22°$.

so that Eq. (7-3-15) becomes

$$\frac{dV}{dt} = -\beta B V_{re}^2 \sin \phi_{re}\, e^{-\beta h} e^{2Be^{-\beta h}} \qquad (7\text{-}3\text{-}17)$$

In Eq. (7-3-17), both B and $\sin \phi_{re}$ are negative and therefore dV/dt will be negative, showing that the E/V is decelerating, as is to be expected. The units of dV/dt are m/s^2; it is a common practice to describe accelerations in terms of Earth-surface g_0 using the symbol n, where $n = (dV/dt)/g_0$.

Equation (7-3-17) is somewhat complicated but at this point we are primarily interested in the magnitude of the maximum deceleration ($n_{\max}$) and the altitude at which it occurs. Setting the derivative of dV/dt with respect to h equal to zero results in

$$\frac{d}{dh}\left(\frac{dV}{dt}\right) = B\beta^2 V_{re}^2 \sin \phi_{re}\, e^{-\beta h} e^{2Be^{-\beta h}} (2eBe^{-\beta h} + 1) = 0 \qquad (7\text{-}3\text{-}18)$$

This is also a complicated equation but fortunately, we only need to set the terms within the parentheses equal to zero to satisfy this equation and the maxima–minima condition. Doing so and solving for h, the altitude where the maximum deceleration occurs, yields

$$h_{n_{\max}} = \frac{\ln(-2B)}{\beta} = \frac{1}{\beta}\ln\left[-\frac{\rho_0 g_0}{(BC)\beta \sin \phi_{re}}\right] \qquad (7\text{-}3\text{-}19)$$

It is interesting to see in Eq. (7-3-19) that the *altitude* for $n_{\max}$ is independent of the magnitude of V_{re} and *is a function only of the ballistic coefficient and the entry angle*.

Example 7-3-3

The E/V of Example 7-3-2 has BC = 94.4 Pa (1.97 lb/ft^2) and V_{re} = 8000 ft/s.
(a) Find the altitude at which the deceleration reaches its maximum when $\phi_{re} = -22°$.
(b) Do part (a) for a steeper entry angle, $\phi_{re} = -45°$ and compare with part (a).
(c) Increase the BC to 5000 Pa (104.5 lb/ft^2) and compare the altitude for $n_{\max}$ (with $\phi_{re} = -22°$) with that of (a).
(d) If V_{re} is increased to 11,000 m/s, will the altitudes change?

Solution

(a) $$B = \frac{1.226 \times 9.81 \times 7254}{2 \times 94.4 \times \sin(-22°)} = -1234$$

and

$$h_{n_{\max}} = 7254 \ln[-2(-1234)] = 56{,}660 \text{ m} = 56.66 \text{ km}$$

(b) With $\phi_{re} = -45°$, the magnitude of B decreases to -653.7, and h for $n_{\max}$ also decreases from 56,660 m (185,800 ft) to 52,050 m (170,700 ft), a decrease of 4610 m (15,100 ft).

(c) With BC = 5000 Pa (104.4 lb/ft^2) and $\phi_{re} = -22°$, B is equal to -23.30 and h for $n_{\max}$ becomes 27,800 m (91,360 ft), which is on the order of half of the altitude found with the much lower value of BC (94.4 Pa).

(d) Changing V_{re} will not affect any of the altitudes found above.

Figure 7-3-4 shows the maximum deceleration altitude as a function of the ballistic coefficient (in kPa) for several entry angles. Notice the knee in the curves as BC increases as well as the decreased effect of increasing the entry angle for the larger values. If the BC were plotted on a logarithmic scale, the altitude curves would be straight lines.

Returning to Eq. (7-3-17), the expression for dV/dt, substituting the condition for the altitude for $n_{\max}$ [$\exp(-\beta h) = -1/2B$] yields an expression for $n_{\max}$, namely,

$$n_{\max} = \frac{1}{g_0}\frac{dV}{dt} = \frac{\beta V_{re}^2 \sin \phi_{re}}{2eg_0} \qquad (7\text{-}3\text{-}20)$$

where $e = 2.718$, the natural logarithm base. It is very interesting to note here that the *maximum deceleration is independent of the ballistic coefficient* and is a function only of the entry angle and the square of the entry velocity.

To find the velocity at $n_{\max}$, the condition for the altitude at that point is substituted into Eq. (7-3-12), the velocity equation, to obtain

$$V_{n_{\max}} = V_{re} e^{Be - \ln(-2B)} \qquad (7\text{-}3\text{-}21)$$

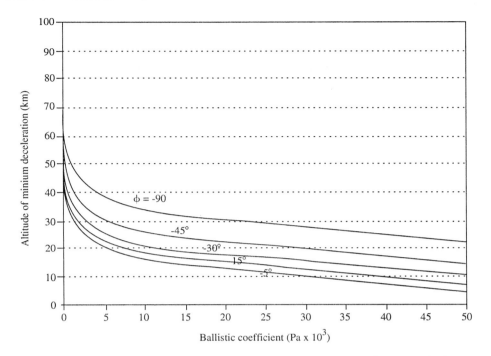

Figure 7-3-4. Altitude of maximum deceleration as a function of the BC for several values of ϕ_{re}.

which can be reduced to

$$V_{n_{\max}} = V_{re} e^{-0.5} = 0.606 V_{re} \qquad (7\text{-}3\text{-}22)$$

which shows that at $n_{\max}$ *the velocity is a function only of the entry velocity*, a somewhat surprising conclusion.

Example 7-3-4

An E/V has an entry angle of $-22°$ and an entry velocity of 8000 m/s.
(a) Find the maximum deceleration (g_0) and the associated altitude and velocity for BC = 5000 Pa (104.4 lb/ft^2).
(b) With BC = 5000 Pa, find the velocity and deceleration at an altitude of 18 km (59,020 ft).
(c) Do part (a) for a more slender and more pointed shape [i.e., for BC = 50,000 Pa (1044 lb/ft^2)].

Solution

(a) $$n_{\max} = \frac{(8000)^2 \sin(-22°)}{2 \times 7254 \times 9.81 \times 2.718} = -62.0 g_0$$

Using Eq. (7-3-13), B is found to be -23.23, so that, from Eq. (7-3-19),

$$h_{n_{\max}} = 27{,}860 \text{ m} = 27.86 \text{ km} = 91{,}360 \text{ ft}$$

and

$$V_{n_{max}} = 0.606 \times 8000 = 4848 \text{ m/s}$$

(b) With $V_{re} = 8000$ m and $B = -23.23$, V can be found from Eq. (7-3-12) to be

$$V = 8000 e^{-23.23 e^{-1.378 \times 10^{-4} \times 18,000}} = 1144 \text{ m/s}$$

From Eq. (7-3-17),

$$\frac{dV}{dt} = -131.8 \text{ m/s}^2 = -13.4 g_0$$

(c) With the BC increased to 50,000 Pa (1044 lb/ft^2), n_{max} is unchanged at $-62.0 g_0$, as is the velocity at 4848 m/s. The altitude, however, decreases inasmuch as B is now -2.323, so that

$$h_{n_{max}} = 7254 \ln(4.646) = 11,140 \text{ m} = 11.14 \text{ km} = 36,530 \text{ ft}$$

Figure 7-3-5 is a plot of the deceleration along a ballistic entry for a suborbital entry velocity of 7300 m/s and an entry angle of $-22°$ with three values for the ballistic coefficient. Note that n_{max} ($-51.6 g_0$) is unaffected by the increase in the BC, but that its altitude decreases as the BC is increased, and the E/V penetrates deeper into the atmosphere. Increasing or decreasing ϕ_{re} would have the same effect as increasing or decreasing the BC. Increasing V_{re} would increase n_{max} and its associated velocity but would not affect the corresponding altitude.

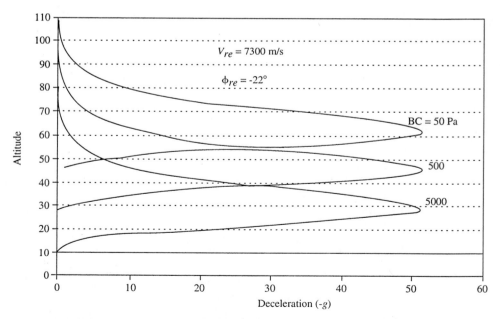

Figure 7-3-5. Deceleration along an entry trajectory as a function of altitude for several values of the BC with $V_{re} = 7300$ m/s and $\phi_{re} = -22°$.

Sec. 7-3 Ballistic Entry

The n_{max} of $-62g_0$ in Example 7-3-4 seems to be very large—and it is. However, n_{max} can be much larger than that. For example, increasing V_{re} to 11,000 m/s, keeping $\phi_{re} = -22°$, results in an n_{max} of $-117g_0$. Leaving V_{re} at 11,000 m/s and increasing ϕ_{re} to 90° (a vertical entry) increases n_{max} to $-313g_0$, a staggering figure. Figure 7-3-6 shows how n_{max} varies with entry velocity and entry angle and emphasizes the importance of keeping the entry angle small if n_{max} is to remain within a reasonable limit so that living occupants and equipment and the E/V itself might survive.

The maximum deceleration experienced by a human being (with survival but with some damage) strapped to a rocket-powered sled on a track was on the order of $12g_0$. The toleration of humans to decelerations is a function of magnitude and duration; decelerations in excess of $12g_0$ can be survived if the duration is sufficiently brief. Ballistic entry was used in the Mercury and Gemini programs, and the lift-to-drag ratio for the Apollo E/V was very small (<0.5). Consequently, the control of the entry angle was of the utmost importance.

Example 7-3-5

With the assumption that the maximum allowable deceleration for a manned entry is $-6g_0$, find the maximum allowable entry angle for the following entry velocities:
(a) 7300 m/s
(b) 8000 m/s
(c) 11,000 m/s

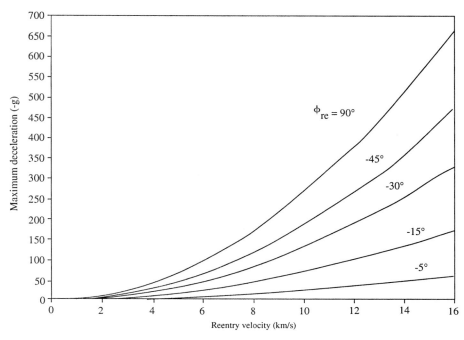

Figure 7-3-6. Maximum deceleration as a function of the reentry velocity for several values of ϕ_{re}.

Solution (a) Rearranging Eq. (7-3-20) to solve for ϕ_{re} produces

$$\phi_{re} = \sin^{-1}\frac{-2n_{max}g_0 e}{\beta V_{re}^2} = \sin^{-1}\frac{-2.313 \times 10^6}{V_{re}^2}$$

With $V_{re} = 7300$ m/s,

$$\phi_{re,max} = -2.50°$$

(b) With $V_{re} = 8000$ m/s,

$$\phi_{re,max} = -2.08°$$

(c) With $V_{re} = 11{,}000$ m/s,

$$\phi_{re,max} = -1.10°$$

Example 7-3-6

If the entry angle is $-5°$, find the maximum deceleration for an Earth entry at the entry velocities of Example 7-3-5.

Solution Applying Eq. (7-3-20), as in the previous example:

With $V_{re} = 7300$ m/s, $n_{max} = -12.0g_0$.
With $V_{re} = 8000$ m/s, $n_{max} = -14.4g_0$.
With $V_{re} = 11{,}000$ m/s, $n_{max} = -27.3g_0$.

Although the relationship between time and altitude (or density ratio) during the linear deceleration phase can be found by using Eq. (7-2-3) ($dh/dt = V \sin \phi_{re}$) and Eq. (7-3-12) (the expression for V in terms of h), the integration is a bit complicated and will not be performed here inasmuch as the exact time along this straight-line portion of the entry trajectory is not of great importance. Suffice it to say that the time is a function of BC, ϕ_{re}, and V_{re} and is relatively short; for example, if the linear trajectory were extended to the surface of the Earth for an entry at $-22°$ and 7300 m/s, the total entry time would be less than 1 minute.

The approximate distance (range) traveled during this time can easily be found by extending the straight-line approximation to the surface and combining Eqs. (7-2-3) and (7-2-4) to obtain

$$dS = V \cos \phi \, dt = \cot \phi \, dh \tag{7-3-23}$$

which, with a constant $\phi = \phi_{re}$, is a simple relationship between range and altitude as a function of ϕ. As was the case with the time, the range is relatively short. For an entry angle of $-22°$ and an entry altitude of 76.25 km, the range is on the order of 185 km (100 nmi), 2.9% of the Earth's radius.

Only when the BC values and entry angles are very large will the straight-line trajectory extend to the planetary surface with the E/V still decelerating as it impacts. The usual case is for the straight-line trajectory to undergo a gravity-turn transition into a vertical trajectory at terminal velocity with the gravitational force balanced by the drag force. With the assumption that the deceleration is small with respect to these forces, it is possible to obtain closed-form expressions for this

Sec. 7-3 Ballistic Entry

vertical portion, which can then be joined with the straight-line solutions if one wished to obtain an approximate solution for the entire entry trajectory. It should be noted that the lower the BC and the lower the entry angle, the earlier the vertical transition will occur and the greater the reduction in the range.

Although the examples in this section pertain to entry into the atmosphere of Earth, the relationships and expressions developed can be used to investigate entry into any planetary (or lunar) atmosphere, provided that there is some knowledge of the planetary characteristics. Table 7-3-1 lists relevant characteristics for Venus and Mars; it must be emphasized that these characteristics are estimates only and do not necessarily represent the most current estimates. They do, however, give a feeling for the relationship to the values for the Earth, which are included in Table 7-3-1 for purposes of comparison. For example, this simple table indicates that the atmospheres of Earth and Venus have similar distributions, with the latter being more dense (high drag devices such as parachutes will be more effective on Venus). Mars, on the other hand, is characterized by a slower decrease in density with altitude than the other two planets and has a much lower density, decreasing the effectiveness (and increasing the size) of drag devices used for braking.

Figure 7-3-7 compares the variation of the velocity ratio (V/V_{re}) with altitude for entry into the three planetary atmospheres with a $W/C_D A$, the ballistic coefficient, equal to 500 Pa (10.44 lb/ft^2) and an entry angle of $-22°$. Figure 7-3-8 shows the variation of the deceleration (in Earth g_0 units) with altitude for the same E/V

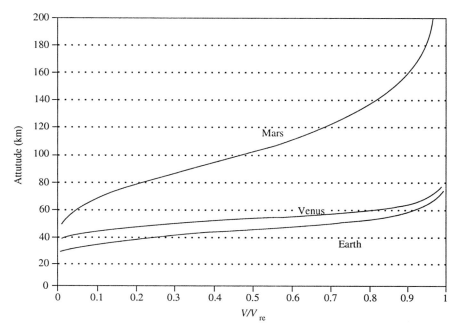

Figure 7-3-7. Velocity along a reentry trajectory in three planetary atmospheres.

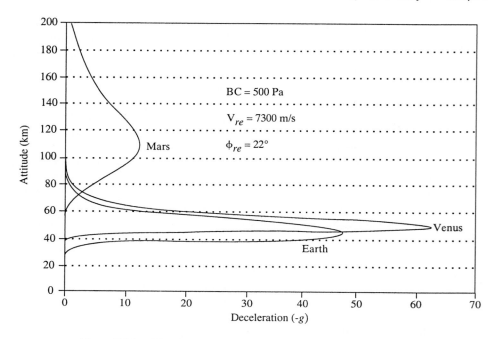

Figure 7-3-8. Maximum deceleration in three planetary atmospheres.

and entry angle with an entry velocity of 7300 m/s. We see that the curves for Venus and Earth are very similar because of the similar atmospheres, with the higher density of Venus manifesting itself in the slightly higher values of the maximum deceleration and its altitude. The lower density and slower variation of density with altitude of Mars, on the other hand, produce a slower variation of velocity with altitude and a much lower maximum deceleration at a higher altitude.

Since the gravitational attractions and the planetary radii of Venus and Earth are of the same order of magnitude, so are their escape velocities and the corresponding maximum decelerations, approximately $325g_0$ for both with a 90° entry angle. Mars, though, with its lower gravitational attraction, would have a lower escape velocity and a corresponding maximum deceleration on the order of only $20g_0$.

Let us return to the entry vehicle itself and use two simple examples to illustrate some of the configuration considerations for a ballistic E/V and to demonstrate the effect(s) of changing the ballistic coefficient while holding the weight constant.

Example 7-3-7

The total weight of an unmanned E/V is 9810 N (1000 kg or 2205 lb), including the structure and all equipment. The desired BC is 5000 Pa (104.4 lb/ft^2) and the selected half-cone angle δ is 25°. The preliminary shape is to be a conical nose section followed by a cylindrical afterbody, if one is needed, as sketched in Fig. 7-3-9, where the total length is $L = l' + l$. If possible, the specific gravity of the E/V is to be 0.5 or more (i.e., a desired density $\geq$ 499.8 kg/m^3).

Sec. 7-3 Ballistic Entry

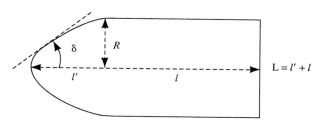

Figure 7-3-9. E/V schematic.

(a) Find A, the cross-sectional area, and the diameter D and radius R of the E/V.
(b) Find L, the total length, and V_T, the total volume, including the volume of the cone and of the afterbody. If the desired density cannot be achieved, determine the maximum value that can be obtained.

Solution

(a) $$C_D = 2\sin^2(25°) = 0.357$$

$$\frac{W}{A} = (BC) \times C_D = 5000 \times 0.357 = 1785 \text{ N/m}^2$$

$$A = \frac{9810}{1785} = 5.496 \text{ m}^2 \qquad D = 2.64 \text{ m} \qquad R = 1.32 \text{ m}$$

(b) Looking at the conical nose section (the nose cone), we find that

$$l' = \frac{R}{\tan\delta} = 2.836 \text{ m}$$

and the volume of the cone, V_c, is

$$V_c = \frac{l'A}{3} = 5.196 \text{ m}^3$$

Since the total volume, V_T, is equal to the total mass divided by the vehicle density, with the desired density V_T should be

$$V_T = \frac{m}{\rho} = \frac{1000}{499.8} = 2.0 \text{ m}^3$$

Since this value is considerably less than the volume of the nose cone itself, it is not realistic. Consequently, $l = 0$, and L must be equal to $l' = 2.836$ m; furthermore, the desired density cannot be attained. Therefore, this E/V is a cone with no afterbody, a cross-sectional area of 5.946 m², a diameter of 2.64 m, and a length of 2.836 m. With the weight fixed at 9810 N, it has the density

$$\rho = \frac{m}{V_c} = \frac{1000}{5.195} = 192.5 \text{ kg/m}^3$$

which corresponds to a specific gravity of 0.192.

Example 7-3-8

Keep the weight at 9810 N and the half-cone angle at 25° but increase the BC to 50,000 Pa.

(a) Find the cross-sectional area and diameter of this E/V and compare with the results of Example 7-3-7(a).

(b) Use the density found in Example 7-3-7 (192.5 kg/m³) to find the total volume and length.
(c) Do part (b) with a specific gravity of 0.5 (ρ = 499.6 kg/m³).
Solution (a) Using the relationships of Example 7-3-7 yields

$$\frac{W}{A} = 17{,}850 \text{ N/m}^2$$

$$A = 0.549 \text{ m}^2 \quad D = 0.8365 \text{ m} \quad R = 0.418 \text{ m}$$

The cross-sectional area and diameter of this E/V are considerably less than those of the E/V with the smaller BC.

(b) Since R is less than that of Example 7-3-6, so will l' and V_c be smaller.

$$l' = 0.897 \text{ m} \quad \text{and} \quad V_c = 0.164 \text{ m}^3$$

Using the density of the preceding example,

$$V_T = \frac{1000}{192.5} = 5.195 \text{ m}^3$$

Since $V_c = 0.164$ m³, the volume of the afterbody, V_a, will be 5.031 m³ and is equal to lA. Therefore, $l = 9.164$ m and the total length of the E/V is 10.06 m, considerably longer than that of the preceding example.

(c) With the higher density of 499.8 kg/m³, $V_T = 2.0$ m³ and $V_c = 1.837$. Therefore, $l = 3.346$ and $L = 4.243$ m, less than half of the length of the less dense E/V.

These two examples show that increasing the ballistic coefficient while keeping the weight constant lengthens and slenderizes the E/V and that increasing the density makes the E/V shorter and more compact, as would be expected. Although *caliber* is commonly used to denote the diameter of a bullet or gun barrel, it is also a measure of the slenderness of a body when defined as the ratio of the length to the diameter: The larger the caliber, the more slender the body. Applying this definition to the E/Vs of these two examples, the caliber of the pure cone with the lower density is 1.07, that of the lower-density cone plus cylindrical afterbody is 12.0, and that of the higher-density E/V is 5.07.

When entry angles become smaller than about −5°, as in Example 7-3-5, this straight-line analysis loses its validity and relevance and the entry trajectories become curved, the decelerations become smaller, and the times become larger. As the entry angle approaches zero, the trajectory becomes what is often referred to as an *orbital decay entry*, in which a spacecraft in a circular orbit gradually loses energy because of atmospheric drag (which is initially small at the orbital altitude in the outer fringes of the atmosphere) and spirals toward the planetary surface until it enters a terminal vertical phase. It can be shown that the maximum deceleration with orbital decay is on the order of $-8g_0$ for all BCs.

We conclude this section on ballistic (direct) entry with the observation that the gas dynamic drag on the E/V is the principal deceleration mechanism. Since the drag is directly proportional to the product of the atmospheric density and the square of the velocity (the dynamic pressure), so is the deceleration. As the E/V approaches the planetary atmosphere, its velocity is at a maximum and the atmospheric density is at a minimum. As the vehicle penetrates the atmosphere, the atmospheric density increases rapidly, increasing the drag, reducing the velocity, and initiating deceleration. Since the deceleration is the product of two variables, one increasing and one decreasing, there will be an inflection point of maximum deceleration (see Fig. 7-3-5) where the velocity is decreasing more rapidly than the density is increasing. (With vehicles with a very large BC and entry angle, this inflection point might well be beneath the planetary surface.) Although the maximum deceleration, which can be very large and which increases as the entry velocity and entry angle increase (and as ρ_0 and g_0 increase), is independent of the characteristics of the E/V, the lighter vehicles decelerate at the higher altitudes and enter the curved terminal phase earlier and consequently have a shorter range than the heavier vehicles.

The entry trajectory can be further modified by the judicious use of lift, and we shall see in the next section that a little lift can be a very useful tool.

7-4 LIFTING ENTRY

Lift can be used to:

1. Increase the width of the entry corridor (see Fig. 7-2-4b)
2. Significantly reduce the decelerations experienced by an E/V
3. Enlarge the landing footprint, thus relaxing the deorbit and entry corridor requirements for the guidance and control system for a specified touchdown location
4. Provide additional entry trajectory options, such as skipping trajectories
5. Execute nonpropulsive plane changes with aerodynamic turns

Although the dynamic and kinetic equations of Section 7-2 are valid for lifting entry, closed-form solutions and analytic expressions that can give insight into the interaction between the trajectories and physical characteristics and parameters require approximations and assumptions, as was the case with ballistic entry.

An important lifting-entry trajectory is the *equilibrium glide*, which is a relatively flat glide in which the gravitational force is balanced by the combination of the lift and centrifugal forces. In addition to the "small" angle assumptions ($\sin \phi \cong 0$ and $\cos \phi \cong 1$) with respect to the elevation angle ϕ, it is further assumed that ϕ is changing slowly so that $d\phi/dt$ can be neglected (to a first approximation ϕ can be assumed to be constant) and that the lift-to-drag ratio is ≥ 0.5 or so.

Equations (7-2-2a) and (7-2-2b) can now be written as

$$\frac{dV}{dt} = -\frac{D}{m} = -\frac{\rho_0 C_D A}{2m} \sigma V^2 \qquad (7\text{-}4\text{-}1)$$

$$\frac{V^2}{r} = -\frac{L}{m} + g = -\frac{\rho_0 C_L A}{2m} \sigma V^2 + g \qquad (7\text{-}4\text{-}2)$$

where C_L and C_D are the lift and drag coefficients of the E/V; they are not necessarily constant, are interrelated, and are a function of the angle of attack of the vehicle.

Since Eq. (7-4-2) is algebraic, it can be solved directly for V as a function of σ, the atmospheric density ratio, and thus of h, the altitude. At this time it is convenient to introduce the lift-to-drag ratio (L/D) of the vehicle, which can be used to relate C_L and C_D as follows:

$$\frac{L}{D} = \frac{C_L}{C_D} \quad \text{and, therefore,} \quad C_L = \frac{L}{D} C_D \qquad (7\text{-}4\text{-}3)$$

With $W = mg_0$ and Eq. (7-4-3), Eq. (7-4-2) can be rewritten as

$$V^2 = \frac{gr}{1 + [(L/D)(C_D A/W)(\rho_0 g_0/2)\sigma r]} \qquad (7\text{-}4\text{-}4)$$

Since $(gr)^{1/2}$ is the circular-orbit velocity V_{cs} at r, $gr = V_{cs}^2$. Since a planetary atmosphere is thin with respect to the planetary radius, g and r (and V_{cs}) within the atmosphere are essentially constant and the surface values can be used, namely, g_0, r_0, and V_{cso}. (In the Earth's atmosphere, there is a 1% difference between r_{re} and r_0, a 2% difference between g_0 and g_{re}, and only a 0.5% difference in the orbital velocities.) With these assumptions, Eq. (7-4-4) can be solved to obtain an expression for the velocity in terms of the density ratio σ,

$$\boxed{\frac{V}{V_{cs}} = \frac{1}{\sqrt{1 + [(L/D)(C_D A/W)(\rho_0 g_0 r_0/2)\sigma]}}} \qquad (7\text{-}4\text{-}6)$$

where V_{cs} is used to denote V_{cso} and σ, the atmospheric density ratio, is approximated by the previously discussed exponential relationship

$$\sigma = e^{-\beta h} \qquad (7\text{-}4\text{-}6)$$

Examining Eq. (7-4-5), we see the familiar ballistic coefficient of direct entry (BC) along with the L/D ratio. Defining a *lifting ballistic coefficient* (LBC) as

$$\text{LBC} = \frac{W}{C_D A (L/D)} = \frac{\text{BC}}{L/D} \qquad (7\text{-}4\text{-}7)$$

Eq. (7-4-5) can now be written as an explicit function of h with the LBC as a major parameter.

$$\boxed{\frac{V}{V_{cs}} = \frac{1}{\sqrt{1 + [\rho_0 g_0 r_0/2(\text{LBC})]e^{-\beta h}}}} \qquad (7\text{-}4\text{-}8)$$

Example 7-4-1

A lifting body entering the Earth's atmosphere in an equilibrium glide with a small elevation angle has a $W/C_D A = 5000$ Pa. Find the LBC and the nondimensional velocity at an altitude of 50 km for the following values of L/D:

(a) 1.0
(b) 1.5

Sec. 7-4 Lifting Entry

(c) 2.0
(d) 2.5
(e) 3.0

Solution (a) With $L/D = 1.0$,

$$\text{LBC} = \frac{W/C_D A}{L/D} = \frac{5000}{L/D} = 5000 \text{ Pa}$$

At $h = 50$ km, with $\beta = 0.1378$ km^{-1},

$$\frac{\rho_0 g_0 r_0}{2(W/C_D A)} e^{-50\beta} = 7.785$$

and

$$\frac{V}{V_{cs}} = \frac{1}{\sqrt{1 + 7.785(L/D)}}$$

With $L/D = 1.0$, $V/V_{cs} = 0.337$.
  (b) With $L/D = 1.5$, LBC = 3333 Pa and $V/V_{cs} = 0.281$.
  (c) With $L/D = 2.0$, LBC = 2500 Pa and $V/V_{cs} = 0.246$.
  (d) With $L/D = 2.5$, LBC = 2000 Pa and $V/V_{cs} = 0.220$.
  (e) With $L/D = 3.0$, LBC = 1667 Pa and $V/V_{cs} = 0.206$.

This example shows that increasing the L/D decreases the LBC and shifts the deceleration to higher altitudes. This effect is confirmed in Fig. 7-4-1, which shows the variation of velocity with altitude for four values of the lifting ballistic coefficient (LBC). When compared with Fig. 7-3-3, a similar plot for direct entry, we see similar

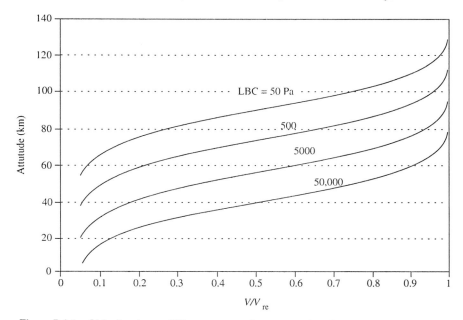

Figure 7-4-1. Velocity along a lifting reentry trajectory as a function of altitude for several values of the LBC.

shapes for the velocity curves but with deceleration occurring at lower altitudes when the L/D is zero (a ballistic entry).

An expression for the deceleration at any point along the trajectory in terms of V/V_{cs} can be found by first rearranging Eq. (7-4-2) to obtain

$$\frac{L}{m} = g - \frac{V^2}{r} = g - \frac{gV^2}{gr} = g - \frac{gV^2}{V_{cs}^2} = g\left[1 - \left(\frac{V}{V_{cs}}\right)^2\right] \tag{7-4-9}$$

Since

$$\frac{D}{m} = \frac{L}{m(L/D)} \tag{7-4-10}$$

and since, from Eq. (7-4-1), $dV/dt = -D/m$,

$$\boxed{n = \frac{1}{g_0}\frac{dV}{dt} = -\frac{[1 - (V/V_{cs})^2]}{L/D}} \tag{7-4-11a}$$

where n is the deceleration expressed in g_0 units and $(V/V_{cs})^2$ is given by Eq. (7-4-8), which can be substituted into Eq. (7-4-11a) to obtain another expression for the deceleration, this time as a function of h directly. This expression can be written as

$$\boxed{n = \frac{1}{g_0}\frac{dV}{dt} = \frac{-1}{(L/D) + [2(W/C_D A)e^{+\beta h}/\rho_0 g_0 r_0]}} \tag{7-4-11b}$$

Figure 7-4-2 is a plot of the deceleration as a function of the altitude for several values of the LBC and for two values of L/D. This figure and the deceleration expressions show that the deceleration increases continuously along the trajectory as the altitude decreases and that there is no maximum deceleration per se (no inflection point as with direct entry). The deceleration does, however, approach an asymptotic maximum that is simply the inverse of the lift-to-drag ratio:

$$\boxed{n_{max} \cong \frac{-1}{L/D}} \tag{7-4-12}$$

Consequently, we see that the decelerations for E/Vs with only modest L/D ratios are significantly less than those for direct-entry (ballistic) E/Vs; a little lift does a lot.

Example 7-4-2

For the lifting E/V of Exercise 7-4-2 (BC = 5000 Pa), find the value of the deceleration at 50 km, using the values of V/V_{cs} and L/D from that example, as well as n_{max}.
Solution (a) With $L/D = 1$, $V/V_{cs} = 0.337$ and using Eq. (7-4-11a), $n = -0.886g$'s. From Eq. (7-4-12), $n_{max} = -1.0g$.
(b) With $L/D = 1.5$ and $V/V_{cs} = 0.281$, $n = -0.614g$ and $n_{max} = 0.666g$.
(c) With $L/D = 2.0$ and $V/V_{cs} = 0.246$, $n = -0.530g$ and $n_{max} = -0.50g$.
(d) With $L/D = 2.5$ and $V/V_{cs} = 0.220$, $n = -0.381g$ and $n_{max} = -0.40g$.
(e) With $L/D = 3.0$ and $V/V_{cs} = 0.206$, $n = -0.319g$ and $n_{max} = -0.333g$.

Sec. 7-4 Lifting Entry

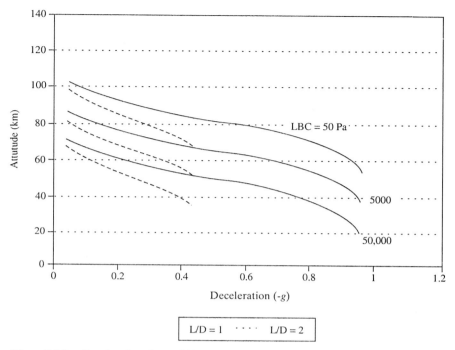

Figure 7-4-2. Deceleration along a lifting reentry trajectory for several values of the LBC and two L/D ratios.

The time along the entry trajectory as a function of the corresponding velocity can be found by rewriting Eq. (7-4-11a) as

$$dt = \frac{-(L/D)\, dV}{g[1 - (V/V_{cs})^2]} \tag{7-4-13}$$

Integrating from the velocity at the altitude of interest until the velocity is zero, the approximate time to touchdown from that altitude (and velocity) can be found from

$$\Delta t = \frac{1}{2}\sqrt{\frac{r_0}{g_0}\frac{L}{D}} \ln \frac{1 + (V/V_{cs})^2}{1 - (V/V_{cs})^2} \tag{7-4-14}$$

Equation (7-4-14) shows that the time in the entry trajectory, in the atmosphere, is directly proportional to the lift-to-drag ratio. Although this expression for time of flight is approximate, as are all of the values obtained from approximate solutions such as these, it provides comparative values. Figure 7-4-3 is a plot of Δt as a function of V/V_{cs} and L/D within the Earth's atmosphere.

Example 7-4-3

For the E/V of Exercise 7-4-2 (BC = 5000 Pa):
(a) Find the time to "touchdown" from a point where $V/V_{cs} = 0.80$ ($V = 6325$ m/s) for $L/D = 1.0, 1.5, 2.0, 2.5,$ and 3.0.

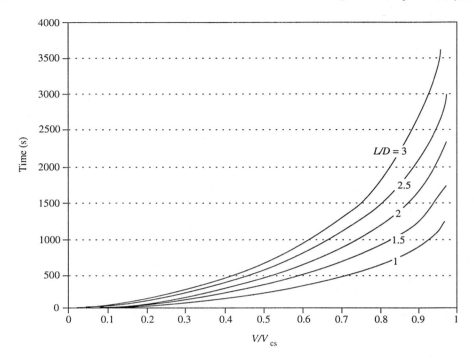

Figure 7-4-3. Reentry time as a function of the velocity along the trajectory for several values of L/D.

(b) Using the results of Example 7-4-2, find the time for each value of L/D from $V/V_{cs} = 0.8$ to an altitude of 50 km.

Solution (a) Substituting into Eq. (7-4-14) yields

L/D	TIME (s)	TIME (min)
1.0	611	10.2
1.5	917	15.3
2.0	1,223	20.4
2.5	1,528	25.5
3.0	1,834	30.6

(b) First, find the time from 50 km to touchdown and then subtract from the values of part (a).

L/D	V/V_{cs}	$t_{touchdown}$ (s)	Δt (s/min)
1.0	0.337	92.0	519/8.65
1.5	0.281	95.7	821/13.7
2.0	0.246	97.4	1126/18.8
2.5	0.220	97.7	1430/23.8
3.0	0.206	102.7	1731/28.8

Sec. 7-4 Lifting Entry

With the small elevation angles associated with the equilibrium glide, the horizontal range of the E/V can be found from the approximation that $dS/dt \cong V$, so that $dS = V\,dt$. Solving Eq. (7-4-11a) for dt in terms of V and dV and substituting for dt produces

$$dS = \frac{-(L/D)V\,dV}{g[1 - (V/V_{cs})^2]} \qquad (7\text{-}4\text{-}15)$$

Integrating from any point on the trajectory, where V is known, to touchdown yields an expression for the range from that point.

$$\boxed{S = -\frac{r_0}{2}\frac{L}{D} \ln\left[1 - \left(\frac{V}{V_{cs}}\right)^2\right]} \qquad (7\text{-}4\text{-}16)$$

Therefore, the range, as was the case with the time of flight, is directly proportional to the lift-to-drag ratio. This distance is traveled in the direction of the entry velocity (in the vertical plane) and is known as the *down-range distance*.

Figure 7-4-4 shows the variation of the maximum down-range distance, expressed in planetary radii, with V/V_{cs} for several L/D ratios. The combination of the higher initial velocities and higher L/D values can result in quite large down-range distances, much larger than any attained with direct entry (no lift). Although our equilibrium glide analysis has implied a constant L/D, the L/D ratio can be varied to touch down at distances less than the maximum.

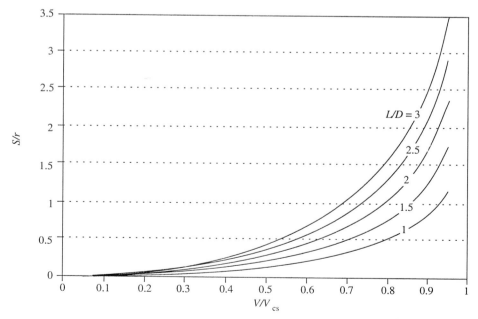

Figure 7-4-4. Downrange distance as a function of the velocity along the trajectory for several values of L/D.

An additional benefit of lift is the ability to make turns within the atmosphere, turns that can be used to reach a touchdown point not in the vertical plane (and to modify the duration of the entry). The out-of-plane distance is known as the *cross-range distance* and in conjunction with the down-range distance forms the *footprint* of the E/V. Without writing the equations for out-of-plane (lateral) flight, it can be shown that the maximum cross (lateral) range occurs when the *bank angle* Φ_{bank} (not to be confused with the elevation or entry angle) takes on an optimum value (Φ_{opt}), which is a function of L/D, namely,

$$\Phi_{opt} \cong \cot^{-1}\sqrt{1 + 0.106\left(\frac{L}{D}\right)^2} \qquad (7\text{-}4\text{-}17)$$

The corresponding maximum cross range is

$$Y_{max} \cong \frac{r_0}{5.2}\left(\frac{L}{D}\right)^2 \frac{1}{\sqrt{1 + 0.106(L/D)^2}} \qquad (7\text{-}4\text{-}18)$$

Equation (7-4-18) shows the variation of the maximum cross range (in planetary radii) and of the optimum bank angle as a function of the L/D ratio. To provide a footprint that will cover the globe (allow the E/V to land at a location of choice), Y/r_0 needs to be equal to $\pi/2$ rad (90°) and Eq. (7-4-18) shows that the required hypersonic L/D is on the order of 3.5. Designing an E/V that can handle the heating problem(s) associated with such an L/D is not a simple task, however.

A lifting vehicle must consider the *aerodynamic load factor*, which is defined as the lift-to-weight ratio (L/W) and has the dimensions of acceleration expressed in g_0 units. Although the standard symbol for the load factor is n, LF will be used to denote the load factor to avoid confusion with the n used for deceleration. Since the load factor is inversely proportional to the cosine of the bank angle, the load factor (g_0) in an optimum turn is

$$LF = \frac{1}{\cos \Phi_{opt}} \qquad (7\text{-}4\text{-}19)$$

Example 7-4-4

Find and tabulate the values of the optimum bank angle, the maximum cross range, and load factor for L/D ratios of 1.0, 2.0, 3.0, and 3.5.

Solution

L/D	Φ_{opt} (deg)	Y_{max}/r_0	LF (g_0)
1.0	43	0.183	1.37
2.0	40	0.698	1.30
3.0	36	1.238	1.23
3.5	33.4	1.554	1.20

In addition to acquiring a cross-range capability, a steady-state turn can be used to reduce the descent time (and down-range distance) by effectively reducing the

Sec. 7-4   Lifting Entry

lift-to-drag ratio by a factor equal to $\cos \Phi_{bank}$. This may be a desirable maneuver to reduce the heating of the E/V during a critical portion of the entry.

It should be pointed out that the introduction of lift to reduce the deceleration and to provide a maneuvering capability also introduces the need to consider the dynamic stability and control of the E/V. The stability analysis is similar to that for conventional aircraft, with the added problem that at hypersonic velocities with low atmospheric densities, there is virtually no aerodynamic damping. If the oscillations in response to a command or a disturbance are to be controlled by the pilot (rather than by a flight control system), the vehicle must be designed so that frequency of any oscillation is low (i.e., the period of the oscillation is high).

As the lifting E/V penetrates the sensible atmosphere, it must function as a conventional aircraft, but to date without power. The flight regime of a lifting E/V covers a broad spectrum, from low touchdown speeds at sea level to hypersonic velocities at the outer fringes of the atmosphere.

Before leaving the equilibrium-glide entry, it should be pointed out that the approximate solutions above imply that the E/V is at a circular orbital velocity at the initiation of the entry. This means that an E/V returning from a high Earth orbit or from outside the SOI will be slowed down prior to entering the glide perhaps by a propulsive ΔV, an aerocapture maneuver using atmospheric braking, a series of grazing trajectories, or a combination thereof.

The equilibrium glide entry gives insight into the characteristics and benefits of lifting entry and approximates the entry of the Space Shuttle (Space Transportation System). It is, however, not the only lifting entry trajectory. There are others of varying degrees of interest and usefulness that lend themselves to approximate closed-form solutions (and others that do not). Several of these are described briefly below with relevant comments where applicable.

The *skip entry* trajectory is a series of joined segments consisting of a penetration into the atmosphere (possibly with a turn), followed by an exit from the atmosphere at a lower velocity (the skipping phase) and a return to an elliptical Keplerian orbit that leads to another penetration, followed by another skip, and so on, until the desired conditions are reached for a descent to the planetary surface. During the skipping phase within the atmosphere, the aerodynamic lift is the dominant force, being much greater than the difference between the gravitational force and the centrifugal force ($mg - mV^2/r_0 \cong 0$). It can be shown that the exit elevation angle is equal to the negative of the entry angle ($\phi_{ex} = -\phi_{re}$). This relationship holds for each atmospheric leg in the overall trajectory so that the elevation angles can be considered to be constant and equal to ϕ.

The relationship between the exit and entry velocities for each skipping segment is given by

$$\frac{V_{ex}}{V_{re}} = \exp\left(\frac{-2\phi}{L/D}\right) \qquad (7\text{-}4\text{-}20)$$

where ϕ is a reasonably large angle, similar to those of ballistic entries; the larger the angle (and the lower the L/D), the greater the velocity reduction. If, for example, the entry (and exit) angle ϕ is $-22°$ (0.384 rad) and the L/D is 2.0, the exit velocity will be 68% of the entry velocity; if the $L/D = 1$, then $V_{ex}/V_{re} = 0.46$.

A single skip followed by a return to a modified Keplerian orbit rather than another atmospheric penetration could be used to execute an aeroassisted transfer or an aerocapture maneuver.

Another lifting trajectory of possible interest is the *negative L/D glide*, which could be used to keep an E/V within the atmosphere (to widen the entry corridor) for reasonable entry velocities and large elevation angles or for large entry velocities with a wide range of elevation angles. In contrast to the equilibrium glide with positive L/D, the elevation angle does not remain constant along a negative-lift trajectory and there is an altitude at which the deceleration reaches a maximum and then decreases. This maximum deceleration is a function of the ballistic coefficient ($W/C_D A$), the L/D, and the entry angle.

The last category of entry trajectories to be mentioned includes those trajectories with *small positive or negative L/D's* (-0.5 to $+0.5$) and *small entry angles* ($\leq 5°$). This category is applicable primarily to orbital decay trajectories (to include the ballistic decay case) and is characterized by low ($\leq 8g_0$) maximum decelerations which decrease in magnitude and increase in altitude with an increase in the magnitude of L/D and are influenced by the magnitude of the ballistic coefficient. Although we have considered the edge of the sensible atmosphere to be at an altitude of 75 km (41 nmi) or so, there is sufficient density and drag at higher altitudes to cause an orbit to decay eventually.

This section can be concluded with the simple statement that lift reduces the decelerations experienced by an E/V and increases its footprint, maneuverability, and duration of entry. However, as we shall see in the next section, this increase in duration introduces problems associated with the heating encountered during entry.

7-5 THE HEATING PROBLEM

An atmospheric entry is successful when an E/V and its contents arrive undamaged at a planetary surface, having survived the decelerations and heating associated with dissipating the large amount of kinetic energy possessed by the E/V at the time of entry. With a discussion of the deceleration problem and its possible solutions behind us, we can turn our attention to the heating problem and its possible solutions. We must not, however, be unaware of or ignore the impact of a heating problem solution upon the deceleration magnitude and profile. *The two entry problems, deceleration and heating, are linked and must be jointly solved.*

The kinetic energy of an E/V is dissipated by transformation into thermal energy (heat) as the E/V decelerates. The magnitude of this thermal energy is such that if all of the heat were transferred to the E/V, the E/V could be severely damaged and conceivably even completely vaporized. Fortunately, not all of the thermal energy is transferred to the surface of the E/V. It is the fraction that is transferred that is of interest to the designer (and of special interest to the occupants).

Sec. 7-5 The Heating Problem

This *thermal conversion fraction* (denoted by f) is a function of the shape (the BC), the altitude, and the velocity. At very high altitudes (above approximately 105 km) in the free-molecule flow region the thermal energy developed appears directly at the surface of the E/V and the value of the fraction is high; it can even approach unity. As the E/V descends and enters the continuum region with a laminar boundary layer, the fraction decreases as the thermal energy is transferred from the hot gas behind the shock wave to the surface of the E/V by conduction and convection. During this phase the fraction can decrease to values as low as 0.01 at altitudes on the order of 30 km (16 nmi) but then increases during the transition to a turbulent boundary layer, after which, in the absence of radiative transfer, the fraction again decreases.

Figure 7-5-1 illustrates the nature of the variation of f with altitude for a particular E/V configuration and set of entry conditions. It shows the increase in f during the transition from laminar to turbulent flow and the subsequent decrease in the absence of any radiative transfer. There may, however, be radiation from the gas to the E/V surface at the lower (denser) altitudes in the case of very blunt bodies and for large entry velocities. The dashed line in Fig. 7-5-1 shows that with such radiation the conversion fraction continues to increase after the transition to turbulent flow.

The surface heating rate depends on this fraction and on the rate of kinetic energy loss by the E/V (i.e., on the nature of the trajectory). For example, a lightweight (small BC) lifting E/V will decelerate slowly in the upper atmosphere and will have a low heat transfer rate despite a large conversion fraction because the kinetic energy loss is low. On the other hand, an E/V with a large BC on a ballistic entry that decelerates rapidly in the lower atmosphere will have a larger surface heating rate despite the lower conversion fraction.

Furthermore, the total heat input depends on both the surface heat transfer rate and the duration of the heating. If the conversion fraction were constant, the total amount of heat transferred would be independent of the type of entry inasmuch as the same fraction of the total initial kinetic energy would be converted into body heat. The actual variation of the fraction offers the opportunity to minimize the total heat by locating the maximum kinetic energy loss where the fraction is low, at the lower altitudes. Unfortunately, this is also the region of relatively high heat transfer rates and the designer is faced with the choice of minimizing the total heat transferred or of minimizing the heat transfer rates (by a slower deceleration at higher altitudes, where the fraction is higher).

The rate of change of the kinetic energy of the E/V into thermal energy in the atmosphere is equal to the product of the drag force and the velocity. The *total heat input* will be denoted by the symbol Q and the *instantaneous surface heat transfer rate* by q, where $q = dQ/dt$.[2]

The instantaneous surface heat transfer rate q can be found from

$$q \equiv \frac{dQ}{dt} = fDV = \frac{f\rho_0 \sigma C_D\, SV^3}{2} \qquad (7\text{-}5\text{-}1)$$

[2] This is not the dynamic pressure of Section 6-7.

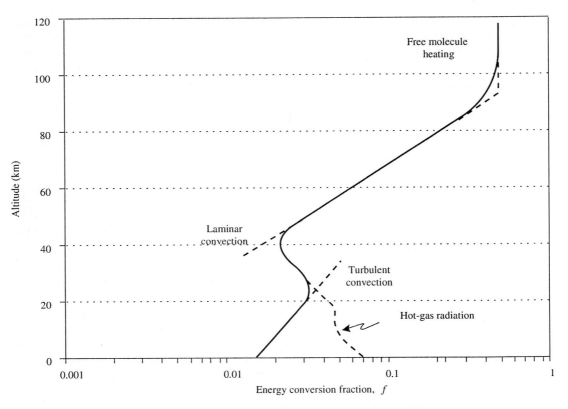

Figure 7-5-1. Typical plot of the energy conversion fraction as a function of altitude (and velocity) for a particular configuration and set of entry conditions.

where S is the relevant surface area for the E/V configuration of interest and is not necessarily the cross-sectional area of the ballistic coefficient. Ignoring for the moment the variation of f with altitude, Eq. (7-5-1) shows that q (the heat transfer rate) is proportional to the product of the atmospheric density ratio σ and the cube of the velocity

$$q \sim \sigma V^3 \qquad (7\text{-}5\text{-}2)$$

where V is the instantaneous velocity as shown in the expressions of Section 7-4. Since σ increases along the entry trajectory and V decreases, their product will reach a maximum at some point along the trajectory, as was the case with the deceleration. (Since the deceleration is proportional to the product of σ and V^2, the altitudes and velocities at its maximum value will be different.) The *maximum heating rate*, q_{max}, is an important design consideration along with the *average heating rate*, q_{ave}.

The other important design consideration is the total heat transferred to the E/V, where

$$Q_{total} \equiv Q = \int q\, dt = q_{ave} \Delta t \qquad (7\text{-}5\text{-}3)$$

Sec. 7-5 The Heating Problem

As mentioned above, the structure and contents of the E/V must be protected during atmospheric entry. Remembering that weight is at a premium, the three current techniques for handling the heat that reaches the surface of the E/V are:

1. The *heat sink*, an absorption technique, whereby the heat capacity of a mass (the heat sink) is used to absorb the heat with a resultant rise in the temperature of the mass. This is essentially a brute-force solution and the energy balance equation is

$$Q = mC_v \Delta T \qquad (7\text{-}5\text{-}4)$$

where m is the mass of the heat sink, C_v the specific heat at constant volume, and ΔT the temperature rise, which must be below the melting temperature of the material. The choice of material is obviously important, and the mass and weight can be very high. The ATLAS (and THOR) ballistic missile(s) used a "heat shield" in excess of 907 kg (2000 lb) of very pure, precisely machined copper. Beryllium, with a high thermal diffusivity, high thermal capacity, and high density, is also a heat shield candidate.

2. *Ablation*, also an absorption technique in which (in contrast to the heat sink) material is destroyed; the outer surface of the E/V chars, melts, and vaporizes. The energy balance equation is

$$Q = h_v \Delta m \qquad (7\text{-}5\text{-}5)$$

where h_v is the heat of vaporization and Δm is the mass lost by vaporization (ablation). The act of vaporization absorbs heat at the surface by virtue of the phase change and the gaseous product modifies the velocity and temperature profiles in the boundary layer, thus reducing the incoming transfer of heat (effectively reducing the conversion fraction f). Again the choice of material is important and requires one with a large heat of vaporization so as to minimize the weight of the ablating material to be carried aboard the E/V. Graphite and phenolic compounds are typical candidates for ablating surfaces.

3. *Radiation* from the surface of the E/V, a rejection scheme in which the temperature of the surface is allowed to increase to the point of melting, to as high a temperature as possible so as to increase the radiation to the point where the outgoing heat transfer rate balances the incoming rate. The equilibrium equation is

$$q_{\text{out}} = q_{\text{in}} \cong \varepsilon \nu T_w^4 \qquad (7\text{-}5\text{-}6)$$

where ε is the emissivity of the surface, ν the Stephan–Boltzmann constant, and T_w the surface temperature, which is assumed to be much larger than the gas temperature. Refractory (ceramic) materials with a high melting temperature coated with a high emissivity substance are the current candidates. Unfortunately, the coatings tend to lose their effectiveness prior to reaching the melting point of the refractory, whose strength also tends to fall off rapidly at high temperatures.

These protective schemes can be used individually or in combination as appropriate to the entry trajectory and E/V configuration. As implied by the energy balance equations above, the two absorption schemes are constrained by the total heat load, whereas radiation is limited by the heat transfer rate and the allowable wall temperature. If insulation is required to protect the contents of the E/V, all three approaches are penalized by the additional weight and the low thermal efficiency of insulating material. Other schemes, such as active cooling, have been considered and rejected for various reasons, primarily weight and complexity, that appear to be valid at this time.

Unfortunately, the equations and relationships involved in the thermal analysis and dynamics of entry are more complicated than those governing the velocity and deceleration, interact with the latter, and do not lend themselves to approximations and closed-form expressions similar to those developed for the latter. Consequently, we shall restrict ourselves to a discussion of the qualitative thermal behavior and parametric relationships for ballistic and lifting entry with remarks as to the suitability and appropriateness of the three thermal protective systems described above.

With *ballistic heating* ($L/D = 0$), convection is the primary mechanism for the transfer of the heat from the atmosphere to the E/V. With the substitution of the velocity profile expression for direct entry [Eq. (7-3-12)] into Eqs. (7-5-1) and (7-5-3) and with appropriate substitutions for the conversion fraction f in terms of skin friction and heat transfer coefficients, integration shows the relationship of the total heat input to the entry and vehicle parameters, namely

$$Q \sim V_{re}^2 \sqrt{\frac{W}{C_D A}} \frac{1}{\sin \phi_{re}} \qquad (7\text{-}5\text{-}7)$$

We see that the total heat input is directly proportional to the square of the entry velocity and the square root of the ballistic coefficient BC and inversely proportional to the square root of the entry angle. Increasing the entry velocity and the BC increases the heat input, whereas increasing the entry angle, on the other hand, reduces the total heat input. This reduction in heat with an increase in the steepness of the descent may seem to be contradictory until we realize that a steep descent reduces the duration of the entry and that the total heat is directly proportional to the duration.

It can also be shown that the maximum heat transfer rate, q_{max}, can be related to the entry and vehicle characteristics by the proportionality

$$q_{max} \sim V^3 \sqrt{\frac{W}{C_D A} \sin \phi_{re}} \qquad (7\text{-}5\text{-}8)$$

where V in Eq. (7-5-8) is the velocity for q_{max} and is equal to

$$V_{q_{max}} = e^{-(1/6)} V_{re} \cong 0.846 V_{re} \qquad (7\text{-}5\text{-}9)$$

Since the velocity for n_{max} was approximately $0.606 V_{re}$, we see that the maximum heating rate occurs at a higher velocity (and prior to) the point of maximum deceleration. Therefore, the maximum heating rate will occur at a higher altitude, where

$$h_{q_{max}} = \ln(3) h_{n_{max}} \cong 1.10 h_{n_{max}} \qquad (7\text{-}5\text{-}10)$$

Sec. 7-5 The Heating Problem

Equation (7-5-8) shows that the maximum surface heat transfer rate is directly proportional, as is the total entry heat transferred, to the magnitude of the entry velocity (the cube in this case rather than the square) and to the square root of the ballistic coefficient ($W/C_D A$). However, the maximum heat transfer rate is directly proportional to the square root of the sine of the entry angle. As a consequence, we see that *increasing the BC or the entry velocity increases both the total heat and the maximum heat transfer rate, whereas increasing the entry angle decreases the total heat transferred but increases the maximum heat transfer rate.*

A small BC implies a high-drag E/V with a low value of W/A. A large C_D value implies a large half-cone angle (on the order of 45°), which is consistent with a large cross-sectional area for a given weight. This results in a "blunt body" that is hemispheric in the limit and will produce a strong standing shock (bow) wave with a substantial stand-off distance. Between the shock wave and the nose of the E/V there will be a region of highly ionized and dissociated gas that absorbs thermal energy and reduces the stagnation temperature at the E/V, thus, in effect, decreasing the energy conversion fraction f. (This region of hot gas behind the bow wave produces luminosity and electromagnetic effects that interfere with communication and tracking for a short period of time; this is *entry blackout*.) Therefore, an E/V with a small BC entering the atmosphere at orbital (or suborbital) velocities at small angles that decelerates slowly at the higher altitudes will have laminar flow with a total heat load and maximum transfer rate that can be handled by a heat sink.

This (a blunt body with a low $W/C_D A$) was the conservative approach used with the ATLAS, the first ICBM, and with the Gemini and Apollo manned-flight programs. The Apollo entry capsules, although blunt bodies, were configured with a small L/D capability (of the order of 0.25) by offsetting the center of gravity. Since blunt-body missile nose cones (as E/Vs were called in the early days) reached a terminal velocity at high altitudes, they were vulnerable to defensive measures and susceptible to wind effects during the remaining portion of the trajectory. In order that the E/V might penetrate deeper into the atmosphere before reaching its terminal velocity, thus reducing the exposure distance and time, the half-cone angle was decreased to make the E/V more slender and increase the BC. The subsequent increases in both the total heat and the heat transfer rate because of the configuration changes and the transition to turbulent flow could not be handled satisfactorily by the heavy and thermally inefficient heat sinks that were replaced by ablating surfaces. The E/V was no longer a blunt body, and it was at this time that the term *nose cone* was dropped and replaced by *reentry vehicle* (R/V) and now by *entry vehicle* (E/V).

With *lifting heating* ($L/D \neq 0$), as might be expected, the proportional relationships for the total heat and the maximum heat transfer rate are modified by the lift-to-drag ratio (L/D). The relationship for the total heat input becomes

$$Q \sim V_{re}^2 \sqrt{\frac{L}{D} \frac{W}{C_D A} \frac{1}{\sin \phi_{re}}} \qquad (7\text{-}5\text{-}11)$$

and for the maximum surface heat transfer rate is now

$$q_{max} \sim V^3 \sqrt{\frac{W}{C_D A} \frac{\sin \phi_{re}}{L/D}} \qquad (7\text{-}5\text{-}12)$$

It can be shown that the velocity at q_{max} is

$$V_{q_{max}} \cong 0.816 V_{re} \qquad (7\text{-}5\text{-}13)$$

and that the altitude for q_{max} is a function of the lift coefficient and is determined by the flight profile.

Equations (7-5-11) and (7-5-12) are the respective expressions for ballistic heating so modified by the L/D that the total heat is increased and the maximum heat transfer rate is reduced for $L/D > 1$. As with ballistic entry, there is a trade-off between the reduction in the heat load and the increase in the heat transfer rate for a reduction in the entry angle. This trade-off can be shown conveniently by dividing Eq. (7-5-11) by Eq. (7-5-12) to obtain the rule of thumb that

$$q_{max} \sim \frac{1}{Q_{total}} \qquad (7\text{-}5\text{-}14)$$

This qualitative trade-off and Eq. (7-5-12) indicate that the reduction in the maximum heat transfer rate might be sufficient to allow for the rejection of the heat by radiation from the surface and yet keep the surface temperature below its melting limit. We must not forget, however, that the total heat input is equal to the time integral of the instantaneous heat transfer rate, which is equal to the product of q_{ave} and Δt, the time of descent, and that it is possible with a lifting entry to have a lower heat transfer rate and a higher total heat input than a ballistic entry. As mentioned in Section 7-4, the time in which the E/V spends in a region of high heat transfer rate can be shortened (thus reducing the total heat transferred) by entering and holding a steady-state turn, which decreases the vertical lift and thus increases the rate of descent and reduces the entry time; the E/V spirals down through the region of concern. A bank angle of 60°, for example, will reduce the heat entering the E/V by a half; turns are used during Space Shuttle entry to reduce the entry time and heat transfer.

With a continuous equilibrium-glide entry, it is possible to construct an entry corridor for a lifting E/V in which the upper boundary is a flight mechanics boundary established by the condition that the lift plus the centrifugal force must be greater than or equal to the weight; in other words, above this boundary the E/V does not have enough velocity (is too slow) to maintain an equilibrium glide. The lower boundary is a thermal boundary established by the maximum heat transfer rate and the maximum allowable surface temperature; below the boundary the surface of the E/V becomes too hot and thermal damage may occur to the surface, the structure, and possibly the contents of the E/V. An entry corridor for an E/V in a continuous entry might resemble that shown in Fig. 7-5-2.

Although the lower (thermal) boundary in Fig. 7-5-2 is generally described as a radiation boundary, with the implication that radiation is the only mechanism for

Sec. 7-5   The Heating Problem

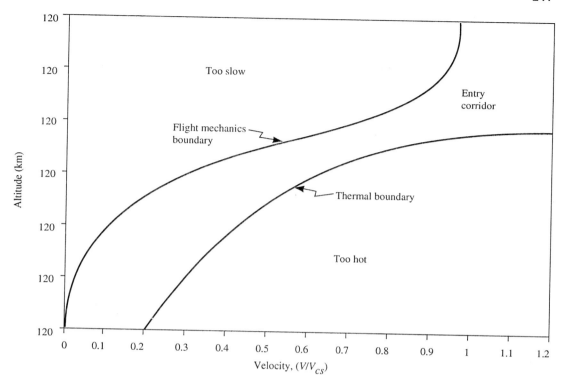

Figure 7-5-2. Example of a continuous-glide lifting-entry corridor.

handling the heating in a lifting entry, that is not necessarily the case. Although the heat sink could conceivably be used, it does not lend itself to lifting entry because of its weight and low thermal efficiency. Ablation, on the other hand, with its high thermal efficiency, is a strong candidate for lifting entry and a serious competitor to radiation. The DynaSoar, a lifting entry R&D vehicle of the early 1960s (canceled prior to first flight), was designed to use ablation exclusively and initially ablation was to be used with the Space Shuttle, at least on the leading edges. Since ablated surfaces cannot be reused without refurbishing or replacement, a reusable vehicle with a short turn-around cycle is better served by a radiative surface, such as the ceramic tiles now being used by the Space Shuttle.

If neither reuse nor rapid turnaround are requirements, the lower boundary in Fig. 7-5-2 could well represent an ablation boundary or possibly a combination radiation–ablation boundary with, for instance, ablation at leading edges (and any other hot spots) and radiative tiles on cooler surfaces. Turnaround refurbishing of ablated material could be a matter of design whereby ablation components, such as leading edges, could be replaced easily and quickly.

Although the Space Shuttle has wings, lifting entry vehicles do not have to be winged in order to generate lift. They might well be *lifting bodies*, which generate lift without wings. Since one surface of a lifting body is usually flat, such vehicles are sometimes referred to as half-cone-angle vehicles, or, if the upper surface is flat, as flying bathtubs. Although lifting bodies are not operational, various configurations have been tested for many years, since the mid-1960s, and may find a use one day.

The objective of aerocapture and aeroassist maneuvers is the reduction of the velocity of a spacecraft, which may or may not be an E/V, by atmospheric penetration and the dissipation of kinetic energy through atmospheric drag (*aerobraking*). Aerocapture implies the use of aerobraking to transform an entering hyperbolic trajectory into an elliptical orbit so that the spacecraft remains within the planetary SOI, is captured. Aeroassist implies reduction of the velocity of an orbit by aerobraking prior to entering a lower-energy orbit or possibly a landing trajectory. There is also the possibility of a plane change by an aerodynamic turn within the atmosphere; the capability for such a turn obviously requires a lifting configuration, which in turn introduces large weight penalties if such a capability is not already present.

Although these aeroassist concepts are attractive and appear to save propulsive weight, they do have associated costs and penalties and pose the dilemma of how to obtain a large velocity reduction (a high deceleration) and a significant plane change without a large weight penalty for thermal protection and lifting capability. Restricting the penetration to the higher altitudes, where the density is low and its variations can be unpredictable, calls for a low ballistic coefficient (a large C_D and a low W/A), which may not be inherent in the design of the spacecraft to accomplish its primary mission. A deep penetration may increase the ΔV obtained but will exacerbate the heating problem; in a current exploratory program the additional weight required for thermal protection is approaching that needed for an all-propulsive maneuver, and the complexity added to the E/V configuration is also discouraging.

There are many schemes for making aeroassisted maneuvers effective with the hopes of keeping the cost increments in weight, complexity, and dollars within reasonable limits and less than the penalties associated with all-propulsive maneuvers. Such schemes include the inflatable ballute (a deployable drag device) with and without forward-firing propulsive thrust, various deployable drag devices, a lifting brake, and new lifting configurations. The potential benefits of aeroassist and aerocapture are there but have yet to be attained.

7-6 CLOSING REMARKS

When a spacecraft enters a planetary atmosphere, it becomes subject to aerodynamic forces and its trajectory is no longer force-free or Keplerian; the spacecraft (or missile warhead) becomes an entry vehicle. The aerodynamic drag is the primary atmospheric force; it decelerates the vehicle and, in doing so, transforms its kinetic energy into thermal energy. Although all of the deceleration is felt by the E/V, only a fraction of the heat is transferred to the vehicle. The magnitude of the deceleration

Sec. 7-6   Closing Remarks

and of the thermal energy entering the E/V is determined by the characteristics of the atmosphere and of the vehicle and by the nature of the entry trajectory. Lift can be, and is, used to modulate the trajectory and the interaction of the E/V with the atmosphere.

With respect to the *deceleration in a ballistic (no lift) entry*, its maximum value (which can be very large, on the order of hundreds of g_0) is independent of the E/V configuration and is directly proportional to the square of the entry velocity and to the sine of the entry elevation angle. (It is also a function of the temperature and gas constant of the planetary atmosphere and of the surface acceleration of the planet.) The altitude at which the peak deceleration occurs, however, is a strong function of the configuration (the $W/C_D A$) and of the entry angle. A blunt-body E/V with a low ballistic coefficient ($W/C_D A$) entering at a small angle will have a relatively low peak deceleration at a high altitude. A slender E/V with a large ballistic coefficient and a large entry angle will penetrate deeply into the atmosphere and undergo a large deceleration. The final portion of a ballistic trajectory is vertical, requiring the use of drag devices or propulsion for a "soft" impact; a conventional horizontal landing is not possible.

Deceleration with lift, on the other hand, is significantly lower and has no maximum or peak value as such. Being inversely proportional to the lift-to-drag ratio (L/D), the maximum deceleration can be less than one g_0 in an equilibrium glide. Lift also introduces a maneuvering capability that can be used to modulate the trajectory and its duration, enlarge the landing footprint, both down-range and cross-range, and to land the E/V horizontally with a conventional touchdown. Lifting entries are characterized by lower decelerations and longer entry times than those of ballistic entries.

The major considerations in handling *transfer of the thermal energy* associated with the deceleration to the entry vehicle are the total amount of heat transferred and the rate (both average and maximum) at which it is transferred. The three current thermal protection schemes are the heat sink and ablation, both of which absorb heat, and radiation, which rejects heat. With the absorption schemes, the objective is to minimize the total heat input, thus favoring the shorter-duration ballistic entries. With radiation, on the other hand, since the objective is to minimize the transfer rates (and surface temperatures), the longer-duration and lower-deceleration lifting entries are appropriate. There is a caveat, however; minimizing the transfer rate implies increasing the duration of the entry, which since the total heat input is the product of the average transfer rate and the duration could result in an excessive total heat load.

Aeroassisted and aerocapture maneuvers have been presented in a simplistic manner to show the concepts along with their potential advantages and disadvantages. The trajectories and their execution, however, are not simple and impose stringent requirements on the performance of the navigation and guidance schemes to position the spacecraft prior to entering the atmosphere and on the control scheme for executing the atmospheric trajectory so as to accomplish the desired results. These added requirements affect the configuration and weight of the vehicle to the

PROBLEMS

7-1. A spacecraft enters the Earth's atmosphere on a hyperbolic orbit with $V_{re} = 13{,}000$ m/s.
   (a) Find the specific kinetic energy.
   (b) If $h_{re} = 76.25$ km, find the specific potential energy with respect to the surface of the Earth.
   (c) What is the total specific energy with respect to the surface of the Earth?
   (d) What are the percentages of the KE and PE of the total energy?
   (e) Is it reasonable to neglect the potential energy in the entry approximation?

7-2. Do Problem 7-1 for a spacecraft returning from a space station (at 300 nmi) with $V_{re} = 7976$ m/s.

7-3. Using the data for the Earth's atmosphere from Table 7-3-1:
   (a) Find σ at the nominal entry altitude (76.25 km).
   (b) What percentage of the planetary radius is this "sensible atmosphere"? Does this justify the reference to a "thin" or "shallow" atmosphere?
   (c) Find the altitude and density at which $\sigma = 0.001$.

7-4. Using the data for Venus from Table 7-3-1:
   (a) If a σ value of 3×10^{-5} is used as the criterion for the edge of the sensible atmosphere, what are the corresponding altitude and density? What percentage of the planetary radius is h?
   (b) How does the h of part (a) compare with the h_{re} of 76.25 km for Earth?
   (c) Find h and ρ when $\sigma = 0.001$.

7-5. Do Problem 7-4 for Mars.

7-6. An E/V in the Earth's atmosphere has BC = 5000 Pa (104.4 lb/ft) and $V = 7300$ m/s. Find the drag force per unit mass (D/m) and compare it with the gravitational force per unit mass (g) at:
   (a) 100 km
   (b) 76.25 km
   (c) 50 km

7-7. Do Problem 7-6 with a higher-drag E/V [i.e., BC = 500 Pa (10.44 lb/ft^2)].

7-8. An OTV with BC = 50 Pa (1.044 lb/ft^2) enters the sensible atmosphere at 76.25 km on a ballistic aeroassisted trajectory with $V = 10{,}350$ m/s and $\phi = -3°$.
   (a) Find the Keplerian (drag-free) perigee altitude and velocity.
   (b) Find the actual velocity (after atmospheric deceleration) at perigee as well as the ΔV due to drag.

Chap. 7   Problems

(c) Compare the velocity of part (b) with the circular orbital velocity at that altitude. Is the OTV trapped within the atmosphere?
(d) Is the perigee altitude of part (a) above or below the altitude for n_{max}?

7-9. Do Problem 7-8 with the BC increased to 5000 Pa.

7-10. Do Problem 7-8 with BC = 50 Pa and V = 10,350 m/s at 76.25 km but with ϕ increased to $-25°$.

7-11. An early ICBM had a blunt-body nose cone with BC $(W/C_D A)$ = 500 Pa (10.44 lb/ft²), V_{re} = 7000 m/s, and ϕ_{re} = $-30°$.
(a) Find n_{max} and the corresponding altitude and velocity.
(b) At 30 km, what are the values of dV/dt (m/s² and g_0) and V?

7-12. Do Problem 7-11 with ϕ_{re} = $-5°$ along with part (c), the deceleration and velocity at 50 km.

7-13. Do Problem 7-12 with the BC increased to 10,000 Pa (209 lb/ft²).

7-14. A spacecraft with a BC = 1000 Pa enters the atmosphere of Mars on a ballistic trajectory at an altitude of 300 km with V = 6200 m/s and ϕ = $-15°$.
(a) Find V and dV/dt (in m/s², Earth g, and Mars g).
(b) Find $(dV/dt)_{max}$ and n_{max} and the corresponding V and h.
(c) At h = 75 km, find V, dV/dt, and n.

7-15. Do Problem 7-14 for a Venus entry and compare with answers of Problem 7-14, if available.

7-16. Do Problem 7-14 for an Earth entry and compare the results with those of Problems 7-14 and 7-15 if those results are available.

CHAPTER 8

Orbital Elements and Earth Tracks

The TITAN III-C booster. (Courtesy of NASA.)

8-1 INTRODUCTION

In our consideration of two-body orbits and trajectories we have focused our attention on the motion of the smaller body in the plane of the trajectory and have ignored the orientation of the plane with respect to the rest of space. Furthermore, we have looked at each trajectory from the viewpoint of a hypothetical observer located at the primary focus (at the center of the attracting body), where orbits appear to be simply circles and ellipses. Real observers, however, are not located at the center of the attracting body and are more likely to be on the surface of a rotating planet, such as the Earth.

In this chapter we first see what additional information is required to locate a moving body in a three-dimensional space (i.e., the complete set of orbital elements) and in the process of doing so define several coordinate systems of interest and relevancy.

In the section following we look at the general problem of determining the characteristics of a geocentric orbit from a set of observations, the orbit determination problem, and introduce the need for coordinate transformations.

Then in the next section we examine the ground tracks of geocentric satellites, the projections of the orbits on the surface of a rotating Earth, with a discussion of the effects of various orbital parameters on the shape and appearance of the ground tracks along with a few words on the effects of the Earth's oblateness.

In the closing remarks in the last section we summarize the effects of changes in the orbital parameters on the ground tracks and take a look at the relationships among certain orbital elements and launch windows.

8-2 THE ORBITAL ELEMENTS

Let us rewrite the vector two-body equation of motion as

$$\ddot{\mathbf{r}} + \frac{\mu}{r^3}\mathbf{r} = 0 \qquad (8\text{-}2\text{-}1)$$

This is a two-degree-of freedom equation with time as the independent variable. If the inertial position vector **r** and the inertial velocity vector **V** of the moving body are known at any particular instant of time, called the *epoch*, the location of the body and the characteristics of the orbit are known and the location of the satellite at any other instant of time can be predicted.

In component form, Eq. (8-2-1) is represented by six scalar equations (or three second-order differential equations) with six constants of integration that are known individually as the *orbital elements*. The set of orbital elements at a particular epoch is referred to as the *ephemeris* of the moving body.

The polar equation that we have been using to describe a planar trajectory or orbit can be written as

$$r(t) = \frac{a(1 - \epsilon^2)}{1 + \epsilon \cos \nu} \tag{8-2-2}$$

In this equation there are two constants, the semimajor axis a and the eccentricity ϵ, that size and shape the trajectory, respectively. In addition, there is the true anomaly ν that locates the moving body along the prescribed trajectory at a particular instant of time (the epoch); ν obviously is not a constant since it is a function of time. These three quantities, a, ϵ, and ν (with which we are already familiar), define the planar orbit and constitute three of the six orbital elements; three additional elements are needed to locate the orbital plane in a three-dimensional space. Before these elements can be defined, it is necessary to establish appropriate reference systems.

There are a number of possible coordinate systems, some with more than one name and all nonrotating and nonaccelerating (inertial). In each right-hand system it is necessary to have a fundamental (a reference) plane, one with which the planar trajectory can be oriented, as well as an origin and a principal direction in the fundamental plane to serve as a reference direction.

One such reference system is the *heliocentric–ecliptic coordinate system*, as sketched in Fig. 8-2-1. The origin is at the center of the Sun and the fundamental plane, containing the X and Y axes, is the *ecliptic*, which is the plane containing the heliocentric orbit of the Earth. The principal direction X is toward the vernal equinox (the extension of the line joining the center of the Earth and the Sun on the first day of Spring, March 21),[1] Y is perpendicular to X (in the plane of the ecliptic), and Z is perpendicular to plane of the ecliptic.

A second widely used coordinate system is the *geocentric–equatorial coordinate system*, sketched in Fig. 8-2-2. The origin is at the center of the Earth, the fundamental plane is the equatorial plane, X is in the direction of the vernal equinox, Y is perpendicular to X and in the equatorial plane, and Z is perpendicular to the equatorial plane in the direction of the North Pole. $\bar{I}, \bar{J}$, and $\bar{K}$ are the nonrotating unit vectors and will be used in later sections to designate the X, Y, and Z axes.

---

[1] Since the vernal equinox once pointed toward Aries (the Ram), γ is often used to designate the direction of the vernal equinox.

Sec. 8-2 The Orbital Elements

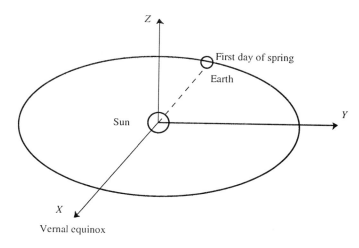

Figure 8-2-1. Heliocentric–ecliptic coordinate system.

The *celestial sphere* is an expansion of the geocentric–equatorial coordinate system to form a fictitious sphere of infinite radius with a celestial equator that includes the equatorial plane of the Earth and with a North Celestial Pole that lies along the North Pole of the Earth. The celestial sphere is used in conjunction with two angles, known as the *right ascension* α and the *declination* δ, to locate an object in space as sketched in Fig. 8-2-3.

Another coordinate system that you might encounter or wish to use is the *perifocal coordinate system*. The origin is at the center of the attracting body, the fundamental plane is the orbital plane with the X-axis toward periapsis, and the Z-axis is perpendicular to the orbital plane along the specific angular momentum vector **H**. In addition to the fundamental plane, there needs to be an additional reference plane, which can be the ecliptic in the case of heliocentric orbits or the equatorial plane in the case of planetocentric orbits. This coordinate system can be convenient in describing planetary orbits.

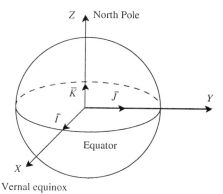

Figure 8-2-2. Geocentric–equatorial coordinate system.

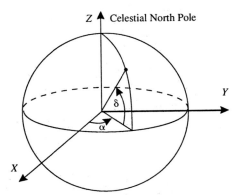

Figure 8-2-3. Celestial sphere and right ascension-declination coordinate system.

Returning to the orbital elements, the six classical orbital elements are generally considered to be a, ϵ, T, i, Ω, and ω. The elements a and ϵ are the familiar semimajor axis and eccentricity that size and shape a trajectory or orbit, respectively. Although v can be used to specify the location of a body at a particular time, it is convention to use T, the *time of periapsis passage*, to establish a reference time for elliptical orbits from which the location of the smaller body can be located at any other time (epoch); T is obviously undefined for circular orbits. This set of three elements (a, ϵ, and T) establishes the planar orbit and the location of the body in the orbit at any particular instant of time.

The set of the remaining three elements, i, Ω, and ω, establishes the orientation of the orbit in space. These elements are defined in Fig. 8-2-4 using a geocentric–equatorial coordinate system.

The orbital element i is the *inclination* of the orbital plane with respect to the equatorial plane. It is also the angle between the polar or $\bar{K}$-axis and the angular momentum vector **H**. This inclination angle is measured in a counterclockwise direction, ranging in value from 0° to +180°. If $i = 0°$ (or +180°), the orbit is an

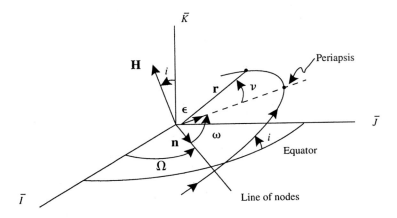

Figure 8-2-4. Orbital plane, its orientation and elements.

Sec. 8-2 The Orbital Elements

equatorial orbit; if $i = +90°$, the orbit is a polar orbit. If $i < 90°$, the satellite has an eastwardly direction (counterclockwise with respect to the polar axis) and is described as a *direct or prograde orbit*. If $i > 90°$, the satellite has a westwardly or backward motion and is called a *retrograde orbit*.

The intersection of the orbital plane and the equatorial plane forms the *line of nodes*. The *ascending node* is the point where the satellite crosses the equatorial plane (and the line of nodes) from below and the *descending node* is the point where the satellite crosses the line of nodes from above. The element with the symbol Ω is the (celestial) *longitude of the ascending node* and is the angle in the equatorial plane measured in a counterclockwise direction from the vernal equinox ($\bar{I}$ in Fig. 8-2-4), which is fixed in space, to the ascending node, always to the east so that $0° \le \Omega \le 360°$. As an aside, geographic longitude is measured to the east and west from a rotating Greenwich reference so that the celestial and geographic longitudes coincide once a day. If the ascending node is at the vernal equinox (the $\bar{I}$-axis), $\Omega = 0°$; if $\Omega = 90°$, the ascending node is at the $\bar{J}$-axis. Ω is undefined for an equatorial orbit.

The sixth orbital element is ω, the *argument of the periapsis*, which is the angle from the ascending node to the periapsis (perigee for geocentric orbits) in the direction of motion and in the orbital plane. When $0° < \omega < 180°$, periapsis is in the northern hemisphere and the satellite is at its northernmost point when $\omega = 90°$. When $180° < \omega < 360°$, periapsis is in the southern hemisphere with the satellite at its southernmost point when $\omega = 270°$. When $\omega = 0°$, periapsis is at the ascending node, and when $\omega = 180°$, periapsis is at the descending node. As might be expected, ω is undefined for circular and equatorial orbits.

These six orbital elements are not necessarily unique. For example, the semimajor axis a may be replaced by p, the semilatus rectum, and the time of periapsis passage T may be replaced by the *true anomaly at epoch* v_0, which is self-defined. Furthermore, in the case of certain orbits, some of the classic elements become undefined and cannot be used to define the location of the spacecraft, thus requiring the use of different or modified orbital elements.

If an inclined orbit is circular ($\epsilon = 0$), there is no periapsis and T is undefined. In this case, T can be replaced by u_0, the *argument of latitude at epoch*, which is the angle in the orbital plane between the ascending node and the satellite. It can be used with elliptical orbits if so desired, in which case

$$u_0 = \omega + v_0 \qquad (8\text{-}2\text{-}3)$$

With an elliptical equatorial orbit ($i = 0$), there is no ascending node from which to measure ω. Periapsis, however, can be located by the use of Π, the *longitude of periapsis*, which is the angle in the equatorial plane measured from the $\bar{I}$ vector to periapsis. With nonequatorial elliptical orbits, there is an ascending node from which ω can be measured and

$$\Pi = \Omega + \omega \qquad (8\text{-}2\text{-}4)$$

In the case of a circular equatorial orbit ($i = \epsilon = 0$), there is no periapsis nor is there an ascending node. The orbital element of relevance in this case is l_0, the *true longitude at epoch*, which is the angle in the equatorial plane between the $\bar{I}$ axis and the position vector $\mathbf{r}_0$. In the general case of an inclined elliptical orbit,

$$l_0 = \Omega + \omega + \nu_0 = \Pi + \nu_0 = \Omega + u_0 \qquad (8\text{-}2\text{-}5)$$

Although these orbital elements were defined in terms of a geocentric–equatorial coordinate system, they are applicable to a heliocentric–ecliptic coordinate system or to a planetocentric–equatorial coordinate system, as evidenced by the use of the term *periapsis* in the definitions.

With respect to time, there is sidereal time and solar time. A *sidereal day* comprising 24 sidereal hours is the time required for the Earth to rotate once about its axis with respect to the distant stars. A *mean solar day* comprising 24 hours, as we measure them, is the time for the Earth to rotate once about its axis with respect to the Sun. Since the Earth itself is orbiting about the Sun, it takes longer to complete a rotation with respect to the Sun than it does with respect to the stars. Consequently, a sidereal day and other units of sidereal time are shorter than corresponding units of solar time. For example, 1 sidereal day is equal to approximately 23.934 mean solar hours.

8-3 ORBIT DETERMINATION

Even with a restricted two-body problem (no drag and a spherical Earth) and with perfect error-free tracking data (not a good assumption) it is not a simple problem to find the orbital elements (define the orbit and locate the spacecraft) from a set of initial conditions that must be obtained from physical observations and the problem becomes much more difficult when the assumptions are relaxed. In this section only the basic approaches are described and discussed briefly with the objective of indicating the scope of the problem and the nature of the possible solutions.

It is necessary to introduce two new vectors, $\mathbf{n}$ and $\boldsymbol{\epsilon}$, that are used to determine the values of certain relevant angles that are also orbital elements. The first of these is the *ascending node vector* $\mathbf{n}$, which is the cross product of $\bar{K}$ and $\mathbf{H}$:

$$\mathbf{n} = \bar{K} \times \mathbf{H} \qquad (8\text{-}3\text{-}1)$$

($\bar{I}, \bar{J}$, and $\bar{K}$ are unit vectors.) As shown in Fig. 8-2-4, $\mathbf{n}$ is aligned with the line of nodes and is positive in the direction of the ascending node. Since Ω, the longitude of the ascending node, is the angle between $\bar{I}$ and $\mathbf{n}$, Ω can be found from the expression

$$\boxed{\Omega = \cos^{-1} \frac{\bar{I} \cdot \mathbf{n}}{n}} \qquad (8\text{-}3\text{-}2)$$

Although we have treated the eccentricity ϵ as a scalar, it actually is a vector as can be seen by returning to Section 2-4, where the relationship that $\boldsymbol{\epsilon} = \mathbf{B}/\mu$ was established. B is the magnitude of the vector $\mathbf{B}$, which was defined in Eq. (2-4-4) as

$$-\mathbf{B} = (\mathbf{H} \times \mathbf{V}) + \frac{\mu}{r}\mathbf{r} \qquad (8\text{-}3\text{-}3)$$

Sec. 8-3 Orbit Determination

or, with its sign changed,

$$\mathbf{B} = (\mathbf{V} \times \mathbf{H}) - \frac{\mu}{r}\mathbf{r} \quad (8\text{-}3\text{-}4)$$

Since **B** is a vector, so is the eccentricity $\boldsymbol{\epsilon}$. With **H** in Eq. (8-3-3) replaced by $(\mathbf{r} \times \mathbf{V})$ and with the expansion of the triple product $\mathbf{V} \times (\mathbf{r} \times \mathbf{V})$, the eccentricity vector $\boldsymbol{\epsilon}$ can be written as

$$\boldsymbol{\epsilon} = \frac{1}{\mu}\left[\left(V^2 - \frac{\mu}{r}\right)\mathbf{r} - (\mathbf{r} \cdot \mathbf{V})\mathbf{V}\right] \quad (8\text{-}3\text{-}5)$$

where $\boldsymbol{\epsilon}$ is in the orbital plane, has its origin at the center of the attracting body, and points toward periapsis. Since the true anomaly v is the angle between $\boldsymbol{\epsilon}$ and the radius vector **r**,

$$\boxed{v = \cos^{-1}\frac{\boldsymbol{\epsilon} \cdot \mathbf{r}}{\epsilon r}} \quad (8\text{-}3\text{-}6)$$

Other angular orbital elements can be found in a similar manner. Since the inclination i is the angle between $\overline{K}$ and **H**,

$$\boxed{i = \cos^{-1}\frac{\overline{K} \cdot \mathbf{H}}{H}} \quad (8\text{-}3\text{-}7)$$

The argument of the periapsis ω is the angle between **n** and $\boldsymbol{\epsilon}$ so that

$$\boxed{\omega = \cos^{-1}\frac{\mathbf{n} \cdot \boldsymbol{\epsilon}}{n\epsilon}} \quad (8\text{-}3\text{-}8)$$

Since the argument of latitude at epoch u_0 is the angle between **n** and **r**,

$$\boxed{u_0 = \cos^{-1}\frac{\mathbf{n} \cdot \mathbf{r}}{nr}} \quad (8\text{-}3\text{-}9)$$

Let us now consider the problem of determining the orbital elements when somehow the inertial position vector **r** and the inertial velocity vector **V** are known at a particular instant of time (the epoch). In the geocentric example to follow, we use the canonical units of Section 5-8 in which 1 DU = r_E = 6.378 × 10^6 m, 1 TU = 806.8 s, and 1 SU = 1 DU/TU = 7905 m/s. Consequently, μ_E = 1 DU^3/TU^2.

Example 8-3-1

Find the orbital elements of a geocentric satellite whose inertial position and velocity in a geocentric–equatorial coordinate system were determined at 1500 GMT (Greenwich Mean Time) on July 23, 1991 to be

$$\mathbf{r} = -0.8\overline{I} + 0.6\overline{J} + 0.5\overline{K}$$
$$\mathbf{V} = -0.4\overline{I} - 0.8\overline{J} + 0.6\overline{K}$$

Solution We first need to find the magnitudes of r and $\mathbf{V}$.

$$r = |\mathbf{r}| = \sqrt{(-0.8)^2 + (0.6)^2 + (0.5)^2} = 1.12 \text{ DU}$$

$$V = |\mathbf{V}| = \sqrt{(-0.4)^2 + (-0.8)^2 + (0.6)^2} = 1.08 \text{ DU/TU}$$

Now to find the specific energy E,

$$E = \frac{V^2}{2} - \frac{\mu}{r} = -0.310 \text{ DU}^2/\text{TU}^2 < 0 \quad \text{(ellipse or circle)}$$

Since $E = -\mu/2a$, the semimajor axis can be found to be

$$\boxed{a = \frac{-\mu}{2E} = 1.61 \text{ DU} > 0} \quad \text{(ellipse or circle)}$$

Rewriting Eq. (8-3-5) yields

$$\boldsymbol{\epsilon} = \frac{1}{\mu}\left[\left(V^2 - \frac{\mu}{r}\right)\mathbf{r} - (\mathbf{r}\cdot\mathbf{V})\mathbf{V}\right]$$

$$(\mathbf{r}\cdot\mathbf{V}) = (-0.8 \times -0.4) + (0.6 \times -0.8) + (0.5 \times 0.6) = 0.14$$

With the substitution of the other values,

$$\boldsymbol{\epsilon} = 0.274\mathbf{r} - 0.14\mathbf{V}$$

Substituting the expressions for $\mathbf{r}$ and $\mathbf{V}$ given in the statement of the problem, the eccentricity vector can be written as

$$\boldsymbol{\epsilon} = -0.163\bar{I} + 0.276\bar{J} + 0.053\bar{K}$$

Therefore, the magnitude of the eccentricity vector is

$$\boxed{\epsilon = 0.32} \quad \text{The orbit is an ellipse.}$$

The specific angular momentum $\mathbf{H}$ can be found from its definition and the knowledge of r and $\mathbf{V}$.

$$\mathbf{H} = \mathbf{r} \times \mathbf{V} = \begin{vmatrix} \bar{I} & \bar{J} & \bar{K} \\ -0.8 & 0.6 & 0.5 \\ -0.4 & -0.8 & 0.6 \end{vmatrix} = 0.76\bar{I} + 0.28\bar{J} + 0.88\bar{K}$$

and the magnitude of $\mathbf{H}$ is

$$H = 1.19 \text{ DU}^2/\text{TU}$$

Knowing $\mathbf{H}$, it is now possible to find the inclination i from

$$\boxed{i = \cos^{-1}\frac{\bar{K}\cdot\mathbf{H}}{H} = \cos^{-1}\frac{0.88}{1.19} = 42.6°}$$

Since $i < 90°$, the elliptical orbit is a direct (prograde) orbit.

The ascending node vector $\mathbf{n}$ is necessary for the determination of Ω and is

Sec. 8-3 Orbit Determination

$$n = \overline{K} \times \mathbf{H} = \begin{vmatrix} \overline{I} & \overline{J} & \overline{K} \\ 0 & 0 & 1 \\ 0.76 & 0.28 & 0.88 \end{vmatrix} = -0.28\overline{I} + 0.76\overline{J}$$

and $n = |\mathbf{n}| = 0.81$. Therefore, Ω, the longitude of the ascending node, is

$$\Omega = \cos^{-1} \frac{\mathbf{n} \cdot \overline{I}}{n} = \cos^{-1} \frac{-0.28}{0.81} = 110°$$

With $\boldsymbol{\epsilon}$ and $\mathbf{n}$ known, the argument of the periapsis can be found from

$$\omega = \cos^{-1} \frac{\mathbf{n} \cdot \boldsymbol{\epsilon}}{n\epsilon} = \cos^{-1} \frac{0.255}{0.81 \times 0.32} = 10.3°$$

The true anomaly at epoch, ν_0, is found to be

$$\nu_0 = \cos^{-1} \frac{\boldsymbol{\epsilon} \cdot \mathbf{r}}{\epsilon r} = \cos^{-1} \frac{0.322}{0.36} = 26.8°$$

Although not an orbital element the period of the orbit is of interest and is

$$P = 2\pi \sqrt{\frac{a^3}{\mu}} = 12.84 \text{ TU} = 10{,}360 \text{ s} = 173 \text{ min} = 2.88 \text{ h}$$

Although ν_0 and the time of the observation (the epoch) can be used as a reference for the prediction of future satellite locations, the time of periapsis passage, T, is the more customary orbital element and can be found by determining the time from periapsis passage to the true anomaly at epoch and subtracting it from the time of epoch. The eccentric anomaly u at anomaly is first found from Eq. (4-3-9),

$$u = \cos^{-1} \frac{\epsilon + \cos \nu}{1 + \epsilon \cos \nu} = \cos^{-1} \frac{1.21}{1.28} = 19.0° = 0.332 \text{ rad}$$

and the mean anomaly M from Eq. (4-3-17).

$$M = u - \epsilon \sin u = 0.332 - 0.32 \sin(0.332) = 0.228 \text{ rad}$$

Note that ν, u, and M at epoch all lie between 0 and π rad (on the same side of the major axis) as they should. From Eq. (4-3-19),

$$t = \frac{P}{2\pi} M = 0.466 \text{ TU} = 375.9 \text{ s} = 6.26 \text{ min}$$

Therefore, the time of periapsis passage on July 23, 1991 (the eighth orbital element) occurred at

$$T = 1500 - 6.26 = 1454 \text{ GMT}$$

As a matter of interest, let us find the other orbital elements, the ones that are used when one or more of the classic elements might be undefined. The argument of latitude at epoch u_0 can be found from Eq. (8-3-9) or from the relationship that

$$u_0 = \omega + \nu_0 = 10.3 + 26.8 = 37.1°$$

The longitude of periapsis Π is

$$\Pi = \Omega + \omega = 110 + 10.3 = 120.3°$$

and the true longitude at epoch l_0 is

$$l_0 = \Omega + u_0 = 110 + 37.1 = 147.1°$$

Let us conclude this example by finding h_p, V_p, h_a, and V_a in both canonical units and real units with $a = 1.61$ DU $= 10.26 \times 10^6$ m, $r = 1.12$ DU $= 7.143 \times 10^6$ m, and $V = 1.08$ SU $= 8537$ m/s. At perigee

$$r_p = a(1 - \epsilon) = 1.61(1 - 0.32) = 1.095 \text{ DU} = 6.98 \times 10^6 \text{ m}$$

$$h_p = r_p - r_E = 1.095 - 1 = 0.095 \text{ DU} = 6.06 \times 10^5 \text{ m} = 329 \text{ nmi}$$

$$V_p = \frac{H}{r_p} = \frac{1.19}{1.095} = 1.087 \text{ SU} = 8590 \text{ m/s}$$

At apogee

$$r_a = a(1 + \epsilon) = 2.125 \text{ DU} = 13.55 \times 10^6 \text{ m}$$

$$h_a = 2.125 - 1 = 1.125 \text{ DU} = 7.175 \times 10^6 \text{ m} = 3900 \text{ nmi}$$

$$V_a = \frac{1.19}{2.125} = 0.560 \text{ SU} = 4427 \text{ m/s}$$

This concludes Example 8-3-1, which started with a knowledge of the inertial position and velocity vectors that originated at the center of the attracting body but ignored the source of that knowledge. If the data were provided by a tracking radar, it is obviously impossible to locate the radar at the center of the Earth. Rather, the radar will be located at a point (the *tropos*) on the surface of the rotating Earth and the radar data will be taken in the *topocentric–horizon coordinate system*. We now have a new problem, that of transforming data from a rotating (noninertial) coordinate system to the inertial coordinate system with its origin at the center of a nonrotating coordinate system. This is the *coordinate transformation problem*, which in this case involves consideration of the angular velocity of the data-gathering coordinate system and the position, velocity, and acceleration of the tropos, all with respect to inertial space.

When two coordinate systems have the same origin, as is the case with perifocal and geocentric–equatorial coordinate systems, only appropriately sequenced rotations are required to transform from one system to the other. Coordinate transformations are best done with matrix algebra and are not discussed further in this book since detailed treatments are available in other texts; Escobal (listed in the References), for example, has an appendix that lists "thirty-six basic coordinate transformations."

Example 8-3-1 treats the case of orbit determination with knowledge of the position and velocity vectors, which knowledge provides six independent pieces of data. Another source of orbital data might be optical sightings that provide only two pieces of independent data, the right ascension and declination angles, from each observation. As a consequence, at least three observations are required to obtain the prerequisite six pieces of data. Requiring a minimum of instrumentation, use of the right ascension and declination (i.e., angles only) is a technique favored by many astronomers to determine an orbit. Three methods for obtaining the inertial position and velocity vectors from these noninertial data (angles only) are the *method of Gauss*, which is very suitable for data obtained from short arcs and which provides a solution that satisfies the orbit at all three data points (dates); the *method of Laplace*, which obtains a solution by a series of successive differentiations, requiring less computational effort than the Gaussian technique but providing accurate representation for only the center observation (date); and the *double r-iteration method*, which is newer and can be used for angular data spread over large arcs, even over a number of revolutions.

Orbit determination can also be carried out with a knowledge of two position vectors and the time interval between the two observations. This is an application of Lambert's theorem and is known as the *Gauss problem* (see Section 4-5 and Example 4-5-2). There are various methods for obtaining a solution, all requiring iteration and some with solutions that are not unique.

The orbit determination problem in real life is further complicated by what are referred to as the perturbations (deviations or variations) in the orbits; these perturbations can be predictable (deterministic) or random (stochastic). Examples of perturbation sources are the nonsphericity of the attracting body (e.g., the oblateness of the Earth), the nonhomogeneity of the attracting body, the presence of atmospheric drag, no matter how small, and the presence of other attracting bodies. Depending on their source, these variations can be secular, with ever-increasing or ever-decreasing changes, or periodic, with either short-period or long-period variations. There are two major classes of schemes for treating perturbations, *general (absolute) perturbations* and *special perturbations*. Special perturbation schemes involve direct numerical integration of the equations of motion with the perturbing accelerations included, usually over a limited time interval, whereas the general perturbation schemes involve the development of infinite series, or possibly a closed expression, in order to make an analytical prediction of the satellites's position and velocity. Among the special perturbation schemes are Cowell's method, Encke's method, Hansen's method, and the variation of parameters or elements.

8-4 SATELLITE GROUND TRACKS

Before examining the relationships between the orbital elements and the projections of geocentric orbits on the surface of a rotating Earth (the ground traces or tracks), let us consider some of the factors influencing the choice of the orbital altitude: namely, the visibility of the Earth's surface, the duration of the mission, and whether the satellite is manned or unmanned.

The altitude of a circular geocentric satellite determines the amount of the Earth's surface that is visible at any instant of time (the field of view), an important consideration for reconnaissance and communications satellites that must see and be able to be contacted by line-of-sight communications techniques. In Fig. 8-4-1a, the tangents to the Earth's surface from the satellite define the field of view from the satellite; the subtended half-angle ϕ is

$$\phi = \cos^{-1} \frac{r_E}{r} \qquad (8\text{-}4\text{-}1)$$

where $r = r_E + h$. Since the area of the spherical segment of the Earth's surface under the satellite's cone of view is equal to $2\pi r_E d$, where $d = r_E(1 - \cos \phi)$, and the entire surface area of the Earth is equal to $4\pi r_E^2$, the field of view (FOV) of the satellite expressed as a percentage of the Earth's surface can be written as

$$\text{FOV} = 50(1 - \cos \phi) = 50\left(1 - \frac{r_E}{r}\right) \qquad (8\text{-}4\text{-}2)$$

Although the area of visibility does increase as the altitude increases, as one would expect, it is not linear, nor does it increase without limit, approaching 50% in the limit as r goes to infinity (see Fig. 8-4-1b). A circular orbit with an altitude of 175 nmi (322 km) will have a field of view of 2.4%, whereas that of a geosynchronous orbit will be 42.45%. As a consequence, two communication satellites in geosynchronous orbits (GEOs) will not provide complete global coverage; a third satellite is required to cover the remaining gap. (One current constellation contains five geosynchronous satellites.) If the altitude is increased so as to reach the edge of the sphere of influence (500,000 nmi), the field of view will increase to the order of 49.65%.

Although in Chapter 7 the edge of the sensible atmosphere was assumed to be on the order of 75 km (41 nmi), there is sufficient atmospheric drag above that altitude [out to the order of 552 km (300 nmi) or more] to cause low Earth circular orbits to decay, thus imposing a finite lifetime on LEO satellites, and to circularize ellipses with low Earth orbit perigees by lowering the apogee altitude prior to orbit decay. Long-duration low-Earth-orbit satellites therefore require periodic propulsive firings to maintain their original orbits, a process of station keeping or orbit trim.

If higher altitudes are chosen to minimize drag effects, the satellite enters the Van Allen radiation belt at altitudes on the order of 736 km (400 nmi). This radiation belt comprises an inner zone of high-energy protons and an outer zone of large quantities of high-energy electrons with both protons and electrons existing throughout the belt. The intensity of the inner zone peaks at altitudes on the order of 3680 km (2000 nmi) and that of the outer zone peaks at altitudes on the order of 27.600 km (15,000 nmi), with the effects of the belt extending out to altitudes on the order of 55,200 km (30,000 nmi) or more. Since these high-energy particles can seriously damage living beings, sensors, and electronic equipment, the Van Allen radiation belt is an area to be avoided, which it normally is, or rapidly traversed, as was the case with the Apollo missions. Although a higher altitude for the Hubble Space Telescope would have been preferable from a data-gathering viewpoint, 607 km (330

Sec. 8-4 Satellite Ground Tracks

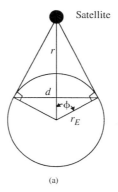

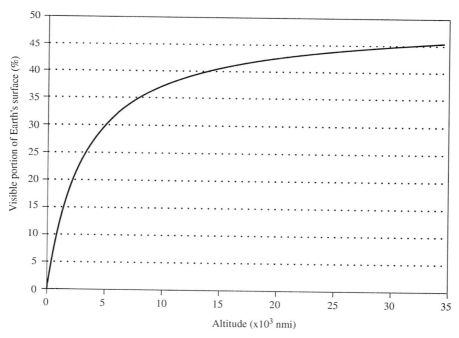

Figure 8-4-1. Field of view (FOV) of a geocentric satellite: (a) schematic; (b) percent coverage.

nmi) was chosen to avoid the radiation belt and potential damage to sensors and electronics. Although the geosynchronous communications satellites are above the regions of high intensity, they are affected to some degree or another by radiation from the belt.

We are now ready to examine satellite ground tracks (using Mercator projection maps) and to relate them to the orbital parameters. With a nonrotating Earth, the projection of a geocentric orbit would be a great circle whose northernmost and southernmost latitudes would be equal to the inclination i. At the completion of an orbit the satellite would be over its original starting point, having traveled through 360° of longitude. In Fig. 8-4-2, track A represents the great circle trace of a circular orbit with an inclination of 50° over a nonrotating Earth. The track starts at the ascending node at 100° W longitude and ends its first orbit (revolution) at point A at 260° E longitude for a longitudinal distance of 360°. The tracks of subsequent orbits would fall exactly on top of this first orbit.

With the Earth rotating to the east with an angular rotation of the order of 15° per hour, the point on the Earth at which the first trace started will have moved to the east during the orbit so that the second orbital track will be displaced to the west, as shown in Fig. 8-4-2, where B is the same orbit as A but with a rotating Earth. The two orbits start at the same point, the common ascending node, but immediately begin to separate with the longitude of the completion point of B's first orbit (the second ascending node) displaced approximately 23° to the west, at point B. Each subsequent track will be displaced by the same amount of longitude, creating a family of tracks as shown in Fig. 8-4-3.

It is possible to use these ground track data to determine the period and semimajor axis a of the orbit in question. The period of the orbit can be found from the expression

$$P \text{ (hours)} = \frac{\text{longitude displacement (deg)}}{\text{Earth's rotational rate (deg/h)}} \cong \frac{\Delta L}{15} \quad (8\text{-}4\text{-}3)$$

where the Earth's rotational rate is approximately equal to 15 deg/h (360°/24 h) and where ΔL is the longitudinal displacement (in degrees) of a trace during one orbit. Since the displacement of the track during one orbit was 23°, the period of the satellite creating track B can be found to be

$$P \cong \frac{23}{15} = 1.533 \text{ h} = 92.0 \text{ min} = 5520 \text{ s}$$

Knowing the approximate period, the expression for the period of an ellipse or circle can be rearranged to give an expression for the semimajor axis, namely

$$a = \left[\mu\left(\frac{P}{2\pi}\right)^2\right]^{1/3} \quad (8\text{-}4\text{-}4)$$

where the P is expressed in seconds. The semimajor axis of track B is on the order of 6.75×10^6 m or 1.058 DU.

Figure 8-4-3 shows several sequential tracks of the circular satellite of

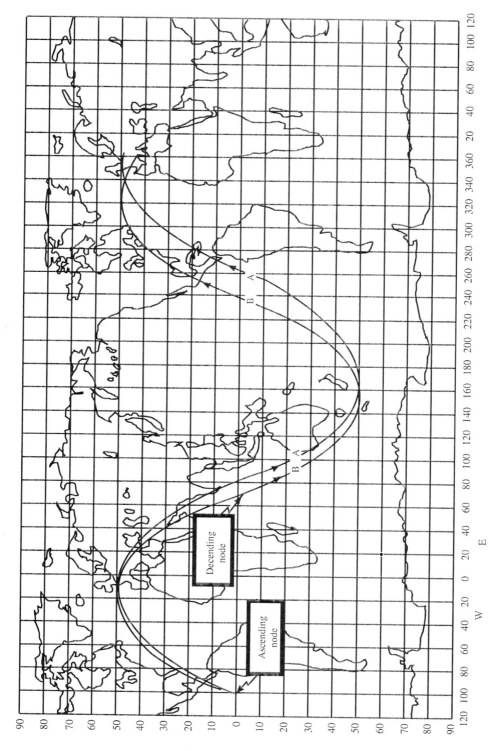

Figure 8-4-2. Effect of rotating Earth on an orbital ground track.

261

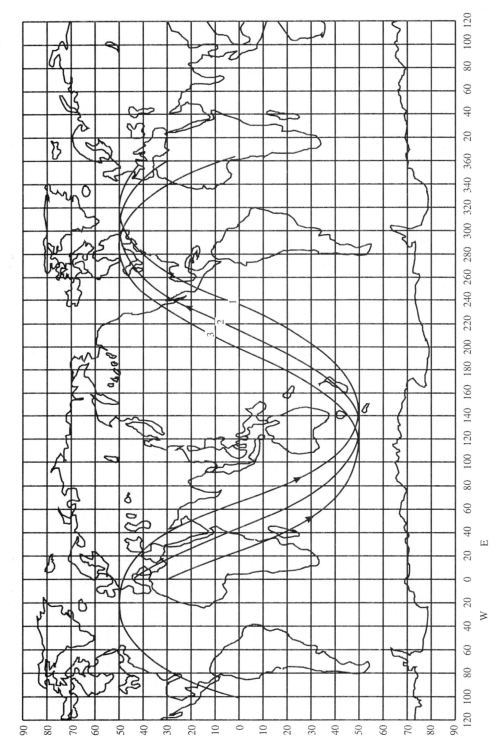

Figure 8.4-3. Several ground tracks for the same geocentric orbit.

Sec. 8-4 Satellite Ground Tracks

Fig. 8-4-2. Note the progressive displacement of the ascending node (and the traces) with each orbit, a displacement that is directly proportional to the period. Also note the symmetry of the traces with respect to both the equator and the lines of longitude through the maximum and minimum latitudes. This total symmetry (which will be related to the estimation of eccentricity and the location of the perigee later in this section) indicates that the orbit is indeed circular, as was stated earlier. Since the orbit is circular, the semimajor axis is the radius of the circle, and the altitude of this particular orbit is 370 km or 201 nmi; this is a low Earth orbit (LEO).

Before examining the effects of increasing the semimajor axis, let us discuss the implications of the inclination i. As mentioned earlier in this section, the values of the maximum northerly and southerly latitudes reached by a direct ground track are identically equal to the value of the inclination; for a retrograde orbit, the maximum and minimum values of the latitudes are equal to $(180° - i)$. A very interesting point with regard to the inclination is that the _inclination of an orbit cannot be less than the latitude of the launching site_ and that this minimum inclination is achieved only with a due east launch. Although spherical trigonometry is required to prove this statement, a look at Fig. 8-4-3 shows that an inclination of 50° is required to reach a latitude of 50°, either above or below the equator, implying that a launch with a lower inclination is not possible from a launch site at 50° latitude. Because of this limitation, a direct launch into an equatorial orbit requires a launch site located directly on the equator. Any other launch site requires a plane change of at least the value of the launch site latitude, which is one reason that Arianespace located its launch center 5° north of the equator and why the Russians, with launch sites above the 45° parallel, use highly eccentric communications satellites with periods of 12 h rather than synchronous equatorial satellites.

Figure 8-4-4 shows the ground tracks of four orbits that are identical with the exception of the inclination. They all have identical periods (and identical longitudinal displacements ΔL), common ascending and descending nodes, but different values for the maximum latitude reached, which is always the inclination of the orbit. Orbit A has an inclination of 10°, orbit B an inclination of 30°, orbit C an inclination of 50°, and orbit D an inclination of 85° (its orbit at the high latitudes is distorted because of the Mercator projection; a polar projection would be more suitable). A polar orbit with an $i = 90°$ would pass over both poles and provide coverage of the entire globe.

A launch site on the equator, at a common ascending node for instance, where the latitude is 0°, has the capability of placing a satellite directly into any of these orbits as well as into higher inclination orbits and into an equatorial orbit with $i = 0°$. On the other hand, a launch site at a latitude of 20° can directly enter (without a plane change) only those orbits whose inclination is 20° or higher.

As the semimajor axis a increases, so does the period and so does the longitudinal displacement ΔL between traces or tracks. Although ΔL is always to the west, there are values of the semimajor axis (on the order of 2.6 to 4.2 DU), where the displacement to the west increases sufficiently far to the west that the tracks appear to be displaced to the east. This is the case for the three ground tracks of a

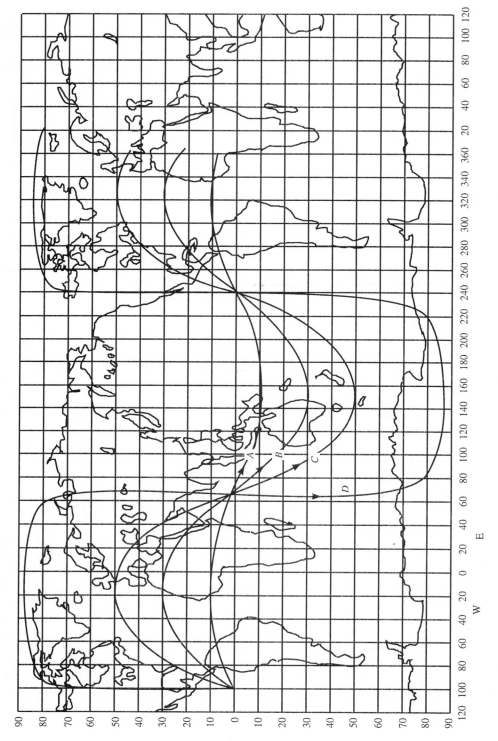

Figure 8-4-4. Effect of different inclination angles on the ground tracks of circular orbits.

Sec. 8-4 Satellite Ground Tracks

circular satellite that are sketched in Fig. 8-4-5. If we examine the trace of the revolution that started at the ascending node at 100° W, we see that the next ascending node is at 90° E for an apparent change of 190° to the east. Since the actual displacement was to the west, the Earth must have rotated through 360° − 190° = 170°, which is the actual ΔL. Consequently, the period of this orbit is

$$\Delta P = \frac{170}{15} = 11.33 \text{ h} = 680 \text{ min} = 40,800 \text{ s}$$

Applying Eq. (8-4-2) with $P = 40,800$ s, the semimajor axis is equal to 2.562×10^7 m = 4.016 DU and the altitude is 19,240 km = 10,460 nmi.

Comparing the traces of Fig. 8-4-5 with those of the lower orbit of Fig. 8-4-3, we see that the ground tracks become more compact as the semimajor axis (altitude) increases; for example, the nodal (equatorial) crossings are closer together. When the semimajor axis increases to 4.166 DU, the period becomes 12 h, the longitudinal displacement is 180° (either west or east as you choose), and the satellite will pass over the same spots on the surface of the Earth ~~twice~~ once each day (successive traces will fall one on the other).

Continuing to increase the semimajor axis increases the period and moves the equatorial crossings closer and closer together until there is only one crossing point and the ground track becomes a single figure-eight pattern that continually retraces itself, as shown in Fig. 8-4-6 for three different inclinations. These are geosynchronous earth orbits (GEOs). If the inclination is 0° (requiring an equatorial launch site or a plane change), the figure eight reduces to a point on the equator and the satellite becomes a geostationary satellite. It is interesting to note that the longitudinal displacement ΔL during each revolution of the Earth by the satellite is 360°. Therefore, the period is $\Delta L/15 = 24$ h, which we already knew, and application of Eq. (8-4-2) shows the semimajor axis to be 42.24×10^6 m = 6.623 DU and the altitude to be 35,860 km or 19,490 nmi, which we also knew.

A further increase in the semimajor axis to 8 DU leads to the ground tracks of Fig. 8-4-7, which are still symmetrical, indicating that $\epsilon = 0$ and that the orbit is still circular. Although the tracks now appear to an observer on the surface of the Earth to be traveling to the west and to make loops at the higher latitudes, the orbit is still a direct circular orbit moving to the east in inertial space. Although the apparent ΔL is 120° W, the actual $\Delta L = 360° + 120° = 480°$ and the period $\cong 480/15 = 32$ h. Notice the straightening of the traces as they cross the equator. This is a characteristic of high-altitude orbits, as is even more evident in the tracks of Fig. 8-4-8 for an a of 10 DU; this is still a direct orbit even though the satellite appears to be traveling to the west. The ground tracks for the moon, a satellite at 60.3 DU with an inclination of 28.5°, are a series of straight lines to the west that take 27.5 days to cover the 57° of latitude.

All of the tracks so far have been for direct orbits, satellites launched in an easterly direction and taking advantage of the Earth's easterly rotation to reduce the required launch velocity. If, however, a satellite were to be launched to the west (with an inclination >90° and ≤180°) and with a period less than 24 h, it would not

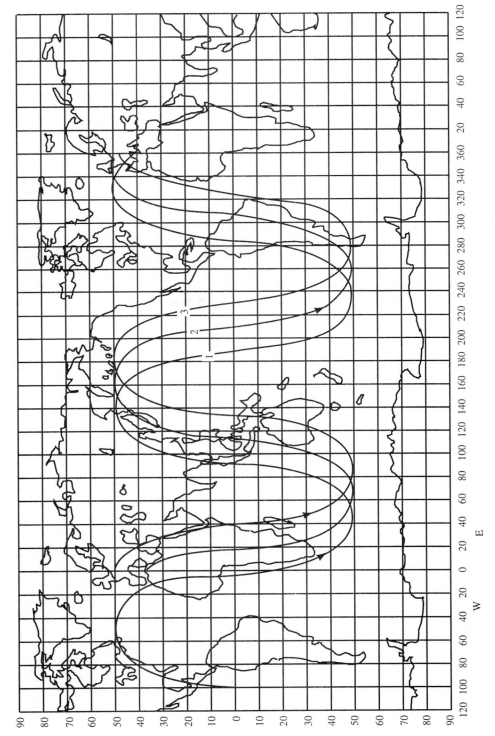

Figure 8-4-5. Effect of increasing orbital altitude until displacement becomes easterly.

266

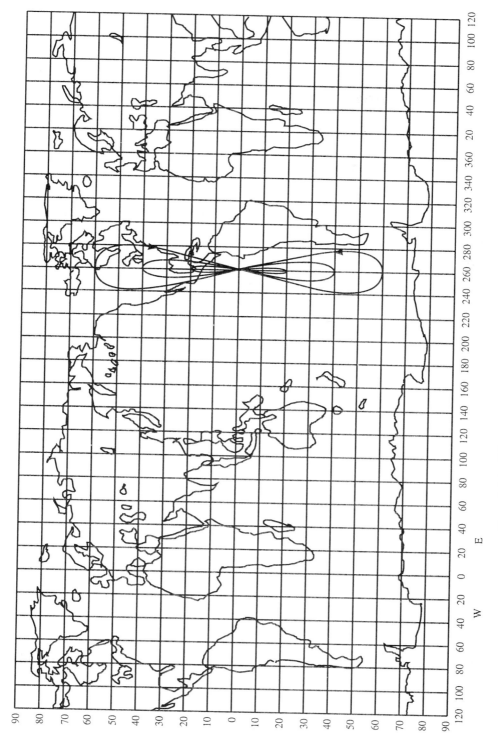

Figure 8-4-6. Effect of inclination angle on geosynchronous orbits.

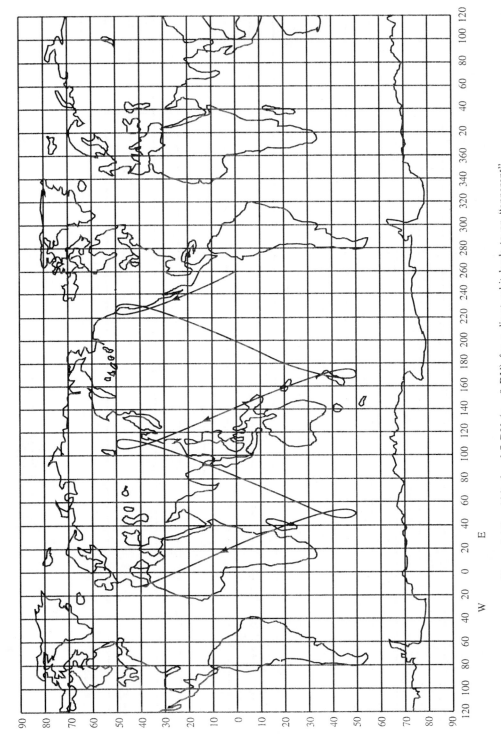

Figure 8.4-7. Altitude of 7 DU ($a = 8$ DU) for a direct orbit leads to an "apparent" retrograde orbit.

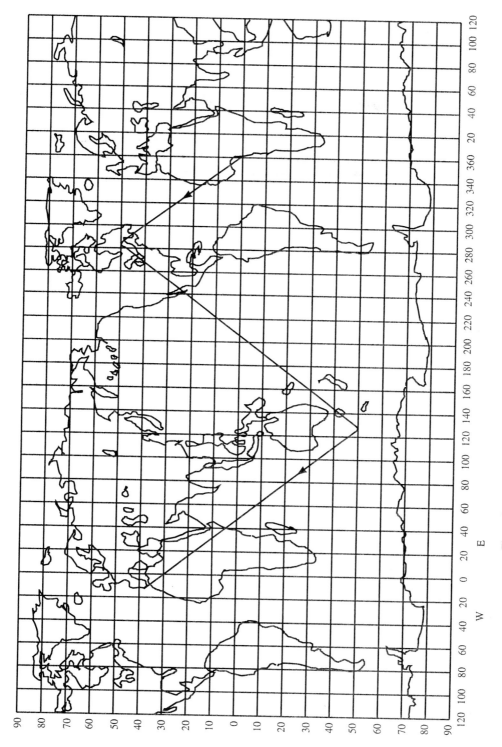

Figure 8-4-8. Direct orbit of Fig. 8-4-7 at an altitude of 9 DU (a = 10 DU).

only travel to the west in a retrograde orbit but would also appear to do so. The period and semimajor axis of retrograde orbits can be found from the longitudinal displacement between traces, which will be to the west as with the direct orbits, and the maximum latitudes will be $180° - i$. Figure 8-4-9 shows the differences and similarities between a high-altitude direct orbit and a low-altitude retrograde orbit; both orbits are circular. Track B, the direct orbit, has an inclination of 50°, an a of 12.1 DU, and a period of 59.3 h, whereas track A, the retrograde orbit, has an inclination of 130°, an a of 4.02 DU, and a period of 11.3 h. Notice that the low-altitude retrograde orbit has the straight-line appearance of high-altitude direct orbits.

A ground track has to pass two symmetry checks to be circular, line symmetry and hinge symmetry; if the ground track fails either test, the orbit is not circular and the eccentricity is not zero. Referring to the tracks of Fig. 8-4-4, the line symmetry check shows that each half of an orbit is symmetrical with respect to a longitude line through the maximum latitude point(s). The hinge symmetry check shows that the portion of the track above the equator can be rotated and superimposed on the portion below the equator. Thus these tracks, and all of the others discussed to this point, pass both symmetry tests and are circular. The track in Fig. 8-4-10 for an orbit with $\epsilon = 0.35$ does not, however, pass the hinge symmetry test, although each lobe does have line symmetry. With line but not hinge symmetry, the argument of the periapsis ω is either 90° or 270°; that is, periapsis is in either the northern or southern hemisphere. With hinge but not line symmetry, ω is either 0° or 180° (i.e., on the equator at either the ascending or descending node). With no symmetry whatsoever, ω can be anywhere else and the task of locating periapsis becomes more difficult. The fact that a satellite travels fastest at perigee, and thus covers more distance over the ground (the track is expanded) in this region, is of help in determining ω and the location of periapsis. Conversely, at the slower speeds in the vicinity of apogee, the Earth has more time to rotate beneath the satellite and the track is compressed. Applying either logic to the track of Fig. 8-4-10 indicates that periapsis is in the southern hemisphere and that $\omega = 270°$. The method of determining the period and semimajor axis from the longitudinal displacement is still valid. For this track with surface coverage of 170°, $\Delta L = 360° - 170° = 190°$, so that the period is 12.67 h and $a = 4.3$ DU. If ω were set equal to 90°, the tracks would be flipped over, perigee would be in the northern hemisphere, and the tracks would be compressed in the southern hemisphere, at apogee.

The two sets of ground tracks in Fig. 8-4-11 have hinge symmetry but not line symmetry; thus $\omega = 0°$ or 180° (both orbits have the same a and ϵ, 4.02 DU and 0.35) and perigee is at either the ascending or descending node. Since the satellite will be moving more quickly to the east at perigee, track A has perigee at the ascending node, where $\omega = 0°$, and track B has an $\omega = 180°$.

Figure 8-4-12 shows three sets of synchronous orbits for different values of ϵ and ω; we know that they are synchronous orbits because the tracks are closed curves and they all have the same inclination of 50°. Track A is that of a circular orbit with $\epsilon = 0$. The second set, tracks B and C with eccentricities of 0.35 and 0.7, respectively,

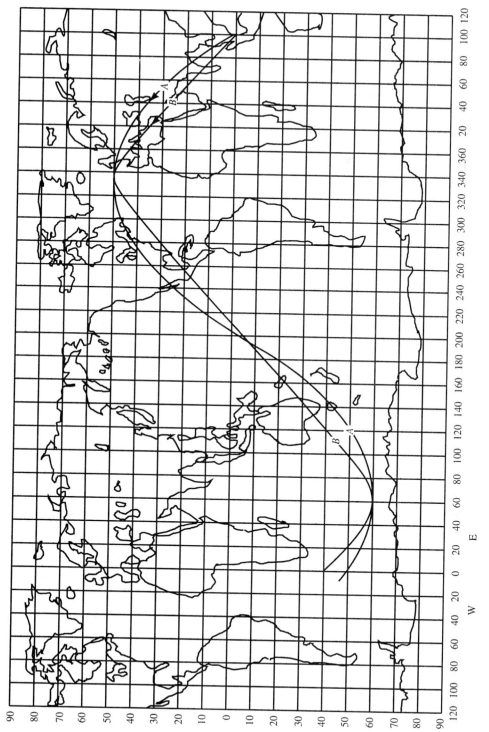

Figure 8-4-9. Track A is a low-altitude retrograde orbit; track B is a high-altitude direct orbit.

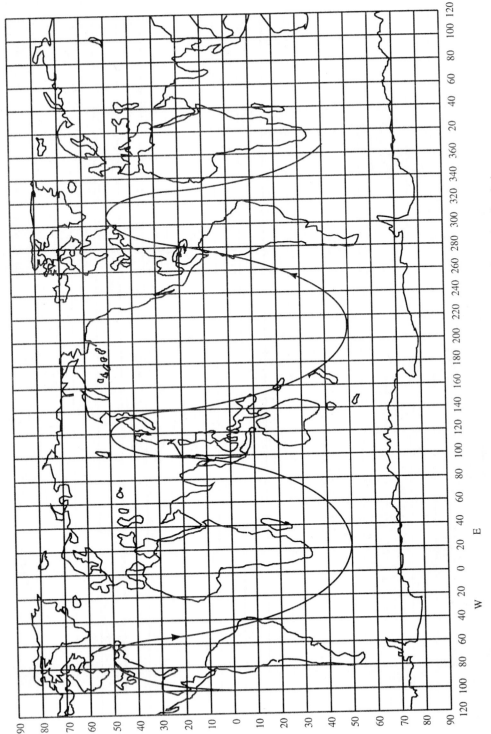

Figure 8.4-10. Direct orbit with $\epsilon = 0.35$ and $\omega = 270°$, periapsis in the southern hemisphere.

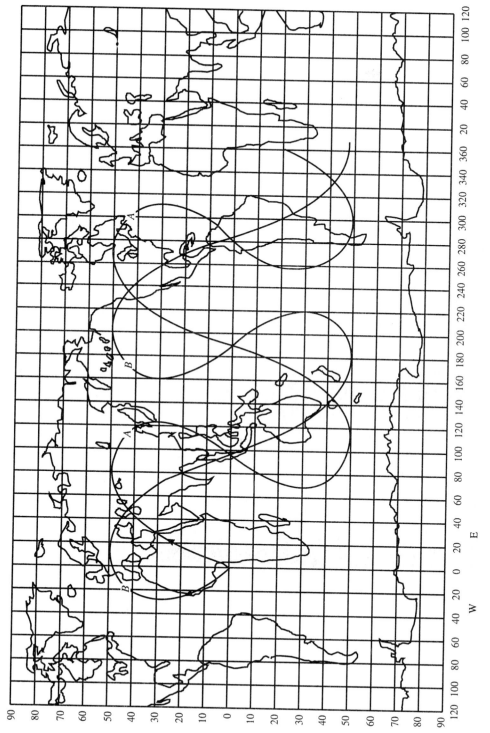

Figure 8-4-11. Two direct orbits with different arguments of periapsis: $\omega_A = 0°$ and $\omega_B = 180°$.

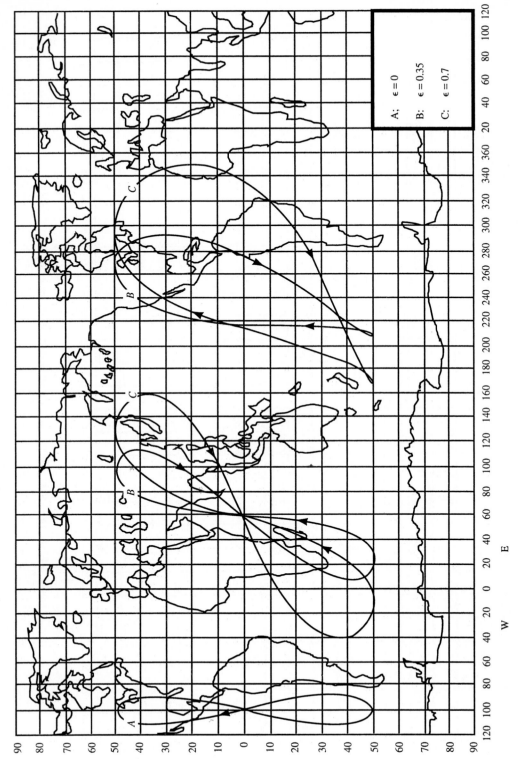

Figure 8-4-12. Three direct orbits with different eccentricities and arguments of periapsis.

have hinge but not line symmetry, so that ω = 0° or 180°, 180° in this case. The third set of tracks, still B and C but with different values of ω, have neither line nor hinge symmetry, so ω is not at any of the four locations mentioned. The narrowing of both tracks (to include a small loop) in the southern hemisphere indicates that apogee is in that region and that perigee is in the northern hemisphere, where the track is wider (i.e., 0° < ω < 180°). For these two tracks, it so happens that ω = 45°.

All of the ground tracks discussed in this section were developed with the assumption of a spherical Earth. The Earth, however, is oblate with an equatorial bulge that results in an average torque on a satellite and an increase in the radial acceleration as the satellite crosses the equator that although small, can perturb the orbits of certain satellites, particularly those at the lower altitudes. The two principal effects are the *regression of the nodes* and the *rotation of the line of apsides* (the major axis).

The nodal regression can be thought of as a precession of the line of nodes because of the applied torques, similar to the precession of the angular momentum vector of a gyroscope. For direct orbits the nodes move toward the west and for retrograde orbits they move toward the east. The rate of precession of the nodes is the time rate of change of Ω, the longitude of the ascending node, which can be expressed (from Kaplan) as

$$\dot{\Omega} = \frac{d\Omega}{dt} = -1.5 n J_2 \left(\frac{r_E}{p}\right)^2 \cos i \qquad (8\text{-}4\text{-}5)$$

where n is the mean motion of the satellite [$n = (\mu/a^3)^{1/2}$], J_2 a dimensionless measure of the oblateness of the Earth with a value of 1.083×10^{-3}, r_E the radius of the Earth, and p the semilatus rectum [$p = a(1 - \epsilon^2)$].

Since the nodal regression rate is directly proportional to the cosine of the inclination and inversely proportional to the square of the semimajor axis, it will be highest for low-altitude equatorial orbits (on the order of −9° per day for low eccentricities) and will go to zero for polar orbits. As the altitude increases, the regression rate decreases and is essentially zero at geosynchronous altitudes. It is interesting to note that this nodal regression mechanism is used to establish orbits for *sun-synchronous satellites* (generally, weather and earth-resource satellites) with a nodal regression of −360° per year (−1° per day) so that the orbit maintains a constant relationship to the Sun. With circular orbits it is a simple matter to establish a suitable relationship between a and i to give the desired regression rate; for sun-synchronous satellites the retrograde orbits are nearly polar and combine large coverage of the Earth's surface with long periods of constant illumination in the same orbit.

The rotation of the line of apsides is given by the time rate of change of the argument of periapsis and can be found from the expression for the average value of $d\omega/dt$ (again from Kaplan):

$$\dot{\omega} = \frac{d\omega}{dt} = 1.5 n J_2 \left(\frac{r_E}{p}\right)^2 (2 - 2.5 \sin^2 i) \qquad (8\text{-}4\text{-}6)$$

From an examination of the final term in parentheses, we see that the major axis (periapsis) will rotate in the direction of the motion of the satellite if the inclination is less than 63.4° or greater than 116.6° and opposite to the direction of motion if the inclination lies between these two values. As with the regression of the nodes, the lower the altitude, the higher the rotation rate.

We also see that there is no rotation of the line of apsides if the inclination is either 63.4° (a direct orbit) or 116.6° (a retrograde orbit); an orbit at these inclinations will remain stationary within its orbital plane. This phenomenon is exploited by the Russians with their Molyniya series of highly eccentric ($\epsilon = 0.74$) communications satellites with 12-h periods and with apogee in the northern hemisphere ($\omega = 270°$) so as to provide the longest possible contact times. If perigee were to migrate into the northern hemisphere, the usefulness of the satellite would be greatly diminished for long periods of time. Consequently, the four Molyniya satellites are placed into orbits with $i = 63.4°$ (one of the two critical inclinations), to preclude any rotation of the major axis.

8-5 CLOSING REMARKS

In this chapter we formally introduced the concept of the orbital elements, identifying a, ϵ, and ν (at epoch) as the orbital elements that define an orbit within the plane of motion[2] and defining three new orbital elements, i, Ω, and ω, that define the relationship of the orbital plane with respect to an inertial reference system. Since there are orbits for which the classic elements are not defined, we introduced additional orbital elements that can be used to describe such orbits.

We then considered the problem of determining the characteristics of an orbit from observed data, data acquired in a noninertial reference frame. Such data sets might be the velocity and position vectors at an epoch, possibly from radar, or two angles (the right ascension and declination) from three or more (optical) observations or two positions and the time between the two positions. We worked an example of the determination of the orbital elements when the inertial position vector and the inertial velocity vector are known. We also showed how an example in Section 4-5 represented a simplified solution to orbit determination from two positions and the time interval between the observations but only discussed in general terms the determination of orbits with only angles known. Orbit determination is not a simple problem, particularly when the complications of the real world must be taken into account.

The ground tracks section illustrated how the simple and straightforward circles and ellipses of the inertial orbits can take on different and unexpected shapes as the orbital elements are varied. Whereas direct circular orbits at low altitudes (smaller values of a) have symmetrical ground tracks that are sinusoidal in shape and appear to travel to the east, increasing the semimajor axis beyond that for a geosynchronous orbit results in tracks that move to the west, become straighter, and even have loops at the maximum latitudes.

[2] As mentioned earlier, T, the time of passage of periapsis, is often used in lieu of the true anomaly at epoch.

Chap. 8   Problems

The ground tracks of elliptical orbits are not symmetrical; the type and degree of asymmetry are determined by the value of ϵ. Increasing ϵ elongates the orbit and increases both the apogee altitude and the time the satellite spends over the Earth below it as the satellite moves more slowly around apogee. The location of the apogee (and perigee) is determined by the value of ω, the argument of the periapsis. Changing ω rotates the line of apsides about the angular momentum vector **H** and thus changes the location of perigee and apogee.

Changing Ω, the longitude of the ascending mode, rotates the plane of the orbit about the polar axis, as we saw during the discussion of the regression of the nodes. The values of Ω and ω for a particular orbit are determined by the nature of the orbit, the location of the launch site, the trajectory of the booster, and the time of the launch. The period of time during which a launch would result in an acceptable trajectory is the launch window for that mission. Launch windows can be expanded by the use of circular parking orbits with the desired i and Ω.

There are several shareware personal computer programs that allow the user to change the orbital parameters and see the effects on the ground tracks. They are enjoyable, informative, and improve one's understanding of the nature of satellite ground tracks and usefulness.

PROBLEMS

8-1. Find the orbital elements and period of a geocentric satellite whose inertial position and velocity vectors in a geocentric–equatorial system were determined at 0100 GMT on January 3, 1992 to be

$$\mathbf{r} = 1.73\bar{I} + 1.00\bar{J}$$
$$\mathbf{V} = -0.27\bar{I} + 0.47\bar{J} + 0.45\bar{K}$$

8-2. Do Problem 8-1 for a satellite whose vectors at 1105 GMT on May 12, 1991 were

$$\mathbf{r} = 0.41\bar{I} + 2.49\bar{J} + 0.61\bar{K}$$
$$\mathbf{V} = -0.35\bar{I} - 0.10\bar{J} + 0.63\bar{K}$$

8-3. Do Problem 8-1 for a geocentric satellite whose vectors at 0913 GMT on July 15, 1993 were

$$\mathbf{r} = -2.45\bar{I} + 0.62\bar{J} + 2.28\bar{K}$$
$$\mathbf{V} = -0.33\bar{I} - 0.47\bar{J} - 0.23\bar{K}$$

8-4. Do Problem 8-1 for a geocentric satellite whose vectors at 1725 GMT on November 12, 1992 were

$$\mathbf{r} = -0.80\bar{I} - 1.29\bar{J} - 0.33\bar{K}$$
$$\mathbf{V} = -0.55\bar{I} + 0.44\bar{J} - 0.36\bar{K}$$

8-5. An object was observed at an altitude of 366 km (198.9 nmi) and 12.5 min later at an altitude of 633 km (344 nmi); the angle between the two observations was 45°. Find the characteristics of the trajectory and the true anomalies at the time of the observations.

8-6. Do Problem 8-5 with observed altitudes of 92,000 km (50,000 nmi) and 7062 km (3838 nmi) with 2.9 h and 96° between the observations.

8-7. Do Problem 8-5 with observed altitudes of 7202 km (3914 nmi) and 11,980 km (6512 nmi) with 14.4 min and 80° between the observations.

8-8. Find the period and FOV (percent of Earth's surface visible) from circular satellites with the following altitudes:
   (a) 300 nmi (552 km)
   (b) 1000 nmi (1840 km)
   (c) 10,000 nmi (18,400 km)
   (d) 40,000 nmi (73,600 km)

8-9. Find the period and FOV (percent of Earth's surface visible) from perigee and apogee of the following geocentric ellipses:
   (a) $a = 1.5$ DU, $\epsilon = 0.3$
   (b) $a = 4$ DU, $\epsilon = 0.7$
   (c) $a = 10$ DU, $\epsilon = 0.1$
   (d) $a = 50$ DU, $\epsilon = 0.2$

8-10. Each of the following ground tracks has an *apparent* longitudinal displacement (ΔL) of 120° per orbit to the west and the inclination shown below. Find the period (hours) and the semimajor axis (meters and DU) and state whether the orbit is direct or retrograde.
   (a) $i = 40°$
   (b) $l = 140°$

8-11. Do Problem 8-10 for an apparent ΔL of 120° to the east.

CHAPTER 9

The Ballistic Missile

The launch the of Peacekeeper Missile, March 21, 1987. (Courtesy of NASA.)

9-1 INTRODUCTION

Although not technically a space vehicle, the ballistic missile is included in this book because it incorporates many of the features of the space flight problem and thus can serve as an interesting tutorial and review. After all, the payload of a missile does travel outside the Earth's atmosphere in a force-free two-body elliptical trajectory whose perigee happens to be below the surface of the Earth. The ballistic missile is also of interest historically, as it was the precursor to space flight systems as we know them today. In addition to being the first systems to accomplish their mission with flight outside the atmosphere, ballistic missiles were used to put the first satellites into geocentric orbits; several missiles are being used today as one or more stages of various space boosters.

Tutorially, the ballistic missile poses the interesting problem of traveling from one point on an ellipse to a second point that is moving with respect to the first (because of the rotation of the Earth). Although the number of possible trajectories through the two points is theoretically unlimited, the choice is restricted by the available energy at the first point (the burnout point), where the engines of the missile's engines are shut down. The missile payload enters the Keplerian orbit at the burnout point and the specific trajectory is determined by the energy and the elevation angle at that point. The motion of the second (the target) point requires consideration of the time of flight in the selection of a trajectory and introduces the need to calculate the lead angle that is associated with a moving target; in other words, the ballistic missile must be aimed.

In this chapter we deal only with the free-fall (Keplerian) portion of the ballistic trajectory, ignoring the powered-boost and reentry[1] phases, which involve aerodynamic and propulsive forces and have been discussed in Chapters 6 and 7. The Earth is assumed to be spherical and homogeneous; the effects of oblateness on the trajectory and the target motion are very complicated. In the next two sections, the Earth is also assumed to be nonrotating, so that the effects of changes in the burnout

---

[1] Reentry and R/V are associated with and appropriate for ballistic missiles.

energy and burnout parameters on the range and time of flight are not compromised by consideration of target motion. Furthermore, without rotation it is possible to obtain closed-form expressions relating the trajectory to the burnout parameters.

In the last section of this chapter the Earth is allowed to rotate so that the realistic problems associated with moving launch and target points can be considered. Interaction between the target motion and the time of flight and the range is such that closed-form expressions and solutions are no longer possible. Although iterative (or graphical) techniques are required for a complete solution, we provide a qualitative insight into this interaction, together with an example of how this interaction would affect one particular trajectory.

9-2 THE RANGE EQUATIONS

There are three phases to the trajectory of the payload of a ballistic missile (normally, a reentry vehicle carrying one or more warheads): the boost phase, the free-fall phase, and the reentry phase. The boost (powered) phase starts with a launch within the atmosphere and ends when the propulsion system shuts down outside the atmosphere at the *burnout point*, leaving the payload with a specific position and velocity, namely, r_{bo} (h_{bo}), V_{bo}, and ϕ_{bo}. The free-fall phase is the portion of a Keplerian ellipse (outside the atmosphere) with its primary focus at the center of the Earth, whose characteristics are determined by the burnout conditions; it starts at burnout and ends at reentry. The reentry phase begins with entry into the Earth's atmosphere and ends with impact or detonation of the warhead.

The total range of a ballistic missile, the distance traveled over the surface of the Earth from the launch site to the target, has three constituents, corresponding to each of the trajectory phases. R_p is the range during the boost phase, R_{ff} the range of the free-flight trajectory, and R_{re} the range covered during entry. With the assumption of a spherical Earth, each of these ranges is the product of the radius of the Earth and the angle between the position vector at the beginning and end of the appropriate phase. For example, if Ψ is the angle (in radians) swept out by r in going from the burnout point to the reentry point,

$$R_{ff} = r_E \Psi \qquad (9\text{-}2\text{-}1)$$

Ψ is the *free-fall range angle*, Γ is the powered (boost) range angle, and Ω is the reentry range angle; all angles are expressed in radians.

Although h_{bo} is outside the atmosphere, on the order of 184 to 368 km (100 to 200 nmi), h_{re} should be lower than h_{bo}, on the order of 75 km (41 nmi), if the edge of the sensible atmosphere is used as the start of reentry and deceleration. Consequently, the portion of an ellipse passing through the burnout and reentry points might not be symmetrical with respect to the major axis of the ellipse. Since the analysis for an asymmetrical trajectory is more complicated and does not provide additional insight into the characteristics and behavior of ballistic missile trajectories, the free-fall phase (and range) will be assumed to start at the burnout point and end at a mirror image point on the other side of the major axis, at a point where $r_{re} = r_{bo}$, $V_{re} = V_{bo}$, and $\phi_{re} = -\phi_{bo}$. Although it is customary to refer to this second point as the reentry point, we should remember that reentry as described in Chapter 7 probably occurs at a lower altitude and that r, V, and ϕ will be slightly different at the actual reentry point from the values at the assumed reentry point. We should also not forget the need for the R/V to have the appropriate angle of attack prior to substantive reentry, zero in the case of a purely ballistic nonmaneuvering R/V.

The geometry of the ballistic missile trajectory is shown in Fig. 9-2-1 with Ψ symmetrical about the major axis and with Ω as the reentry range angle. Consequently, the total range angle Λ is

$$\Lambda = \Gamma + \Psi + \Omega \tag{9-2-2a}$$

and the total range R_T is

$$R_T = r_E \Lambda = R_p + R_{ff} + R_{re} \tag{9-2-2b}$$

Since the determination of R_p and R_{re} is not germane to this chapter, and since their sum is of the order of 5% of the total range, they will be neglected and the total range will be approximated by the free-fall range alone.

$$R_T \cong r_E \Psi = R_{ff}$$

Before developing a range equation for the free-fall range, let us restate the assumptions and approximations. The Earth is spherical, homogeneous, and nonrotating (the effects of rotation are discussed in a subsequent section). The center of the Earth is the origin of a central force field and is located at the primary focus of the ellipse. The free-fall trajectory is symmetrical and elliptical and traces a great circle arc on the surface of the Earth. All the two-body relationships developed and used in the earlier orbital mechanics chapters are applicable.

The range equation, which defines Ψ, the free-fall range angle, as a function

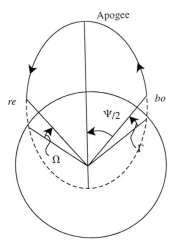

Figure 9-2-1. Geometry of a ballistic missile trajectory.

Sec. 9-2 The Range Equations

of the burnout conditions can be developed algebraically or geometrically. Although the form of the two resulting equations will be somewhat different, each is equally valid.

The algebraic development is based on the solution of the two-body polar equation, namely,

$$r = \frac{p}{1 + \epsilon \cos \nu} \qquad (9\text{-}2\text{-}3a)$$

where

$$p = \frac{H^2}{\mu_E} \quad \text{and} \quad \epsilon = \sqrt{1 + \frac{2EH^2}{\mu^2}} \qquad (9\text{-}2\text{-}3b)$$

The true anomaly ν in Eq. (9-2-3a) is replaced by an auxiliary angle θ, which is measured from apogee rather than perigee, so that

$$\theta = \nu + \pi \qquad (9\text{-}2\text{-}4a)$$

and

$$\Psi = \theta_{re} - \theta_{bo} = -2\theta_{bo} \qquad (9\text{-}2\text{-}4b)$$

Without going through the details of the algebraic development, the resulting range equation can be written as

$$\boxed{\cot \frac{\Psi}{2} = \frac{2}{Q_{bo}} \csc 2\phi_{bo} - \cot \phi_{bo}} \qquad (9\text{-}2\text{-}5)$$

where

$$\boxed{Q_{bo} = \frac{V_{bo}^2 r_{bo}}{\mu} = \frac{V_{bo}^2}{V_{cs}^2}} \qquad (9\text{-}2\text{-}6)$$

In Eq. (9-2-6), V_{cs}, which is equal to $(\mu/r_{bo})^{1/2}$, is the circular orbital velocity at r_{bo}.

Q_{bo}, which is dimensionless and nameless, is an interesting and insightful parameter. It is a measure of the energy of a ballistic missile payload at burnout; actually, it is twice the ratio of the kinetic energy to the potential energy. If $Q_{bo} < 1$, only suborbital flight (missile trajectories) is possible. If $Q_{bo} = 1$, the payload has the capability of circular orbital flight, but only if $\phi_{bo} = 0°$. If $Q_{bo} > 1$, both missile trajectories and orbital trajectories are possible. Finally, if $Q_{bo} \geq 2$, escape from the Earth's SOI is possible.

The geometric development of the range equation, which results in a different form, will now be presented in more detail. Before starting, we need to express Q_{bo} in yet another form. Starting with the vis-viva integral, written at burnout,

$$\frac{V_{bo}^2}{2} - \frac{\mu}{r_{bo}} = \frac{-\mu}{2a} \qquad (9\text{-}2\text{-}7a)$$

Equation (9-7-2a) can be arranged to show that

$$\frac{V_{bo}^2 r_{bo}}{\mu} = 2 - \frac{r_{bo}}{a} = Q_{bo} \tag{9-2-7b}$$

The two right-hand terms can be rearranged to form the relationship that

$$a = \frac{r_{bo}}{2 - Q_{bo}} \tag{9-2-7c}$$

where a is the semimajor axis of the ellipse.

Referring to Fig. 9-2-2, which shows a suborbital elliptical trajectory (a ballistic trajectory), one definition of an ellipse is that

$$2a = r_{bo} + r' \tag{9-2-8a}$$

which, with the substitution of Eq. (9-2-7c), can be rewritten as

$$r' = \frac{r_{bo} Q_{bo}}{2 - Q_{bo}} \tag{9-2-8b}$$

All we need now is an expression for r' as a function of Ψ and ϕ_{bo}. The characteristic of an ellipse that the normal to a tangent (V_{bo}) bisects the angle between r_{bo} and r' means that this angle is equal to twice the elevation angle (i.e., $2\phi_{bo}$). Applying the law of cosines to find $F_1 F_2$ yields

$$(F_1 F_2)^2 = r_{bo}^2 + r'^2 - 2r_{bo} r' \cos 2\phi_{bo} \tag{9-2-9a}$$

Applying the law of cosines for the second time yields the expression for r' that

$$r'^2 = r_{bo}^2 + (F_1 F_2)^2 - 2r_{bo}(F_1 F_2) \cos\frac{\Psi}{2} \tag{9-2-9b}$$

Combining these two equations with Eq. (9-2-8b) and solving for an expression for Ψ results, after some messy manipulation, in another range equation, namely,

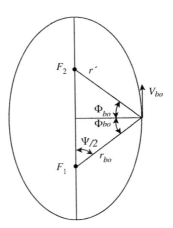

Figure 9-2-2. Geometry for the range angle.

Sec. 9-2 The Range Equations

$$\boxed{\cos\frac{\Psi}{2} = \frac{1 - Q_{bo}\cos^2\phi_{bo}}{\sqrt{1 + Q_{bo}(Q_{bo} - 2)\cos^2\phi_{bo}}}} \qquad (9\text{-}2\text{-}10)$$

Although neither range equation is particularly straightforward, Eqs. (9-2-5) and (9-2-10) are equivalent and either one can be used to find the (free-fall) range angle. It can be shown that when $Q_{bo} < 1$, Ψ will be less than 180°, implying, as before, that the only trajectories possible are those of a ballistic missile with a range less than half the distance around the Earth.

Example 9-2-1

A ballistic missile has the following burnout conditions: $V = 7168$ m/s, $h = 276$ km (150 nmi), and $\phi = 25°$.
(a) Find the value of Q_{bo} and discuss its implications.
(b) Using Eq. (9-2-5), find Ψ and R_{ff}.
(c) Using Eq. (9-2-10), find Ψ and R_{ff}.

Solution

(a)
$$r_{bo} = r_E + h_{bo} = 6.654 \times 10^6 \text{ m}$$

$$Q_{bo} = \frac{V_{bo}^2 r_{bo}}{\mu} = \frac{(7168)^2 \times 6.654 \times 10^6}{3.986 \times 10^{14}} = 0.8575$$

The fact that $Q_{bo} < 1$ indicates that any trajectory will be suborbital, that it will hit the surface of the Earth before completing an orbit, a ballistic missile trajectory. As a matter of interest, the circular orbital velocity at r_{bo} is 7740 m/s, a value greater than V_{bo}, and $(V_{bo}/V_{cs})^2 = 0.8575 = Q_{bo}$.

(b) Using Eq. (9-2-5) yields

$$\cot\frac{\Psi}{2} = \frac{2}{0.8575}\csc(50°) - \cot(25°) = 0.90$$

and

$$\Psi = 96.02° = 1.676 \text{ rad}$$

Therefore,

$$R_{ff} = 6378 \times 1.676 = 10{,}690 \text{ km} = 5809 \text{ nmi}$$

(c) Using Eq. (9-2-11) gives

$$\cos\frac{\Psi}{2} = \frac{1 - 0.8575\cos^2(25°)}{\sqrt{1 + 0.8575(0.8575 - 2)\cos^2(25°)}} = 0.6690$$

and

$$\Psi = 96.01° = 1.676 \text{ rad}$$

Therefore,

$$R_{ff} = 10{,}690 \text{ km} = 5809 \text{ nmi}$$

We see that both forms of the range equation give identical answers, as should be the case.

Although the range equations provide the range for a given set of burnout conditions, a more realistic problem is to know the range (the launch and target positions) and to find the burnout conditions. Since V_{bo} and h_{bo} are dependent on the configuration of the missile (the booster) and the powered trajectory and are relatively restricted as to their values, the burnout elevation angle ϕ_{bo} is the easiest burnout condition to control. Consequently, it would be convenient to have an expression for ϕ_{bo} in terms of the range angle Ψ and Q_{bo}, since Q_{bo} is independent of ϕ_{bo}. Continuing with the trigonometric approach only and referring to Fig. 9-2-3, we find that

$$d = r' \sin \alpha = r_{bo} \sin \frac{\Psi}{2} \tag{9-2-11a}$$

so that

$$\sin \alpha = \frac{r_{bo}}{r'} \sin \frac{\Psi}{2} \tag{9-2-11b}$$

But

$$\alpha = 180° - \left(2\phi_{bo} + \frac{\Psi}{2}\right) \tag{9-2-11c}$$

and

$$\sin \alpha = \sin\left[180° - \left(2\phi_{bo} + \frac{\Psi}{2}\right)\right] = \sin\left(2\phi_{bo} + \frac{\Psi}{2}\right) \tag{9-2-12}$$

With the substitution of Eqs. (9-2-12) and (9-2-8b) (r' as a function of r_{bo} and Q_{bo}), Eq. (9-2-11b) becomes a function of Q_{bo} and Ψ only.

$$\boxed{\sin\left(2\phi_{bo} + \frac{\Psi}{2}\right) = \frac{2 - Q_{bo}}{Q_{bo}} \sin \frac{\Psi}{2}} \tag{9-2-13}$$

Example 9-2-2

A ballistic missile is to have a free-fall range of 5500 nmi (10,120 km). It is capable of achieving a burnout velocity of 7168 m/s at an altitude of 150 nmi (276 km).
(a) Find the appropriate value(s) of ϕ_{bo}.

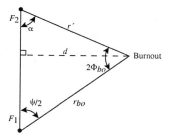

Figure 9-2-3. Geometry for the burnout elevation angle.

Sec. 9-2 The Range Equations

(b) Find the characteristics of the trajectory associated with each ϕ_{bo}: E, a, H, ϵ, h_a, and h_p.

Solution

(a) $\psi = 10{,}120/6378 = 1.587$ $r = 90.92°$
With $r_{bo} = 6.654 \times 10^6$ m, $Q_{bo} = 0.8577 < 1$. Using Eq. (9-2-13), we find that

$$\sin(2\phi_{bo} + 45.46°) = \frac{2 - 0.8577}{0.8577}\sin(45.46°) = 0.9493$$

The arcsine of 0.9493 has two possible values, 71.67° and 108.3°. Consequently, *the burnout elevation angle has two possible values, 13.10° and 31.43°*, a low value (a low trajectory) and a high value (a high trajectory); both trajectories will have a range of 5500 nmi.

(b) The low trajectory ($\phi_{bo} = 13.10°$):

$$E = \frac{(7168)^2}{2} - \frac{3.986 \times 10^{14}}{6.654 \times 10^6} = -3.421 \times 10^7 \text{ m}^2/\text{s}^2$$

$$a = \frac{-\mu}{2E} = 5.825 \times 10^6 \text{ m}$$

$$H = r_{bo} V_{bo} \cos\phi_{bo} = 4.645 \times 10^{10} \text{ m}^2/\text{s}$$

$$\epsilon = \sqrt{1 + \frac{2EH^2}{\mu^2}} = 0.2657 \quad \text{(an ellipse)}$$

$$h_a = a(1 + \epsilon) - r_E = 994.5 \text{ km} = 540 \text{ nmi}$$

$$h_p = a(1 - \epsilon) - r_E = 2100 \text{ km} = -1141 \text{ nmi} \quad \text{(below the surface)}$$

The high trajectory ($\phi_{bo} = 31.43°$):

$$E = -3.421 \times 10^7 \text{ m}^2/\text{s}^2 \quad \text{(unchanged)}$$

$$a = 5.825 \times 10^6 \text{ m} \quad \text{(unchanged)}$$

but

$$H = 4.067 \times 10^{10} \text{ m}^2/\text{s} \quad \text{(lower)}$$

$$\epsilon = 0.5357 \quad \text{(larger, more elliptical)}$$

$$h_a = 2567 \text{ km} = 1395 \text{ nmi} \quad \text{(much higher)}$$

$$h_p = -3673 \text{ km} = -1996 \text{ nmi}$$

Although not immediately apparent from the form of Eq. (9-2-13), this example shows that for $Q_{bo} < 1$, there are two different trajectories with the same range, a high (lofted) trajectory and a low trajectory. The two trajectories have the same energy and semimajor axis but different values of H, ϵ, apogee height, and time of flight (the high trajectory will have a longer time of flight). It is customary, even traditional, at this point to draw the analogy of the water hose with a fixed nozzle velocity. Any spot on the ground within the range of the water can be reached with either a high or a low trajectory by changing the elevation angle of the nozzle.

When $1 < Q_{bo} < 2$, there are still two mathematically possible trajectories.

The low trajectory, however, is not physically possible, as it penetrates the surface of the Earth shortly after burnout. This is illustrated in the following example.

Example 9-2-3

The ballistic missile of the preceding example (5500 nmi range and 150-nmi burnout altitude) has its burnout velocity increased to 8825 m/s.
(a) Find ϕ_{bo} and the relevant characteristics of the high and low trajectories.
(b) Are they both physically realistic?

Solution

(a) $Q_{bo} = 1.3 > 1$ and $\Psi = 90.92°$. From Eq. (9-2-14),

$$2\phi_{bo} + 45.46° = \arcsin 0.3838$$

Since the arcsin 0.3838 has two possible values, 22.57° and 157.4°, then $\phi_{bo} = -11.44°$ or 55.98°. Let us first look at the high trajectory with the positive elevation angle and then at the low trajectory with the negative elevation angle.

The high trajectory ($\phi_{bo} = 55.98°$):

$$E = -2.096 \times 10^7 \text{ m}^2/\text{s}^2 \qquad a = 9.507 \times 10^6 \text{ m}$$
$$H = 3.285 \times 10^{10} \text{ m}^2/\text{s} \qquad \epsilon = 0.8457$$
$$h_a = 11{,}170 \text{ km} = 6070 \text{ nmi}$$
$$h_p = -4911 \text{ km} = -2669 \text{ nmi}$$

The low trajectory ($\phi_{bo} = -11.44°$):

$$H = 5.755 \times 10^{10} \text{ m}^2/\text{s} \qquad \epsilon = 0.355$$
$$h_a = 6504 \text{ km} = 3534 \text{ nmi}$$
$$h_p = -245.6 \text{ km} = -133.5 \text{ nmi}$$

(b) The high trajectory is possible because ϕ_{bo} is positive, indicating that the payload is ascending toward apogee. The low trajectory, however, with a negative ϕ_{bo} has passed apogee and burnout point is actually the reentry point. Shortly after burnout, therefore, the payload would reach the edge of the atmosphere and begin a short-range reentry trajectory followed by surface impact or detonation.

Although only the high trajectory is physically possible when $1 < Q_{bo} < 2$, the range angle Ψ can now be greater than 180°; Ψ can also be less than 180°. In fact, a ballistic missile with $Q_{bo} > 1$ has the capability of delivering a payload anywhere on a great circle, up to a distance all the way around the Earth. With such a capability, if the desired range angle is less than 180°, the target can be reached by either the customary short-way trajectory ($\Psi < 180°$) or a long-way trajectory ($\Psi > 180°$), which travels around the back side of the Earth, as indicated in Fig. 9-2-4.

Example 9-2-4

For the ballistic missile of Example 9-2-3, which has a desired range of 5500 nmi ($\Psi = 90.92°$) to a target, take the long way round to the target ($\Psi = 269.08°$) and find:
(a) ϕ_{bo} and compare it with the ϕ_{bo} of 56.0° for the short-way trajectory.
(b) The trajectory characteristics and compare them with those of the short trajectory of Example 9-2-3.

Sec. 9-2 The Range Equations

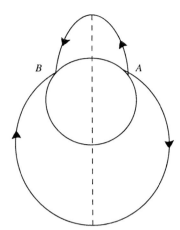

Figure 9-2-4. Short-way and long-way trajectories between A and B.

Solution

(a) With $Q_{bo} = 1.3$ and $\Psi/2 = 134.5°$,

$$\sin(2\phi_{bo} + 134.5°) = \frac{2 - 1.3}{1.3} \sin(134.5°) = 0.384$$

Rejecting the negative value for ϕ_{bo} (the low trajectory),

$$\phi_{bo} = 11.45°$$

which is lower than the ϕ_{bo} of 56.0° for the short way.

(b) As for the trajectory characteristics,

$E = -2.096 \times 10^7 \text{ m}^2/\text{s}^2 \quad a = 9.507 \times 10^6 \text{ m}$ (both the same)

$H = 5.755 \times 10^{10} \text{ m}^2/\text{s}$ (higher)

$\epsilon = 0.3552$ (lower, more circular)

$h_a = 6505 \text{ km} = 3536 \text{ nmi}$ (lower)

$h_p = -248 \text{ km} = -135 \text{ nmi}$ (not as far below the surface)

Since the long-way range is larger (16,270 nmi), the time of flight will also be longer than that of the short-way trajectory.

It is possible to develop equations for the eccentricity ϵ in terms of Q_{bo}, ϕ_{bo}, or Ψ that are more convenient and do not require the calculation of E and H.

Starting with Eqs. (9-2-3) and (9-2-5), after some manipulation one arrives at the expression

$$\boxed{\epsilon = \frac{\sin \phi_{bo}}{\sin(\phi_{bo} + \Psi/2)}} \quad (9\text{-}2\text{-}14)$$

This is an expression for ϵ as a function of ϕ_{bo} and Ψ (burnout elevation angle and range angle).

Starting with the expression for ϵ as a function of E and H and making appropriate substitutions, it can be shown that

$$\epsilon = \sqrt{1 + Q_{bo}(Q_{bo} - 2) \cos^2 \phi_{bo}} \qquad (9\text{-}2\text{-}15)$$

This expression gives ϵ when Q_{bo} and ϕ_{bo} are known.

With the previously developed expression for the semimajor axis as

$$a = \frac{r_{bo}}{2 - Q_{bo}} \qquad (9\text{-}2\text{-}16)$$

it is now a simple matter to calculate a and ϵ from the range and burnout conditions and then determine the remaining trajectory characteristics.

Equation (9-2-15) gives some interesting results for $Q_{bo} = 1$ and $Q_{bo} = 2$. When $Q_{bo} = 1$,

$$\epsilon = \sqrt{1 - \cos^2 \phi_{bo}} = \sin \phi_{bo}$$

Now when $\phi_{bo} = 0°$, $\epsilon = 0$ and the resulting trajectory is a circular orbit with $a = r_{bo}$. When $\phi_{bo} = 90°$, $\epsilon = 1$ and the resulting trajectory is the vertical "straight-line" ellipse of a sounding rocket. When $Q_{bo} = 2$ (the escape condition), $\epsilon = 1$ for all values of the elevation angle at the burnout point and the payload will be on a parabolic escape trajectory.

The time of free-fall flight (TOF) between the burnout and reentry points (bo and re) on the elliptical trajectory can be found from the expressions of Chapter 4. Equation (4-3-21) can be written as

$$\text{TOF} = \sqrt{\frac{a^3}{\mu}} (M_{re} - M_{bo}) \qquad (9\text{-}2\text{-}17a)$$

where M_{re} and M_{bo} are the respective mean anomalies, which in turn are functions of the eccentricity ϵ and the eccentric anomaly u, namely,

$$M_{re} = u_{re} - \epsilon \sin u_{re} \quad \text{and} \quad M_{bo} = u_{bo} - \epsilon \sin u_{bo} \qquad (9\text{-}2\text{-}17b)$$

where u is expressed in radians and can be found from

$$\cos u = \frac{\epsilon + \cos \nu}{1 + \epsilon \cos \nu} \qquad (9\text{-}2\text{-}17c)$$

where ν is the true anomaly.

With a symmetrical trajectory (with the burnout and reentry points on opposite sides of the major axis), the true anomalies are

$$\nu_{bo} = \pi - \frac{\Psi}{2}$$

and

$$\nu_{re} = \pi + \frac{\Psi}{2} \qquad (9\text{-}2\text{-}17d)$$

Sec. 9-2   The Range Equations   291

Consequently, the two eccentric anomalies have identical expressions but are also on opposite sides of the major axis, so that

$$u_{re} = 2\pi - u_{bo}$$

and

$$\sin u_{re} = -\sin u_{bo} \qquad (9\text{-}2\text{-}17e)$$

Combining Eqs. (9-2-17b) and (9-2-17e) yields the relationship that

$$M_{re} - M_{bo} = 2(\pi - u_{bo} + \epsilon \sin u_{bo}) \qquad (9\text{-}2\text{-}18)$$

where the burnout eccentric anomaly is

$$u_{bo} = \cos^{-1} \frac{\epsilon - \cos(\Psi/2)}{1 - \epsilon \cos(\Psi/2)} \qquad (9\text{-}2\text{-}19)$$

With the substitution of Eq. (9-2-18), Eq. (9-2-17a) can now be written as

$$\boxed{\text{TOF} = 2\sqrt{\frac{a^3}{\mu}}(\pi - u_{bo} + \epsilon \sin u_{bo})} \qquad (9\text{-}2\text{-}20)$$

which is a function of ϵ, u_{bo}, and a. We see that Eq. (9-2-20) is simply twice the time of flight from burnout to apogee.

Example 9-2-5

A ballistic missile has a free-fall range of 5500 nmi, $V_{bo} = 7168$ m/s, and $h_{bo} = 150$ nmi (176 km). In Example 9-2-2 we found that there are two acceptable values of ϕ_{bo}, 13.1° and 31.43°.
(a) Find the TOF of the low trajectory, $\phi_{bo} = 13.1°$.
(b) Find the TOF of the high trajectory and compare it with that of part (a).
Solution

(a)
$$r_{bo} = r_E + h_{bo} = 6.654 \times 10^6 \text{ m}$$

$$\Psi = 1.587 \text{ rad} = 90.92°$$

$$Q_{bo} = \frac{(7168)^2 \times 6.654 \times 10^6}{3.986 \times 10^{14}} = 0.8577$$

The low trajectory ($\phi_{bo} = 13.1°$): Equation (9-2-16) can be used to find ϵ:

$$\epsilon = \sqrt{1 + 0.8577(0.8577 - 2)\cos^2(13.1°)} = 0.2657$$

The eccentric anomaly at burnout is

$$u_{bo} = \cos^{-1} \frac{0.2657 - \cos 0.7934}{1 - 0.2657 \cos 0.7934} = 2.136 \text{ rad}$$

We now need the semimajor axis a.

$$a = \frac{6.654 \times 10^6}{2 - 0.8577} = 5.825 \times 10^6 \text{ m}$$

$$\sqrt{\frac{a^3}{\mu}} = 704.2 \text{ s}$$

Therefore, the time of flight is

$$\text{TOF} = 2 \times 704.2(\pi - 2.136 + 0.2657 \sin 2.136) = 1732 \text{ s} = 28.87 \text{ min}$$

(b) The high trajectory ($\phi_{bo} = 31.43$): Q_{bo} and a are identical for both trajectories. This time, Eq. (9-2-14) will be used to find ϵ:

$$\epsilon = \frac{\sin(31.43°)}{\sin(31.43° + 45.46°)} = 0.5354$$

The eccentricity of the high trajectory is larger than that for the low trajectory, as expected for the lofted (more elliptical) high trajectory. The burnout eccentric anomaly is

$$u_{bo} = \cos^{-1} \frac{0.5354 - \cos 0.7935}{1 - 0.5354 \cos 0.7935} = 1.840$$

Therefore, the free-fall time of flight is

$$\text{TOF} = 2 \times 704.2(\pi - 1.84 + 0.5354 \times \sin 1.84) = 2560 \text{ s} = 42.67 \text{ min}$$

The TOF of the high trajectory is 48% longer than that of the low trajectory, again as expected.

Figure 9-2-5 shows the free-fall TOF for various values of the range angle and of Q_{bo} (with $h_{bo} = 166$ nmi = 305 km). Note that the TOF is double-valued for $Q_{bo} < 1$, the smaller value for the low trajectory and the larger value for the high trajectory. For values of $Q_{bo} \geq 1$, the TOF is single-valued since there is no low trajectory per se.

In the next section we develop the conditions for maximizing the range for a given set of burnout variables, determine the minimum-energy trajectory for a specified range, and investigate the sensitivity of the range for changes in the burnout conditions.

9-3 TRAJECTORIES AND SENSITIVITIES

In the preceding section, the range equations were applied to find ranges for specified burnout energies and elevation angles. Figure 9-3-1 is plot of range versus V_{bo} for various values of ϕ_{bo} and a fixed value of h_{bo} (150 nmi). This is a busy plot, particularly at the lower velocities (<7740 m/s) where $Q_{bo} < 1$, and the qualitative relationships among the velocity, elevation angle, and range are not obvious. This figure shows that a range of 6000 nmi can be achieved with several combinations of V_{bo} and ϕ_{bo}: 7200 m/s and 15°, 7400 m/s and 30°, 8000 m/s and 45°, 9800 m/s and 60° (all approximate values), and that there is no possibility of reaching 6000 nmi with $\phi_{bo} = 75°$, no matter how large the burnout velocity.

Table 9-3-1 contains the data for Fig. 9-3-1, and although not as cluttered as the figure, does not show specific correlations. It does indicate, though, that there is an elevation angle that maximizes the range for a particular V_{bo} (Q_{bo}). For instance,

Sec. 9-3 Trajectories and Sensitivities

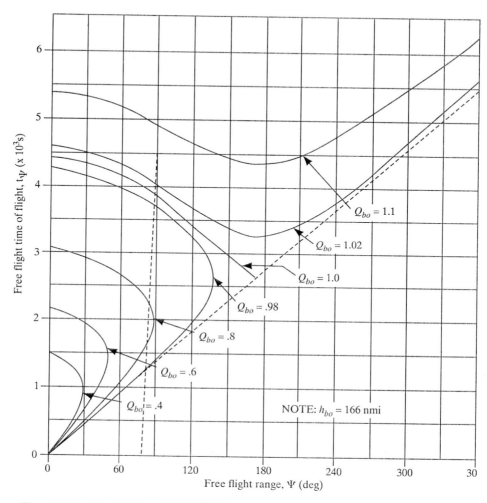

Figure 9-2-5. Free-fall time of flight (TOF) as a function of range angle for several values of Q_{bo}.

with a burnout velocity of 6000 m/s ($Q_{bo} = 0.6010$), the range increases from 2284 nmi to 3066 nmi as ϕ_{bo} is increased from 15° to 30° but decreases monotonically as ϕ_{bo} is further increased, becoming 1077 nmi when $\phi_{bo} = 75°$. One would suspect that the burnout elevation angle that gives the maximum range for this value of Q_{bo} might be of the order of 30°, more or less.

Although the relationships among Ψ_{max}, Q_{bo}, and ϕ_{bo} can be obtained from trigonometry, the conventional approach would be to set the (partial) derivative of Ψ with respect to the principal variable of interest (ϕ_{bo}) equal to zero and solve for the conditions for a maximum (or a minimum). Using either form of the range

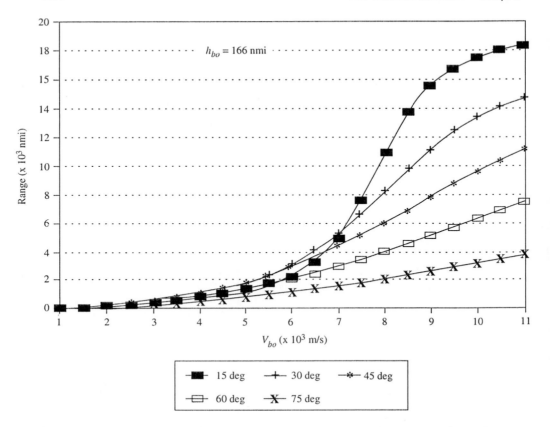

Figure 9-3-1. Ballistic missile range as a function of burnout velocity and elevation angle.

equation [Eq. (9-2-5) or (9-2-10)], setting the partial derivative of Ψ with respect to ϕ_{bo} to zero reduces to

$$\frac{\partial \Psi}{\partial \phi_{bo}} = \frac{2 \sin(\Psi + 2\phi_{bo})}{\sin 2\phi_{bo}} - 2 = 0 \tag{9-3-1}$$

Using the trigonometric identity that

$$\sin x - \sin y = 2 \sin \frac{x-y}{2} \cos \frac{x+y}{2} \tag{9-3-2}$$

Eq. (9-3-1) becomes

$$\sin \frac{\Psi}{2} \cos \frac{\Psi + 4\phi_{bo}}{2} = 0 \tag{9-3-3}$$

Equation (9-3-3) is satisfied when either term is identically zero. Setting

TABLE 9-3-1 RANGE VERSUS V_{bo} AND Φ_{bo}

V_{bo} (m/s)	Q_{bo}	\multicolumn{5}{c}{Range (nmi) for Φ_{bo} (deg):}				
		15	30	45	60	75
1,000	0.0167	29	51	58	50	29
1,500	0.0376	67	116	133	114	65
2,000	0.0668	123	211	239	204	116
2,500	0.1043	200	340	381	321	182
3,000	0.1502	303	507	562	468	263
3,500	0.2045	437	722	786	645	359
4,000	0.2671	615	996	1,060	855	471
4,500	0.3380	851	1,342	1,391	1,099	598
5,000	0.4173	1,173	1,783	1,787	1,380	742
5,500	0.5050	1,624	2,346	2,258	1,700	901
6,000	0.6010	2,284	3,067	2,813	2,060	1,077
6,500	0.7053	3,300	3,988	3,458	2,462	1,269
7,000	0.8180	4,938	5,142	4,196	2,904	1,478
7,500	0.9390	7,522	6,530	5,022	3,386	1,702
8,000	1.0684	10,810	8,077	5,918	3,904	1,942
8,500	1.2061	13,620	9,636	6,854	4,451	2,198
9,000	1.3522	15,450	11,060	7,792	5,021	2,468
9,500	1.5066	16,590	12,250	8,696	5,603	2,751
10,000	1.6693	17,320	13,210	9,534	6,187	3,046
10,500	1.8405	17,820	13,980	10,290	6,762	3,352
11,000	2.0199	18,180	14,580	10,960	7,320	3,665

$\sin(\Psi/2) = 0$ means that Ψ is either 0 or 2π radians, neither of which is of practical interest. Therefore, the condition of interest must occur when

$$\cos\frac{\Psi + 4\phi_{bo}}{2} = 0 \qquad (9\text{-}3\text{-}4)$$

Since $\cos^{-1}(0) = \pi/2$, the relationship between the maximum free-fall range angle Ψ_{max} and the optimum elevation angle at burnout $\phi_{bo,opt}$ becomes

$$\boxed{\phi_{bo,opt} = \frac{\pi - \Psi}{4}} \qquad (9\text{-}3\text{-}5)$$

where both Ψ and $\phi_{bo,opt}$ are expressed in radians. Notice that *the optimum burnout elevation angle is single-valued*.

For a specified range angle Ψ, this particular burnout elevation angle requires the least amount of burnout energy, the smallest value of Q_{bo} (i.e., $Q_{bo,min}$). The relationship between $Q_{bo,min}$ and Ψ can be obtained by substituting Eq. (9-3-5) into either of the range equations and solving for Q_{bo} to obtain

$$\boxed{Q_{bo,min} = \frac{2\sin(\Psi/2)}{1 + \sin(\Psi/2)}} \qquad (9\text{-}3\text{-}6)$$

The third relationship of interest is the one that establishes the maximum range angle for a given Q_{bo} and which can be obtained by solving Eq. (9-3-6) for $\sin(\Psi/2)$. The result is

$$\sin\frac{\Psi_{max}}{2} = \frac{Q_{bo}}{2 - Q_{bo}} \qquad (9\text{-}3\text{-}7)$$

These three boxed equations establish the relationships required to maximize the range for a specified Q_{bo} or to minimize Q_{bo} for a specified range. Therefore, *the maximum-range trajectory is the minimum-energy trajectory* and the terms are interchangeable. Using the definition of Q_{bo}, these maximum-range (minimum-energy) relationships can be expressed in terms of appropriate combinations of V_{bo} and ϕ_{bo}.

Example 9-3-1

An ICBM has a $Q_{bo} = 0.6010$.
(a) Find the maximum range and corresponding burnout elevation angle.
(b) If $h_{bo} = 150$ nmi (276 km), what V_{bo} is required?
(c) Check these values with Table 9-3-1.
Solution
(a) Knowing Q_{bo}, Eq. (9-3-7) can be used to find Ψ_{max}.

$$\frac{\Psi_{max}}{2} = \sin^{-1}\frac{0.601}{2 - 0.601} = 0.4440 \text{ rad}$$

$$\Psi_{max} = 0.8880 \text{ rad} = 50.88°$$

$$R_{max} = 6378\Psi_{max} = 5663 \text{ km} = 3078 \text{ nmi}$$

Using Eq. (9-3-5) gives

$$\phi_{opt} = \frac{\pi - 0.888}{4} = 0.5634 \text{ rad} = 32.29°$$

(b) From the definition of Q_{bo}, with $r_{bo} = 6.654 \times 10^6$ m,

$$V_{bo} = \sqrt{\frac{Q_{bo}\,\mu}{r_{bo}}} = 6000 \text{ m/s}$$

(c) Table 9-3-1 shows a range of 3067 nmi for a $\phi_{bo} = 30°$, which is consistent with the results of part (a): 3078 nmi and 32.29°.

In the next example, we start with a fixed range and determine the optimum burnout angle and minimum required energy ($Q_{bo,min}$).

Example 9-3-2

You are doing a feasibility design of an ICBM with a range of 5500 nmi (10,120 km).
(a) Find the optimum burnout angle.
(b) Find the minimum energy required at burnout, $Q_{bo,min}$.
(c) If h_{bo} is 100 nmi (184 km), find V_{bo}.

Sec. 9-3 Trajectories and Sensitivities

(d) Using the value of Q_{bo} from part (b) and a range of 5500 nmi, find the appropriate value(s) of ϕ_{bo} and compare them with the value found in part (a).

Solution

(a)
$$\Psi = \frac{10{,}120}{6378} = 1.587 \text{ rad} = 90.92°$$

$$\phi_{bo,\,opt} = \frac{\pi - 1.587}{4} = 0.3886 \text{ rad} = 22.27°$$

(b)
$$Q_{bo,\,min} = \frac{2 \sin 0.7935}{1 + \sin 0.7935} = 0.8323$$

(c) With $r_{bo} = 6.562 \times 10^6$ m,

$$V_{bo} = \sqrt{\frac{0.8323\mu}{r_{bo}}} = 7110 \text{ m/s} = 23{,}310 \text{ ft/s}$$

(d) From Eq. (9-2-14),

$$2\phi_{bo} + 0.7935 = \sin^{-1}\left(\frac{2 - 0.8323}{0.8323} \sin 0.7935\right) = \frac{\pi}{2} \text{ rad}$$

so that

$$\phi_{bo} = 0.3886 \text{ rad} = 22.27°$$

Since there is only one value for ϕ_{bo} and it is identical with that for $\phi_{bo,\,opt}$, 5500 nmi is the maximum-range (minimum-energy) trajectory for $Q_{bo} = 0.8323$.

This example illustrates the fact that there is only one value of ϕ_{bo} for a maximum-range (minimum-energy) trajectory. *A maximum-range (minimum-energy) trajectory is unique.* Although there are other requirements, such as entry angle, that may be more important and may dictate the choice of the actual trajectory, this minimum-energy trajectory does establish a lower limit for Q_{bo}. If Q_{bo} is less than $Q_{bo,\,min}$, there is no trajectory that will meet the range requirement. If Q_{bo} is exactly equal to $Q_{bo,\,min}$, there is only one trajectory, the maximum-range (minimum-energy) trajectory. Finally, if Q_{bo} is greater than $Q_{bo,\,min}$ (and < 1), both high and low trajectories are possible.

With a symmetrical free-fall ballistic trajectory, the so-called reentry point may not be the substantive start of reentry and ϕ_{re}, which is equal to $-\phi_{bo}$, and h_{re} and V_{re}, which are equal to the burnout values, may not be the reentry conditions. The following example will indicate the differences in the values at this "reentry point" and at the edge of the sensible atmosphere.

Example 9-3-3

The ballistic missile of Example 9-3-2 has a range of 5000 nmi and the following burnout conditions: $V_{bo} = 7110$ m/s, $h_{bo} = 100$ nmi (184 km), and $\phi_{bo} = 22.27°$.

(a) What are the values at the symmetrical reentry point?
(b) If reentry is considered to start at an altitude of 76.25 km (41.44 nmi), find the new reentry conditions.

(c) How do the reentry conditions at 76.25 km compare with those at the symmetrical reentry point? Is it reasonable to assume reentry at the higher altitude?

Solution

(a) Since the free-fall trajectory is symmetrical, $V_{re} = V_{bo} = 7110$ m/s, $h_{re} = h_{bo} = 100$ nmi (184 km), and $\phi_{re} = -\phi_{bo} = -22.27°$.

(b) The characteristics of the ballistic trajectory must be determined before the "new" reentry conditions can be found. With $Q_{bo} = 0.8323$ and $r_{bo} = 6.562 \times 10^6$ m,

$$a = \frac{6.562 \times 10^6}{2 - 0.8323} = 5.620 \times 10^6 \text{ m}$$

$$E = \frac{-3.986 \times 10^{14}}{2a} = -3.456 \times 10^7 \text{ m}^2/\text{s}^2$$

$$H = V_{bo} r_{bo} \cos\phi_{bo} = 4.318 \times 10^{10} \text{ m}^2/\text{s}$$

$$\epsilon = \sqrt{1 + 0.8323(2 - 0.8323)\cos^2(22.27°)} = 0.3190$$

The eccentricity was found as a matter of interest and not necessity. At 76.25 km

$$V_{re} = \sqrt{2\left(E + \frac{\mu}{r_{re}}\right)} = 7253 \text{ m/s} = 23{,}800 \text{ ft/s}$$

$$\phi_{re} = \cos^{-1}\frac{H}{V_{re} r_{re}} = -22.73°$$

(c) At 76.25 km V_{re} is 2% larger than the symmetrical V_{re} and ϕ_{re} is 2% larger than the symmetrical ϕ_{re}. It seems reasonable, therefore, to assume the start of reentry at the symmetrical point, where $h = h_{bo}$.

The range equation shows the nominal range (angle) to be a function of the three burnout conditions. It would be worthwhile to know how sensitive the range (angle) is to errors in these burnout conditions. This can be done by expanding the range equation in a Taylor's series about the nominal burnout conditions and then truncating the series by assuming that the errors are sufficiently "small" that the higher-order differentials and products are on the order of zero; if the missile is to be effective, the errors must be "small." The truncated Taylor's series expansion about the burnout point can be written in general form with the differentials, such as dR, written as differences ΔR, so that

$$\Delta R \cong \frac{\partial R}{\partial V_{bo}} \Delta V_{bo} + \frac{\partial R}{\partial r_{bo}} \Delta r_{bo} + \frac{\partial R}{\partial \phi_{bo}} \Delta\phi_{bo} \qquad (9\text{-}3\text{-}8)$$

Using either of the range equations and evaluating the partial derivatives at the nominal burnout point, the respective derivatives can be written as

$$\frac{\partial \Psi}{\partial V_{bo}} = \frac{8\mu}{V_{bo}^3 r_{bo}} \frac{\sin^2(\Psi/2)}{\sin 2\phi_{bo}} \qquad \frac{\text{rad}}{\text{m/s}} \qquad (9\text{-}3\text{-}9)$$

$$\frac{\partial \Psi}{\partial r_{bo}} = \frac{4\mu}{V_{bo}^2 r_{bo}^2} \frac{\sin^2(\Psi/2)}{\sin 2\phi_{bo}} \qquad \frac{\text{rad}}{\text{m}} \qquad (9\text{-}3\text{-}10)$$

$$\frac{\partial \Psi}{\partial \phi_{bo}} = \frac{2 \sin(\Psi + 2\phi_{bo})}{\sin 2\phi_{bo}} - 2 \qquad \frac{\text{rad}}{\text{rad}} \qquad (9\text{-}3\text{-}11)$$

Sec. 9-3   Trajectories and Sensitivities   299

These partial derivatives are *sensitivity functions* that indicate the errors in range that would occur for deviations from the nominal (design) burnout conditions. They are often referred to as *influence coefficients* or *error coefficients* and occur in other disciplines and applications. Remember that the burnout values used in these functions are the nominal burnout values.

Let us discuss each sensitivity function in turn and apply it to an illustrative missile that has both a high and a low trajectory, such as the missile in Example 9-2-2.

Considering the velocity error function of Eq. (9-3-9), it is positive for $\phi_{bo} < 90°$, the usual case. Consequently, if the burnout velocity is greater than intended, the payload will overshoot the target, be long. Conversely, a lower velocity will undershoot the target, fall short. The range error for a velocity error is found from

$$\Delta R = r_E \Delta \Psi = r_E \frac{\partial \Psi}{\partial V_{bo}} \Delta V_{bo} \qquad (9\text{-}3\text{-}12)$$

Example 9-3-4

The ICBM of Example 9-2-2 has a design range of 5500 nmi (10,120 km) with $\Psi = 90.92°$ (1.587 rad) and a nominal burnout point where $V_{bo} = 7168$ m/s and $h_{bo} = 150$ nmi (276 km) ($r_{bo} = 6.654 \times 10^6$ m). The low trajectory has $\phi_{bo} = 13.10°$ and the high trajectory, $\phi_{bo} = 31.43°$.

(a) Find the value of the velocity error function for the low trajectory for a $\Delta V_{bo} = 1$ m/s along with the magnitude of the range error.
(b) Do part (a) for the high trajectory.
(c) Which trajectory has the lower error?

Solution

(a) For the low trajectory ($\phi_{bo} = 13.10°$):

$$\frac{\partial \Psi}{\partial V_{bo}} = \frac{8 \times (3.986 \times 10^{14})}{(7168)^3 \times (6.654 \times 10^6)} \frac{\sin^2(45.46°)}{\sin(26.2°)} = 1.497 \times 10^{-3} \frac{\text{rad}}{\text{m/s}}$$

$$\Delta R = 6378 \times (1.497 \times 10^{-3}) \times 1 = 9.55 \text{ km} = 5.19 \text{ nmi}$$

(b) For the high trajectory ($\phi_{bo} = 31.43°$):

$$\frac{\partial \Psi}{\partial V_{bo}} = \frac{8 \times (3.986 \times 10^{14})}{(7168)^3 \times (6.654 \times 10^6)} \frac{\sin^2(45.46°)}{\sin(62.86°)} = 7.428 \times 10^{-4} \frac{\text{rad}}{\text{m/s}}$$

$$\Delta R = 6378 \times (7.428 \times 10^{-4}) \times 1 = 4.738 \frac{\text{km}}{\text{m/s}} = 2.575 \frac{\text{nmi}}{\text{m/s}}$$

(c) The high trajectory has the lower range error for a specified velocity error. Note that a velocity error of 1 m/s is only 1.313×10^{-4} of the nominal burnout velocity of 7168 m/s.

In the plane of the free-fall trajectory, an error in r_{bo} is actually an error in h_{bo}, the burnout altitude. The form of the altitude error function of Eq. (9-3-10) is similar to that of the velocity error function in that it is positive for elevation angles less than 90°, an increase in the height of burnout produces an overshoot of the target. The range error for an altitude error is

$$\Delta R = r_E \Delta \Psi = r_E \frac{\partial \Psi}{\partial r_{bo}} \Delta h_{bo} \qquad (9\text{-}3\text{-}13)$$

Example 9-3-5

For the ICBM of Example 9-3-5:
(a) Find the value of the altitude error function and the range error for the low trajectory with an altitude error of 50 m ($\Delta r_{bo} = \Delta h_{bo} = 50$ m).
(b) Do part (a) for the high trajectory.
(c) Which trajectory has the lower range error?

Solution

(a) For the low trajectory ($\phi_{bo} = 13.1°$):

$$\frac{\partial \Psi}{\partial r_{bo}} = \frac{4 \times (3.896 \times 10^{14})}{(7168)^2 \times (6.654 \times 10^6)^2} \frac{\sin^2 45.46}{\sin 26.2} = 8.065 \times 10^{-7} \frac{\text{rad}}{\text{m}}$$

$$\Delta R = 6378 \times (8.065 \times 10^{-7}) \times 50 = 257.2 \text{ m} = 0.1398 \text{ nmi}$$

For every error of 1 m in the burnout altitude, there will be a 5.144 m error in range.

(b) For the high trajectory ($\phi_{bo} = 31.43°$):

$$\frac{\partial \Psi}{\partial r_{bo}} = 8.065 \times 10^{-7} \frac{\sin 26.2}{\sin 62.86} = 4.0 \times 10^{-7} \frac{\text{rad}}{\text{m}}$$

$$\Delta R = 6378 \times (4.0 \times 10^{-7}) \times 50 = 127.6 \text{ m} = 0.0693 \text{ nmi}$$

For every meter error in the burnout altitude, there will be a 2.55 m error in range.

(c) As with the velocity error, the high trajectory has the lower range error. Note that a Δr_{bo} of 50 m is only 1.82×10^{-3} of the nominal h_{bo} of 276 km, 1820 parts in a million.

The elevation angle error function of Eq. (9-3-13) should be familiar inasmuch as it is the expression of Eq. (9-3-1), which was used to find the value of ϕ_{bo} for maximum range. The range error for an error in the burnout angle is

$$\Delta R = r_E \Delta \Psi = r_E \frac{\partial \Psi}{\partial \phi_{bo}} \Delta \phi_{bo}$$

Example 9-3-6

Do Example 9-3-5 for a $\Delta \phi_{bo} = 1°$.

Solution

(a) For the low trajectory ($\phi_{bo} = 13.1°$):

$$\frac{\partial \Psi}{\partial \phi_{bo}} = \frac{2 \sin(90.2° + 26.2)}{\sin(26.2°)} - 2 = 2.058 \frac{\text{rad}}{\text{rad}} = 0.0359 \frac{\text{rad}}{\text{deg}}$$

$$\Delta R = 6378 \times 0.0359 \times 1 = 229 \text{ km} = 124.5 \text{ nmi}$$

(b) For the high trajectory ($\phi_{bo} = 31.43°$):

$$\frac{\partial \Psi}{\partial \phi_{bo}} = \frac{2 \sin(90.2° + 62.86°)}{\sin 62.86} - 2 = -0.982 \frac{\text{rad}}{\text{rad}} = -0.0171 \frac{\text{rad}}{\text{deg}}$$

$$\Delta R = 6378 \times -0.0171 \times 1 = -109 \text{ km} = -59.27 \text{ nmi}$$

Notice the sign change in the sensitivity function and thus a sign in the range error.

(c) As with the other two cases, the high trajectory has the lower error, but with this error function the sign for the two trajectories is different. Since it is negative for the high trajectory, a positive $\Delta\phi_{bo}$ results in a negative ΔR, an undershoot or a "short." The positive sign for the low trajectory means that a positive $\Delta\phi_{bo}$ will result in an overshoot of the target or a "long."

These influence coefficients clearly indicate how sensitive the range performance of a ballistic missile is to errors in the design burnout conditions. The rule of thumb in the early ballistic missile days was that for an ICBM (intercontinental ballistic missile) with a range of 5500 nmi, an error of 1 ft/s in the burnout velocity (approximately 23,500 ft/s) resulted in a range error of 1 nmi. As an aside of interest, an error of 1 lb in the burnout weight also resulted in an approximate range error of 1 nmi. For the IRBM (intermediate-range ballistic missile) with a 1500-nmi range, the trade-off penalties were much smaller, 1 nmi for an error of 5 ft/s in V_{bo} and 1 nmi for a 5-lb error in the burnout weight.

To add to the visualization of the high sensitivity of an ICBM to the burnout conditions, consider a missile with a range of 5000 nmi using a high trajectory. If the maximum miss distance (range error) is to be ± 1 km, then ΔV_{bo} should be less than 0.3 m/s, Δh_{bo} should be less than 500 m, and $\Delta\phi_{bo}$ should be less than $0.01°$.

In addition to errors in the magnitudes of the burnout conditions, there can be errors in the location of the burnout point (its latitude and longitude) and in the azimuth of the nominal plane of the trajectory. Errors in the latitude and longitude of the burnout point are usually resolved into a downrange displacement error ΔX_{bo} and a cross-range displacement error ΔY_{bo}.

With a down-range displacement error ΔX_{bo}, Ψ is not affected or involved and

$$\Delta R = \Delta X_{bo} \qquad (9\text{-}3\text{-}14)$$

where ΔR represents a down-range miss distance rather than a change in range or the range angle Ψ.

With a cross-range displacement error ΔY_{bo}, the effect is to displace the impact point along a great circle arc. It can be shown that

$$\Delta R \cong \Delta Y_{bo} \cos \Psi \qquad (9\text{-}3\text{-}15)$$

Note that with $\Psi = 90°$ (or $270°$), ΔR becomes small, going to zero in the limit.

A burnout azimuth error $\Delta\beta_{bo}$ is the angular displacement of the trajectory plane from the intended plane and manifests itself as an angular displacement error ΔC (a great circle arc). Applying the spherical law of cosines yields the approximation

$$\Delta C \cong \Delta\beta_{bo} \sin \Psi \qquad (9\text{-}1\text{-}16a)$$

and

$$\Delta R = r_E \Delta C = r_E \Delta\beta_{bo} \sin \Psi \qquad (9\text{-}3\text{-}16b)$$

When $\Psi = 180°$, ΔC approaches zero but when $\Psi = 90°$, ΔC becomes a maximum.

Since the range of most ICBMs is on the order of 5500 nmi (10,000 km), Ψ is on the order of $90°$. Consequently, the effect of a cross-range displacement error is minimized, but the effect of a burnout azimuth error is maximized.

Example 9-3-7

An ICBM with a 6600-nmi (12,140-km) range and a design burnout azimuth angle of 85° has an actual $\beta_{bo} = 86°$ and misses its burnout point with a down-range displacement of -0.35 km (actually an uprange displacement) and a cross-range displacement of 0.67 km to the left. What are the corresponding miss distances (km and nmi) at the target?

Solution The design range angle is

$$\Psi = \frac{12{,}140}{6378} = 1.903 \text{ rad} = 109.1°$$

The downrange miss distance with $\Delta X_{bo} = -0.85$ km is

$$\Delta R = -0.35 \text{ km} = -0.190 \text{ nmi (a short)}$$

The cross-range displacement miss distance with $\Delta Y_{bo} = -0.67$ km is

$$\Delta R = -0.67 \cos(109.1°) = -0.219 \text{ km} = -0.119 \text{ nmi}$$

This is an impact to the left of the target.

Since the burnout azimuth error is $\Delta \beta_{bo} = +1° = +0.01745$ rad, the resulting miss distance is

$$\Delta R = r_E \Delta C = 6378 \times 0.01745 \times \sin(109.1°) = 105.2 \text{ km} = 57.2 \text{ nmi}$$

with impact to the right of the target.

Although the probability of any weapon (gun, bomb, rocket, or missile) hitting a target is generally expressed as the circular error probable (CEP), the footprint of a ballistic missile tends to be elliptical rather than circular with the major axis down-range. The elliptical shape is a consequence of the fact that the errors in the burnout variables, V_{bo}, h_{bo}, and ϕ_{bo} all result in misses up- and downrange and the primary source of the cross-range error is the burnout azimuth error. It is possible to calculate an elliptical error probable (EPE) and convert it into an artificial CEP to be used in a comparison of the effectiveness of ballistic missiles with other delivery systems.

9-4 THE ROTATING EARTH

A ballistic missile trajectory is a trajectory in an inertial reference frame that has its origin at the center of a nonrotating Earth; the burnout velocity, for example, is measured with respect to the center of the Earth and not with respect to its surface. Since the launch and target points are located on the surface of the rotating Earth, they also have velocities with respect to the center of the Earth (i.e., they have their own individual inertial velocities). It is their motion that modifies the ballistic trajectory and introduces the need for a lead angle.

The inertial velocity of the launch point is in an easterly direction and has its maximum at the equator. Depending on the direction of launch, this velocity will either decrease or increase, depending on the direction of launch, the velocity that must be imparted to the missile to attain the necessary burnout velocity. In other

Sec. 9-4 The Rotating Earth

words, the motion of the launch point affects the launch velocity, as has been discussed in Chapter 3.

At the time of launch, the target has a fixed range angle (and range) and azimuth angle with respect to the launch point. These are the angles of the preceding sections and they are constant since the launch and target points are moving together. However, after the missile lifts off and enters its inertial trajectory, the target continues to move with respect to the center of the Earth and is in a different inertial location at the end of flight. In other words, the range angle is no longer a constant function of the initial locations of the launcher and target; the *range angle is a function of the time of flight*. To compensate for the target motion, the burnout elevation and azimuth angles must be adjusted to provide the appropriate lead angle. The burnout velocity and altitude (and thus Q_{bo}) are unaffected.

With the definition that Λ is the *total* range angle and that t_Λ is the *total* time of flight, the expression for Λ, from the law of cosines of spherical trigonometry, is

$$\cos \Lambda = \sin L_0 \sin L_T + \cos L_0 \cos L_T \cos(\Delta N \pm \omega_E t_\Lambda) \qquad (9\text{-}4\text{-}1)$$

In Eq. (9-4-1), L_0 is the latitude of the launch point, L_T the latitude of the target at time of launch, and ΔN the difference in the terrestial longitude of the two points at the time of launch. ΔN must be measured in an easterly direction from the launch point to the target point and can have a value between 0 and 360° (0 and 2π radians).

In Eq. (9-4-1), $\omega_E t_\Lambda$ is a term to correct the difference in longitude (and thus the range angle) for the motion of the target along its parallel of latitude. ω_E is the Earth's rotational angular velocity, which at 15° per hour is 7.272×10^{-5} rad/s or 4.167×10^{-3} deg/s. As a reminder, the inertial velocity of the surface of Earth to the east at the equator is 463.8 m/s (1520 ft/s) to the east.

The correction term $\omega_E t_\Lambda$, has a ± sign in front of it. The choice of the sign is dependent on the direction of launch. If the missile is launched to the east, the target will move away from the burnout point in inertial space, thus increasing the range angle; use the + sign. If launch is to the west, the target will move toward the burnout point, decreasing the range angle; use the − sign.

Since the solution of Eq. (9-4-1) involves an arccosine function, there will be two solutions for Λ. One solution lies between 0 and 180° (0 and π radians) and represents the short-way great circle path to the target. The other solution, which lies between 180 and 360° (π and 2π radians) represents the long-way trajectory. Inasmuch as the long-way trajectory is generally in a westerly direction (requiring a larger launch velocity) and has a longer time of flight (in spite of the negative longitudinal correction term), the short-way path customarily represents a preferred solution.

Without going further, examination of Eq. (9-3-1) clearly shows another effect of the Earth's rotation. Since the range is now a function of the time of flight and since the high and low trajectories have different times of flight, with a specified Q_{bo} *the range of a high trajectory is not equal to the range for a low trajectory* ($\Lambda_{high} \neq \Lambda_{low}$ and $\Psi_{high} \neq \Psi_{low}$). Since the ranges have changed, so will the burnout elevation angles.

Applying the law of cosines of spherical trigonometry again, with the azimuth angle β as the included angle, and solving for β yields

$$\cos \beta = \frac{\sin L_T - \sin L_0 \cos \Lambda}{\cos L_0 \sin \Lambda} \qquad (9\text{-}4\text{-}2)$$

Once the corrected value of Λ is known, it is a simple matter to calculate the new β, where $\Delta\beta$ is the lead angle. Since β is measured in a clockwise direction from a meridian, an increase represents a change in the clockwise direction. Whether β increases or decreases is a function of the range problem and the value of Λ. Since the solution of Eq. (9-4-2) also involves an arccosine function, there will be two solutions for β, which are linked to Λ and the value of $\Delta N \pm \omega_E t_\Lambda$. With Λ between 0 and 180°, β will lie between 0 and 180° if $\Delta N \pm \omega_E t_\Lambda$ also lies between 0 and 180°.

Finding the corrected value of Λ is unfortunately not simple, since Eq. (9-4-1) is transcendental and requires an iterative solution or a graphical solution to determine Λ. Furthermore, since there is an infinite number of trajectories that will pass through the launch and target points, constraints and operational requirements are needed and used to define acceptable trajectories.

An example will be used to demonstrate the use of Eqs. (9-4-1) and (9-4-2) and the interaction between Λ and t_Λ and their influence on β. To simplify the problem without losing any substance, assume that Ψ and Λ are equivalent to avoid the intrusion of the powered and reentry phases. Consequently, t_Λ is the free-fall TOF of the preceding sections. In the example that follows, Q_{bo} is set at 0.80. This is not a good choice; it is on the low side for the launch–target combination of the problem, being too close to the value of $Q_{bo, min}$, but was chosen for convenience as its TOF curve is given in Fig. 9-2-5 and the next-higher value of Q_{bo} is too high.

Example 9-4-1

A launch point is located at 30° N and 60° W and its target is located at 40° N and 40° E. Q_{bo} is to be constant and equal to 0.80 with h_{bo} = 305 km (165.8 nmi). Neglect the powered-boost and reentry phases so that $\Psi \cong \Lambda$.
(a) Neglecting the Earth's rotation find Λ and β for the short-way trajectory.
(b) Using Fig. 9-2-5, find the approximate times of flight (t_Λ) for the high and low trajectories, still with no rotation.
(c) Now include the Earth's rotation and use a combination of iteration and Fig. 9-2-5 to find Λ for the high trajectory. Find β and R.
(d) With $\Psi \cong \Lambda$, find the effect of the Earth's rotation on the burnout elevation angle.
(e) Without the correction for the Earth's rotation, what would have been the miss distance at the target?

Solution

(a)
$$\Delta N = 40 - (-60) = 100°$$
$$\cos \Lambda_0 = \sin(30°)\sin(40°) + \cos(30°)\cos(40°)\cos(100°) = 0.2062$$

and
$$\Lambda_0 = 78.1° \text{ or } 281.9°$$

Sec. 9-4 The Rotating Earth

With $\Lambda = 78.1°$ (1.363 rad), the short-way trajectory,

$$\cos\beta = \frac{\sin(40°) - \sin(30°)\cos(78.1°)}{\cos(30°)\sin(78.1°)} = 0.6369$$

$$\beta = 50.44°$$

and

$$R = 6378 \times 1.363 = 8693 \text{ km} = 4724 \text{ nmi}$$

(b) Approximating the free-fall range angle Ψ by Λ, the approximate times of flight for the short-way trajectory can be found from Fig. 9-2-5 (rather than by calculating the two ϕ_{bo}'s and calculating the respective TOFs). The order of magnitude of t_Λ for the low trajectory is 1800 s (30 min) and 2400 s (40 min) for the high trajectory. The two times are fairly close together because, as can be seen in the figure, Q_{bo} is not much larger than the value for the minimum-energy trajectory.

(c) With $\omega_E = 4.167 \times 10^{-3}$ deg/s and 2400 s as $t_{\Lambda 1}$, the first trial value for the high trajectory.

$$\omega_E t_\Lambda = 10.0° \text{ to the east}$$

Now

$$\cos\Lambda_1 = \sin(30°)\sin(40°) + \cos(30°)\cos(40°)\cos(100° + 10.0°) = 0.0945$$

(notice that the + sign was used with $\omega_E t_\Lambda$ because of the launch in an easterly direction) and

$$\Lambda_1 = 84.58° \quad \text{or} \quad 275.42°$$

If the short-way Λ_1 is correct, $\Lambda_1 - \Lambda_0$ should be approximately equal to $\omega_E t_\Lambda \cos\beta$ (or 6.37°) and

$$84.58 - 78.1 = 6.48°$$

The next step is to find the time of flight for Λ_1 (Ψ_1) = 84.58° and use this value with Λ_1 to solve for Λ_2, to be compared with Λ_1.

Returning to Fig. 9-2-5, the time of flight (within the limits of accuracy with this plot) for this new value of Λ (Ψ) has decreased to 2000 s, so that $\omega_E t_\Lambda = 8.334°$. Solving for Λ_2, its value is found to be 83.5° so that

$$\Lambda_2 - \Lambda_1 = 83.5 - 84.58 = -1.04°$$

The difference is not equal to zero but it is decreasing, indicating that the process is converging. Entering Fig. 9-2-5 with $\Psi = 83.5$ gives the new time to be 2100 s, $\omega_E t_\Lambda = 8.75°$ and $\Lambda_3 = 83.78°$. Now

$$\Lambda_3 - \Lambda_2 = 83.78 - 83.5 = 0.28°$$

This difference of 0.28° is sufficiently small to indicate convergence for the purpose of this example. Therefore, the solution with consideration of the Earth's rotation is

$$\Lambda = 83.78° = 1.462 \text{ rad}$$

$$R = r_E \Lambda = 9325 \text{ km} = 5068 \text{ nmi}$$

$$\beta_{bo} = 46.87°$$

Notice that the Earth's rotation and launch in an easterly direction has increased the range angle (and thus the range) as well as the burnout azimuth angle. The range over the surface of a rotating Earth is 9325 km = 5068 nmi as compared to the nonrotating range of 4724 nmi (a 7.3% increase).

(d) With $\omega_E = 0$, $\Psi = 78.1°$, and $Q_{bo} = 0.80$,

$$2\phi_{bo} + 39.05° = \sin^{-1}\left[\frac{2 - 0.8}{0.8}\sin(39.05°)\right] = 70.91°, 109.1°$$

For the high trajectory,

$$\phi_{bo,\,high} = 35.0°$$

With Earth's rotation, $\Psi = 83.78°$ and Q_{bo} is still 0.80, so that

$$2\phi_{bo} + 41.89° = \sin^{-1}\left[\frac{2 - 0.8}{0.8}\sin(41.89°)\right] = 90°$$

$$\phi_{bo,\,high} = 24.06°$$

The burnout elevation angle has decreased from 35.0° to 24.06° to accommodate the increase in range (angle) associated with the movement of the target away from the payload during the time it takes the payload to reach the target. The azimuth angle β has increased from 43.9° to 46.85° to accommodate the shift in the trajectory plane. The earlier statement that a Q_{bo} of 0.8 may be too low for this mission is substantiated by the fact that the argument of the arcsine in the expression for ϕ_{bo} could well be greater than unity, indicating that there was no elevation angle that would produce the required rotating-Earth range.

(e) The inertial velocity of the target V_T along its parallel of latitude is

$$V_T = 463.8 \cos L_T = 463.8 \cos(40°) = 355.3 \text{ m/s}$$

where 463.8 m/s is the inertial velocity of the surface of a spherical Earth at the equator. If, with no correction for a rotating Earth, the TOF is 2400 s, the miss distance would be

$$\Delta R = 355.3 \times 2400 = 8.527 \times 10^5 \text{ m} = 852.7 \text{ km} = 463.4 \text{ nmi}$$

This is not a trivial miss.

To find the changes in the low trajectory to accommodate the Earth's rotation, it is necessary to repeat the iteration in conjunction with the time of flight plot. Notice that consideration of the Earth's rotation introduces the need to know the absolute locations (latitude and longitude) of the launch and target points, not just the great circle distance between them.

The solution of the rotating Earth range problem of this example was accomplished with a combination of a graph and a manual iteration. It could have been solved by graphical techniques alone by superimposing on the graphs of Fig. 9-2-5 an appropriate plot of Eq. (9-4-1)[2] and finding the point(s) of intersection for the specified Q_{bo}. Two points, the normal situation, would identify the corrected high and low trajectories. A single intersection, at the inflection point of the appropriate

---

[2] The dashed line in Fig. 9-2-5 is the plot for Example 9-4-1.

curve of Fig. 9-2-5, would indicate that the solution is a single minimum-energy trajectory. No intersection(s) means that the specified $Q_{bo} < Q_{bo,\,min}$ and that there is no trajectory for that value of Q_{bo} that would hit the moving target. Rather than a graphical solution, it is obviously possible to use a computer to link the time-of-flight computation to the solution of the moving Earth range angle and to solve them simultaneously by iteration.

The sensitivity of the range of a ballistic missile to minor changes in the burnout conditions emphasizes the importance of the guidance and control (G&C) system, which has the task of directing the missile to the correct burnout point in the presence of atmospheric winds and other perturbations. One guidance scheme is known as "flying the wire," in which the objective is to reach the predetermined burnout point by following a specified boost trajectory. If the missile wanders off this trajectory for any reason, the G&C system returns it to the trajectory and strives to make the actual and design burnout points coincide.

Another scheme is the "velocity to go" guidance and control system, which also has a design trajectory and burnout point but does not attempt to return the missile to this trajectory if it should stray. Instead, the system computes a new burnout point and a new burnout velocity vector. The G&C system then subtracts the current velocity vector to obtain the "velocity to go" vector and devotes its efforts to driving this vector to zero. When the "velocity to go" vector becomes zero, the missile has arrived at the computed-on-the-run burnout point and the propulsion system is shut down.

We close this section and chapter with the comment that an understanding of the complete ballistic missile trajectory, to include boost and reentry, can provide an excellent foundation and basis for understanding how to get into and out of space; it does not help us to get around in space, however.

PROBLEMS

9-1. An IRBM with a design range of 1400 nmi (2760 km) has $h_{bo} = 100$ nmi (184 km).
  (a) Find V_{bo} and Q_{bo} if $\phi_{bo} = 15°$.
  (b) If $V_{bo} = 5500$ m/s, find the possible values of ϕ_{bo}.
  (c) Do part (b) with $V_{bo} = 6200$ m/s.

9-2. A ballistic missile has $Q_{bo} = 0.65$.
  (a) Find V_{bo} for $h_{bo} = 75$ nmi (138 km); for 100 nmi (184 km); for 125 nmi (230 km).
  (b) With $h_{bo} = 100$ nmi and $\phi_{bo} = 25°$, find the range (km and nmi) and the apogee altitude.
  (c) Do part (c) with $\phi_{bo} = 45°$ and compare the results.

9-3. An ICBM with a range of 6000 nmi (11,040 km) has a $Q_{bo} = 1.1$ and an $h_{bo} = 150$ nmi (276 km).
  (a) Find the mathematically possible ϕ_{bo} values for the short-way trajectory. Are both physically possible?
  (b) Find the trajectory characteristics: E, a, H, ϵ, h_a, and h_p.
  (c) Do parts (a) and (b) for the long-way trajectory and compare with the results above.

9-4. An ICBM has a burnout altitude of 125 nmi (230 km), $V_{bo} = 7300$ m/s, and $\phi_{bo} = 24°$.
  (a) Find Q_{bo} and discuss its significance.
  (b) Find the range angle and range and the characteristics of the trajectory.
  (c) For the range of part (b), is there another possible ϕ_{bo} and another trajectory? If so, compare with that of part (b).

9-5. A missile burns out at 175 nmi (322 km) with $V = 8300$ m/s and $\phi_{bo} = 3°$.
  (a) Find the range angle and range along with h_a and h_p.
  (b) What type of a trajectory is that of part (a), and what is the significance of the results of part (a)?

9-6. An IRBM with a range of 1500 nmi (2760 km) has $h_{bo} = 100$ nmi (184 km) and $V_{bo} = 6200$ m/s.
  (a) Find the values of ϕ_{bo} for possible trajectories.
  (b) Find the time of flight (TOF) in minutes for the low trajectory.
  (c) Find the TOF for the high trajectory and compare with that of part (b).

9-7. A ballistic missile has a $Q_{bo} = 0.65$, $h_{bo} = 100$ nmi (184 km), and $\phi_{bo} = 25°$.
  (a) Find V_{bo} and the range.
  (b) Find the TOF (min).

9-8. For the ICBM of Problem 9-3, find the TOFs of all the possible trajectories of that problem.

9-9. The ICBM of Problem 9-4 has $h_{bo} = 125$ nmi (230 km) and $V_{bo} = 7300$ m/s.
  (a) Find the short-way range and time of flight for $\phi_{bo} = 24°$.
  (b) Find the other short-way ϕ_{bo} for the range of part (a) and find its TOF.

9-10. A submarine-launched ballistic missile is being designed for a 6000 km (3260 nmi) range with $h_{bo} = 100$ nmi (184 km).
  (a) Find the minimum value of Q_{bo} to accomplish this mission along with the corresponding value of V_{bo}.
  (b) Find the value of the burnout elevation angle, $\phi_{bo,opt}$.
  (c) Find the TOF.
  (d) Find the minimum-energy range.

9-11. An IRBM is being designed for a 1500-nmi range with $h_{bo} = 90$ nmi.
  (a) Find the minimum value of V_{bo} and the associated Q_{bo}.
  (b) What is the time of flight for this trajectory?

9-12. A ballistic missile has a burnout velocity and altitude of 7680 m/s and 135 nmi (248.4 km).
  (a) Find the maximum range and the associated ϕ_{bo} and V_{bo}.
  (b) Find the minimum-energy range and associated values.

9-13. Do Example 9-4-1 for the low trajectory and compare the results.

CHAPTER 10

Attitude Dynamics and Control

Artist's concept of the Galileo spacecraft after deployment of antennas and booms. The Galileo trajectory includes a Venus flyby (gravity assist maneuver) and two Earth flybys. (Courtesy of NASA.)

10-1 INTRODUCTION

A rigid body has six degrees of freedom: three translational modes and three rotational modes. The translational modes describe the motion of the center of mass of the body and the three rotational modes describe the attitude of the body. In the preceding chapters, the spacecraft (or entry vehicle) was assumed to be a point mass, a rigid body with its mass concentrated at the center of mass, and the attitude was explicitly ignored. There was, however, mention of the need to maintain a zero angle of attack during powered boost through the atmosphere and to acquire the requisite angle of attack (zero for ballistic entries) prior to entry into an atmosphere and the importance of the appropriate direction of the thrusting ΔV's for trajectory modification was stressed.

Implicit in the neglect of attitude in these chapters are the assumptions that the spacecraft (or E/V) is in a state of stable equilibrium along its trajectory and is not subjected to applied torques or disturbances. Neither assumption is necessarily nor usually valid, particularly for spacecraft in planetocentric orbits. Usually, spacecraft are inherently unstable and are subjected to torques engendered by the interaction of the spacecraft with its environment and by movement within the vehicle itself. Although the forces associated with these torques are not normally large enough to affect the trajectory of a spacecraft, the torques themselves can significantly affect the attitude of the spacecraft.

Environmental (external) torques for a planetocentric satellite can arise from solar radiation pressure, planetary gravitational and magnetic fields, and free molecular forces (if the satellite is sufficiently close to a planetary atmosphere). There may also be impact effects from meteorites and dust particles. Torques arising from internal movement can be generated by the shift of payloads and by the movements and actions of occupants as well as by the sloshing of propellants and other liquids.

If a spacecraft in orbit is unstable, it will tumble if disturbed; if it is marginally stable (undamped), it will oscillate about an equilibrium condition; and if it is stable (with external or internal damping), it will return to the equilibrium condition after any transients die out. The rotational response of a spacecraft to a disturbance torque is sometimes referred to as *libration*.

Although an unstable spacecraft is unacceptable, a marginally stable spacecraft may be acceptable if the amplitude and frequency of the oscillations are small; a dynamically stable spacecraft is the design goal. Stabilization techniques may be classified as *passive* or *active*. Since active stabilization devices, such as propulsive thrusters, momentum gyros, and reaction (momentum) wheels, require the expenditure of energy and increase the weight of a spacecraft, they can be expensive; furthermore, if the energy supply is exhausted, the attitude control system fails, the spacecraft tumbles, and the mission is aborted.

In addition to the stabilization problem, there may well be a separate pointing problem, generally an active control problem, in which the spacecraft or a piece of equipment in the spacecraft must be pointed in a specific direction. Consider an Earth-exploration satellite in an LEO with cameras and infrared (IR) sensors that must always be aimed at the surface of the Earth. There may be missions to track the Sun or other celestial bodies and navigational missions in which antennas must maintain or acquire required orientations. There is always the need for proper pointing prior to ΔV thrusting.

Although, generally, this chapter is restricted to spacecraft in geocentric orbits, the treatment can be extended to orbits about other planets and other bodies, such as moons and asteroids, and to interplanetary transfers. The relevant Newtonian equations of motion for a rigid body are developed in the next section to include gyroscopic effects. In the two sections following the equations of motion we discuss single-spin and dual-spin stabilization, respectively, to be followed by a section on gravity-gradient stabilization, a good example of passive stabilization. In the concluding section we discuss the use of active control (and sensors) to stabilize (maintain a reference) and to point. The emphasis throughout this chapter is on the qualitative behavior of a spacecraft and not necessarily on the quantitative details. Consequently, there are no problems at the end of this chapter; there are, however, examples in several of the sections that can be reworked to serve as problems.

This introductory section concludes with the remark that there are no truly rigid bodies, especially lightweight spacecraft, which are characterized by booms, antennas, protruding sensors, and other appendages. Furthermore, spinning bodies with imperfections of elasticity will dissipate energy in a manner that may well change the orientation of the body and even the spin axis, possibly leading to instability.

10-2 THE EQUATIONS OF MOTION

In its most general form, Newton's law governing the linear momentum and translational modes of a continuous system can be written in vector form as

$$\mathbf{F} = \int \mathbf{a}_I \, dm = \frac{d\mathbf{Q}}{dt}\bigg|_I \tag{10-2-1}$$

where $\mathbf{F}$ is the vector resultant of all the applied forces, $\mathbf{a}_I$ is the acceleration of a particle mass dm with respect to inertial space, and the integration is performed over the entire system (mass) of particles. This equation is applicable to all continuous systems, to both rigid and nonrigid bodies. In this equation $\mathbf{Q}$ is the linear momentum of the system and is defined as

$$\mathbf{Q} = \int \mathbf{V}_I \, dm$$

where $\mathbf{V}_I$ is the inertial velocity of the particle dm and the indicated differentiation must be performed with respect to an inertial reference frame.

Figure 10-2-1 shows the relationship of a particle mass dm to two sets of axes: $Oxyz$ is a set of moving axes fixed to the body (the system of particles) of interest and rotating with the body and $O_I X_I Y_I Z_I$ is an inertial reference frame. $\mathbf{a}_I$, the inertial acceleration of the particle, can be found from the expressions

$$\mathbf{a}_I = \frac{d\mathbf{V}_I}{dt} = \frac{d^2 \mathbf{r}_I}{dt^2} \tag{10-2-2}$$

where $\mathbf{V}_I$ is the inertial velocity. $\mathbf{r}_I$, the inertial position vector, can be written as

$$\mathbf{r}_I = \mathbf{r}_o + \mathbf{r} \tag{10-2-3}$$

$\mathbf{r}_o$ is the inertial position vector of the origin of the moving reference frame and in terms of inertial (nonrotating) unit vectors and components is

$$\mathbf{r}_o = X_I \mathbf{i}_I + Y_I \mathbf{j}_I + Z_I \mathbf{k}_I \tag{10-2-4a}$$

where the unit vectors are nonrotating. $\mathbf{r}$ is the position vector of the particle with respect to the origin of the moving reference frame and in terms of the components along the moving reference unit vectors can be written as

$$\mathbf{r} = x\mathbf{i} + y\mathbf{j} + z\mathbf{k} \tag{10-2-4b}$$

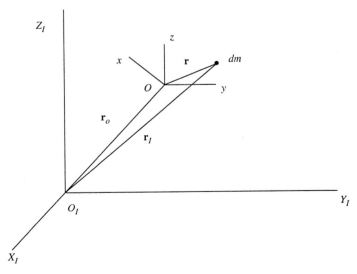

Figure 10-2-1. Relationships of a particle mass to inertial and moving reference frames.

Sec. 10-2 The Equations of Motion

Remembering that the time derivative of an inertial unit vector is zero and that the time derivative of a rotating unit vector is the cross product of the angular velocity and the unit vector itself, the first derivative of Eq. (10-2-3) shows the inertial velocity $\mathbf{V}_I$ to be

$$\mathbf{V}_I = \mathbf{V}_o + \mathbf{V}_r + \boldsymbol{\omega}_m \times \mathbf{r} \tag{10-2-5}$$

where $\mathbf{V}_o$ is the velocity of point O with respect to inertial space, $\mathbf{V}_r$ is the relative velocity of dm with respect to the moving reference frame, and $\boldsymbol{\omega}_m$ is the vector angular velocity of the moving reference frame with respect to the inertial reference frame. Although $\boldsymbol{\omega}_m$ is an inertial angular velocity, it can be resolved into components about the moving axes so that

$$\boldsymbol{\omega}_m = \omega_x \mathbf{i} + \omega_y \mathbf{j} + \omega_z \mathbf{k} \tag{10-2-6}$$

where $\mathbf{i}$, $\mathbf{j}$, and $\mathbf{k}$ are unit vectors of the moving axes.

Differentiation of Eq. (10-2-5) yields an expression for $\mathbf{a}_I$, the inertial velocity of the particle mass, namely,

$$\mathbf{a}_I = \mathbf{a}_o + \mathbf{a}_r + \boldsymbol{\omega}_m \times (\boldsymbol{\omega}_m \times \mathbf{r}) + 2(\boldsymbol{\omega}_m \times \mathbf{V}_r) + \dot{\boldsymbol{\omega}}_m \times \mathbf{r} \tag{10-2-7}$$

In this equation, $\mathbf{a}_o$ is the inertial acceleration of point O, $\mathbf{a}_r$ is the relative acceleration of dm with respect to point O, the next term is the centripetal acceleration, the next term is the Coriolis acceleration, then the angular acceleration.

At this point let us *assume that the body is rigid*, that is, that every particle mass maintains its original position with respect to point O so that

$$\mathbf{V}_r = \mathbf{a}_r = 0$$

Now the expressions for the inertial velocity and acceleration of dm reduce to

$$\mathbf{V}_I = \mathbf{V}_o + \boldsymbol{\omega}_m \times \mathbf{r} \tag{10-2-8a}$$

$$\mathbf{a}_I = \mathbf{a}_o + \boldsymbol{\omega}_m \times (\boldsymbol{\omega}_m \times \mathbf{r}) + \dot{\boldsymbol{\omega}}_m \times \mathbf{r} \tag{10-2-8b}$$

If Eq. (10-2-8b) is substituted into Eq. (10-2-1) and point O is placed at the center of mass (cm) of the body (where $\int \mathbf{r}\, dm = 0$), Eq. (10-2-1) reduces to the familiar form for a point mass,

$$\mathbf{F} = m\mathbf{a}_o \tag{10-2-9}$$

where $\mathbf{a}_o$ is the acceleration of the center of mass of the body with respect to inertial space. With $\mathbf{F}$ restricted to the gravitational force, this is the equation used in Section 2-2 to develop the orbital equations of motion.

The most general form of Newton's law governing the angular momentum and rotational modes of a continuous system can be written in vector form as

$$\mathbf{M} = \int (\mathbf{r} \times \mathbf{a}_I)\, dm = \left.\frac{d\mathbf{H}}{dt}\right|_I \tag{10-2-10}$$

where $\mathbf{M}$ is the resultant of all applied and generated moments and torques, external and internal, and $\mathbf{r}$ is the moment-arm vector, the position vector from the point about which moments are taken to a particle mass. **H** is the angular momentum (moment of linear momentum) of the system and can be expressed as

$$\mathbf{H} = \int (\mathbf{r} \times \mathbf{V}_I) \, dm \qquad (10\text{-}2\text{-}11)$$

With the assumption of a rigid body and with point O located at the cm, substituting the expression of Eq. (10-2-8a) for $\mathbf{V}_I$ into Eq. (10-2-11), performing the indicated integration, and then using the definitions of the moments of inertia, the angular momentum vector can be written as

$$\mathbf{H} = [\mathbf{I}]\boldsymbol{\omega}_m \qquad (10\text{-}2\text{-}12)$$

where $[\mathbf{I}]$ is the symmetric inertia matrix, which is represented by

$$[\mathbf{I}] = \begin{bmatrix} I_{xx} & -I_{xy} & -I_{xz} \\ -I_{xy} & I_{yy} & -I_{yz} \\ -I_{xz} & -I_{yz} & I_{zz} \end{bmatrix} \qquad (10\text{-}2\text{-}13)$$

where I_{xx}, I_{yy}, and I_{zz} are the moments of inertia about the three major axes, respectively, and the remaining elements are the cross products of inertia; the units are kg-m^2.

H in Eq. (10-2-12) is the total angular momentum of the rigid body (the spacecraft) as it rotates about its axes, whereas the H of Chapter 2 is the angular momentum per unit mass of the point mass of the body as it moves along its trajectory. The former might be thought of as the "internal" angular momentum and the latter as the "external" angular momentum of a rigid body; the two are independent of each other. If the body is not rotating ($\omega_m = 0$), the "internal" angular momentum is zero but the "external" is not; it is not affected by the value of ω_m.

Returning to Eq. (10-2-10), an expanded and usable expression can be found either by substituting Eq. (10-2-8b) for $\mathbf{a}_I$ and integrating or by finding the derivative of **H** with respect to inertial space by application of the theorem of Coriolis. In either case, the resulting vector equation is

$$\boxed{\mathbf{M} = \left.\frac{d\mathbf{H}}{dt}\right|_m + \boldsymbol{\omega}_m \times \mathbf{H}} \qquad (10\text{-}2\text{-}14)$$

With **M** and **H** expressed in terms of their components along the body (moving) axes, Eq. (10-2-14) can be written as a set of three first-order differential equations in **H**, namely,

$$\begin{aligned} M_x &= \dot{H}_x + \omega_y H_z - \omega_z H_y \\ M_y &= \dot{H}_y + \omega_z H_x - \omega_x H_z \\ M_z &= \dot{H}_z + \omega_x H_y - \omega_y H_x \end{aligned} \qquad (10\text{-}2\text{-}15)$$

These are known as *Euler's moment equations* and model the attitude motion of a rigid body. The fixed body axes themselves are known as the *Euler axes*. These moment equations can be converted to first-order differential equations in ω by applying Eq. (10-2-12). Although there is no general solution to this set of equations, there are special situations of interest for which there are solutions.

If the spacecraft is torque and moment free ($\mathbf{M} = 0$), then

$$\left.\frac{d\mathbf{H}}{dt}\right|_I = 0 \tag{10-2-16a}$$

which indicates that the angular momentum vector is constant (in both magnitude and direction). However, Eq. (10-2-14) shows that

$$\left.\frac{d\mathbf{H}}{dt}\right|_m = -\boldsymbol{\omega}_m \times \mathbf{H} \tag{10-2-16b}$$

indicating that $\mathbf{H}$, although constant in the inertial reference frame, is not necessarily constant with respect to the body axes.

$\mathbf{H}$ can be represented by its components along the moving axes, namely,

$$\mathbf{H} = H_x \mathbf{i} + H_y \mathbf{j} + H_z \mathbf{k} \tag{10-2-17}$$

Now consider a spacecraft with the major axes defined as the principal axes (the cross products of inertia are all zero), so that

$$H_x = I_{xx}\omega_x \qquad H_y = I_{yy}\omega_y \qquad H_z = I_{zz}\omega_z \tag{10-2-18}$$

With $\mathbf{M} = 0$, the set of equations can be written as

$$I_{xx}\dot{\omega}_x + \omega_y\omega_z(I_{zz} - I_{yy}) = 0$$
$$I_{yy}\dot{\omega}_y + \omega_x\omega_z(I_{xx} - I_{zz}) = 0 \tag{10-2-19}$$
$$I_{zz}\dot{\omega}_z + \omega_x\omega_y(I_{yy} - I_{xx}) = 0$$

Although these equations are nonlinear and coupled, solutions are possible for special configurations such as the axisymmetric spacecraft of the next section.

These equations show that if a rigid body is initially nonrotating (the angular velocities are zero) and torque free, the angular accelerations are also zero and the spacecraft should remain stationary. However, the angular momentum is also zero and the body is not resistant to overturning or tumbling if disturbances, even slight disturbances, do occur. The resistance to disturbances can be greatly increased by spinning the spacecraft about a principal axis, as discussed in the next two sections.

Euler's equations describe the rotation of the body but do not describe its orientation with respect to the inertial frame. The three Euler angles are used to relate the body (Euler) axes to either the inertial axes themselves or to an intermediate set of axes; a useful intermediate axis system might be the local-horizon axes that are parallel to the Earth axes. Its origin is also located at the cm of the body with x' and y' defining the local horizon, which is a plane parallel to the Earth's surface, and with z' along the vertical, perpendicular to both x' and y'. Two coordinate systems with a common origin are related by three angles, by three rotations in a prescribed sequence. To move from the local-horizon axes to the body

(Euler) axes, the first rotation is in the horizontal plane about the z' axis through the angle Ψ with x' and y' shifted to intermediate positions x_1 and y_1, as shown in Fig. 10-2-2a. The second rotation is in the vertical plane about the y_1-axis through the angle θ, moving x_1 out of the horizontal plane to x_2 and rotating z' away from the vertical to z_2, as shown in Fig. 10-2-2b. The third and final rotation is about the x_2-axis, which is the x body axis, through the angle ϕ, moving y_2 out of the horizontal plane to y, another body axis, and rotating z_2 out of the vertical plane to z, the third body axis, as shown in Fig. 10-2-2c. These are the same Euler angles used in aircraft stability and control analyses, where Ψ, θ, and ϕ are the yaw angle, pitch angle, and bank (roll) angle, respectively.

The components of $\boldsymbol{\omega}$, the angular velocity of the body (Euler) axes, can be expressed in terms of the Euler angles and their derivatives as

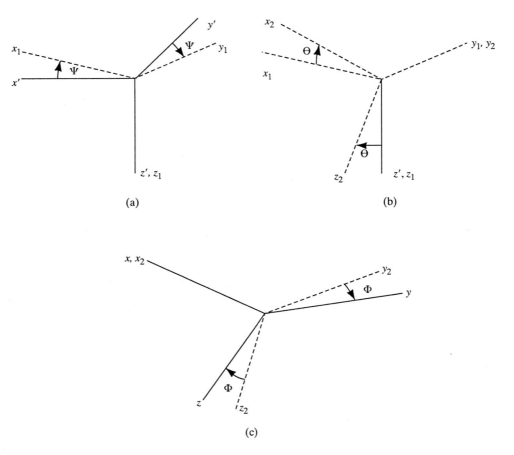

Figure 10-2-2. Sequential rotations to move from local-horizon axes to Euler (body) axes: (a) first about z'; (b) second about y_1; (c) third about x-axis.

$$\omega_x = \dot{\phi} - \dot{\psi} \sin \theta$$
$$\omega_y = \dot{\theta} \cos \phi + \dot{\psi} \cos \theta \sin \phi \quad \quad (10\text{-}2\text{-}20)$$
$$\omega_z = \dot{\psi} \cos \phi \cos \phi - \dot{\theta} \sin \phi$$

It should be noted that the Euler angle derivatives (the Euler rates) are nonlinear and are coupled.

The objectives of this section were to provide sufficient background for the derivation of the Newtonian equations of motion, with a lot of work, for nonrigid bodies, to develop the moment equations for a rotating rigid body (the Euler equations), to define the Euler angles that describe the attitude of the body, and to relate the components of the angular velocity of the body to the Euler rates. There was no mention of spinning the body about an axis or about any concomitant gyroscopic characteristics or behavior. Spinning and gyroscopic effects are discussed in the next two sections.

10-3 SINGLE-SPIN STABILIZATION

A nonspinning rigid body (spacecraft) has no inherent angular momentum and thus no resistance to disturbing and overturning torques, no inherent stability. The simplest way to increase the angular momentum of a spacecraft is to spin the entire spacecraft about a principal axis so that it acquires the rigidity in space that is a characteristic of a free gyroscope. This spin can be obtained by the use of external springs at the time of separation of the spacecraft from its delivery system, as is the case with satellites launched from the cargo bay of a Space Shuttle, by spinning the booster, or by small (micro) thrusters in the satellite itself. The thrusters can be rocket motors (solid, monopropellant, or hybrid) or stored gas jets and are also used to compensate for any decay in the spin rate, to correct any changes in attitude resulting from disturbance torques, and to change the orientation of the spacecraft if required by the mission (the pointing problem).

The most common configuration for a spinning spacecraft is axisymmetric in which two of the principal moments of inertia are identical. If the spacecraft is elongated (cylindrical), it is said to have a *prolate* inertia distribution, whereas if it is flattened (disklike), it is said to have an *oblate* inertia distribution. Any elasticity or flexibility in a prolate configuration spinning about its axisymmetric axis will eventually lead to dynamic instabilities; *oblate configurations are safer*.

Let us consider an axisymmetric spacecraft with $I_{xx} = I_{yy}$ and spinning about the z-axis with its angular velocity ω_z denoted by ω_{sp} to distinguish it from the components of the rotating body angular velocities. Now the torque-free Euler's equations of Eq. (10-2-19) become

$$I_{xx}\dot{\omega}_x + (I_{zz} - I_{xx})\omega_y \omega_{sp} = 0$$
$$I_{xx}\dot{\omega}_y - (I_{zz} - I_{xx})\omega_x \omega_{sp} = 0 \quad \quad (10\text{-}3\text{-}1)$$
$$I_{zz}\dot{\omega}_{sp} = 0$$

The third equation in this set shows that $\dot{\omega}_{sp} = 0$, that ω_{sp} is constant (in the absence of any torques or forces). Consequently, the other two differential equations are now linear and can be written as

$$\dot{\omega}_x + \lambda \omega_y = 0$$
$$\dot{\omega}_y - \lambda \omega_x = 0 \quad (10\text{-}3\text{-}2)$$

where

$$\lambda = \frac{I_{zz} - I_{xx}}{I_{xx}} \omega_{sp} \quad (10\text{-}3\text{-}3)$$

If the first equation in the pair of Eq. (10-3-2) is differentiated and $\dot{\omega}_y$ is replaced by its value from the second equation, the result is the equation of a simple harmonic oscillator,

$$\ddot{\omega}_x + \lambda^2 \omega_x = 0 \quad (10\text{-}3\text{-}4)$$

which has the general solution, in terms of the initial conditions, that

$$\omega_x = \omega_x(0) \sin \lambda t - \frac{\dot{\omega}_x(0)}{\lambda} \cos \lambda t \quad (10\text{-}3\text{-}5)$$

Since $\omega_y = -\dot{\omega}_x/\lambda$,

$$\omega_y = \omega_x(0) \cos \lambda t - \frac{\dot{\omega}_x(0)}{\lambda} \sin \lambda t \quad (10\text{-}3\text{-}6)$$

where $\omega_y(0) = -\dot{\omega}_x(0)/\lambda$.

These two equations describe a *coning* (elliptical) motion of the spin axis about the angular moment vector, which is constant with respect to the inertial reference. This motion represents the *nutation*[1] of the spinning body, the "wobbling" motion of the spin axis that results from initial conditions or impulsive torques. Nutation can be excited by separation disturbances, thruster operations, and even variations in the gravitational torques arising from the oblateness of the Earth.

Since λ represents the nutation frequency, examination of Eq. (10-3-3) shows that the direction of nutation is determined by the relationship between I_{xx} and I_{zz}. If $I_{zz} > I_{xx}$, the body is oblate (disklike) and the motion is in the same (prograde) direction as the spin. If, however, $I_{zz} < I_{xx}$, the spacecraft is prolate (cylindrical), the motion is in the opposite (retrograde) direction. (As mentioned earlier, prolate configurations are susceptible to instabilities induced by elasticity effects.) Since the amplitude of nutation is inversely proportional to the nutation frequency, a high spin rate is desirable, all other things being equal.

Nutation represents the periodic and continuous transfer of energy between the x and y axes and may interfere with the mission of the spacecraft. If there is no dissipation of this energy (as in a force and torque-free orbit), nutation will continue

---
[1] The term *precession* will be reserved for the motion of the spin axis in response to an *applied* torque.

Sec. 10-3 Single-spin Stabilization

indefinitely (as does the Earth about its spin axis) unless artificially damped by a nutation damper. Nutation dampers can be active or passive. The active dampers use thrusters and the passive dampers use tubes, either cylindrical or annular, that are partially filled with a fluid (such as mercury) and provide internal damping. Such fluid dampers can be used only with oblate spinning spacecraft; prolate spinners will transition from spin about the axis of symmetry to tumbling end over end. The damping is negative, since the damper gives the appearance of a nonrigid body.

Now let us look at *precession*, the motion of the spin angular momentum vector in response to an applied torque. Let the total angular momentum $\mathbf{H}_T$ be separated into its nonspin contribution $\mathbf{H}_{nsp}$ and its spin contribution $\mathbf{H}_{sp}$, so that

$$\mathbf{H}_T = \mathbf{H}_{nsp} + \mathbf{H}_{sp} \qquad (10\text{-}3\text{-}7)$$

where $\mathbf{H}_{nsp}$ is the product of the inertia matrix of the spacecraft and $\boldsymbol{\omega}_m$, the angular velocity of the spacecraft, and

$$\mathbf{H}_{sp} = I_{sp}\,\boldsymbol{\omega}_{sp} \qquad (10\text{-}3\text{-}8)$$

where I_{sp} is the moment of inertia about the spin axis and is not to be confused with the specific impulse of Chapter 6 with the same symbology. When the entire spacecraft is spinning, the spin axis will be a body axis. If only an element of the spacecraft is spinning, such as with a gyro or gyrostat, the spin axis will not necessarily be a body axis and another set of reference axes may be required.

With a constant spin velocity and a constant I_{sp}, the local derivative of $\mathbf{H}_{sp}$ with respect to time will be zero. Then if $\mathbf{H}_{sp} \gg \mathbf{H}_{nsp}$, the usual case, the torque equation of Eq. (10-2-14) can be approximated by

$$\boxed{\mathbf{M} \cong \left.\frac{d\mathbf{H}_{nsp}}{dt}\right|_m + \boldsymbol{\omega}_m \times \mathbf{H}_{sp}} \qquad (10\text{-}3\text{-}9)$$

Equation (10-3-9) is often referred to as the *gyroscope equation*. This equation indicates that the source of nutation is the unsteady dynamics of the nonspin angular momentum,[2] the local time derivative of $\mathbf{H}_{nsp}$.

When the transients associated with nutation have died out, the equation remaining, known as the *steady-state gyro equation*, becomes

$$\boxed{\mathbf{M} = \boldsymbol{\omega}_m \times \mathbf{H}_{sp}} \qquad (10\text{-}3\text{-}10)$$

When $\mathbf{M} = 0$, $\boldsymbol{\omega}_m$ is also zero and the spin angular momentum vector $\mathbf{H}_{sp}$ is constant in both magnitude and direction. However, when a torque is applied about a nonspin axis, $\mathbf{H}_{sp}$ must change its direction in such a manner that Eq. (10-3-10) is satisfied. Consider Fig. 10-3-1a, which shows the body axes at the time of application of a

---

[2] Nutation can also occur during the spinup and spindown of the spinning element when $\mathbf{H}_{sp}$ is not constant.

torque about the x-axis ($M_x \mathbf{i}$) with $\mathbf{H}_{sp}$ along the z-axis ($H_{sp} \mathbf{k}$). Equation (10-3-10) becomes

$$M_x \mathbf{i} = \boldsymbol{\omega} \times H_{sp} \mathbf{k} \qquad (10\text{-}3\text{-}11)$$

and the solution is for $\boldsymbol{\omega}_m = \omega_y \mathbf{j}$ (since with the right-hand rule, $\mathbf{j} \times \mathbf{k} = \mathbf{i}$) and $\omega_y = M_x/H_{sp}$. As a consequence of the applied torque, the spin axis rotates about the y-axis with a constant angular velocity in an attempt to align itself with the torque, which in turn moves away from the spin axis, leading to a chase of the torque vector by the spin axis. This movement of the spin axis in response to an applied torque is called *precession*.

In Fig. 10-3-1b, the torque is applied about the y-axis ($M_y \mathbf{j}$), so that $\boldsymbol{\omega}_m = -\omega_x \mathbf{i}$, with $\omega_x = -M_y/H_{sp}$. Once more we see precession, the spin axis in pursuit of the applied torque. If, however, the torque is applied about the spin axis

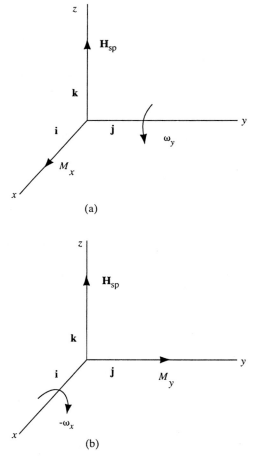

Figure 10-3-1. Precession in response to applied torque: (a) about x-axis; (b) about y-axis.

Sec. 10-3 Single-spin Stabilization

(the z-axis), the only effect is to change the magnitude of $\boldsymbol{\omega}_{sp}$; the direction of the spin axis is unaffected, and there is no precession.

This phenomenon of precession in which the spin vector tries to align itself with the applied torque vector becomes apparent when the vector gyro equation [Eq. (10-3-10)] is written as a set of scalar equations:

$$M_x = \omega_y H_{sp} \quad \text{or} \quad \omega_y = \frac{M_x}{H_{sp}}$$

$$M_y = -\omega_x H_{sp} \quad \text{or} \quad \omega_x = \frac{M_y}{H_{sp}} \quad (10\text{-}3\text{-}12)$$

$$M_z = 0$$

These scalar equations also show that if somehow the spinning body is forced to rotate about a nonspin axis, a torque is developed that opposes the rotation, a torque that is directly proportional to the magnitude of both $\mathbf{H}_{sp}$ and the forcing angular velocity. This reactive torque is known as the *gyro torque* and is the source of the rigidity in space (the resistance to overturning) that is associated with a spinning body or a gyroscope. The larger the spin angular momentum (H_{sp}), the smaller the precession angular velocities.

Spinning an entire spacecraft to stabilize it is essentially a passive technique; the expenditure of energy to maintain a desired spin rate is relatively low. That is the good news. The bad news with respect to planetocentric satellites is the major problem of pointing the antennas and data-gathering sensors that are integral to the spinning spacecraft so as to achieve the desired coverage as the spacecraft orbits the planet along with the secondary problem of handling the rotation of these payload devices.

Since a spin-stabilized satellite is stabilized with respect to inertial space, it strives to maintain its initial attitude as the satellite travels along its orbit. For example, if initially, the spin axis of a geocentric satellite is aligned with the local vertical, so that any antennas or transponders are pointed toward the Earth, it will not remain aligned with the local vertical ($\mathbf{H}_{sp}$ is in the plane of the orbit, whereas $\mathbf{H}$ of the point mass is normal to the plane of the orbit). For example, without compensating torques, the antennas will be pointing directly away from the Earth when the satellite is halfway around its orbit, as shown in Fig. 10-3-2. There is a similar pointing problem if the spin axis is normal to the orbital plane with one of the nonspin axes aligned with the local vertical, which is often the case. $\mathbf{H}_{sp}$, the spin angular momentum, and $\mathbf{H}$, the angular momentum of the orbit, are now both normal to the plane of the orbit.

To keep the fixed sensors pointing toward the Earth, the spinning satellite must precess at an appropriate rate in an appropriate direction; controlled precession requires an applied torque. A geosynchronous communications satellite with a period of 24 h must precess through one revolution each day. Since the period of the orbit decreases with altitude, LEO satellites have higher precession rates and require larger applied torques than do GEOs. (Deep-space probes, on the other

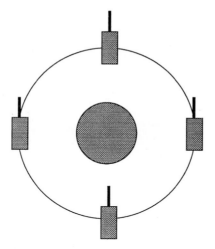

Figure 10-3-2. Effects of spin stabilization on sensor pointing.

hand, only require realignment of the spin axis when necessary to maintain communications contact with the Earth.)

Example 10-3-1

A communications satellite in a circular orbit has a spin rate about its axis of symmetry of 5 r/s and an $I_{sp} = 300$ kg-m². Initially, the spin axis is pointed toward the Earth along **g** and the x-axis is along the velocity vector (tangent to the circular orbit).

(a) If the satellite is at geosynchronous altitude with a period of 24 h, find the required precession rate and the magnitude and direction of the applied torque required.

(b) Do part (a) for a period of 1 h, a LEO.

(c) Which appear to be preferable for such satellites, higher or lower altitudes?

Solution

(a) The spin angular momentum vector is

$$\mathbf{H}_{sp} = I_{sp}\boldsymbol{\omega}_{sp} = 300 \times 5\mathbf{k} = 1500\mathbf{k} \text{ kg-m}^2/\text{s}$$

If x is to remain along the velocity vector so that the spin axis remains pointed to the Earth, it must rotate about the y-axis with a negative angular velocity so that $\boldsymbol{\omega} = -\omega_y \mathbf{j}$, where

$$\omega_y = \frac{2\pi}{24 \times 3600} = 7.272 \times 10^{-5} \text{ r/s}$$

From Eq. (10-3-11),

$$\mathbf{M} = \boldsymbol{\omega} \times \mathbf{H}_{sp} = \omega_y H_{sp}(-\mathbf{j} \times \mathbf{k}) = -\omega_y H_{sp} \mathbf{i}$$

so that the required torque is

$$M_x \mathbf{i} = -\omega_y H_{sp} \mathbf{i}$$

The torque must be applied about the negative x-axis and have a magnitude

$$M_x = 7.272 \times 10^{-5} \times 1500 = 0.1091 \text{ N-m}$$

If, for instance, the torque is to be generated by a pair of thrusters mounted 1.0 m apart

(0.5 m on each side of the axis), each thruster must produce a continuous thrust of 0.05454 N (0.01226 lb).

(b) With a satellite period of 1 h,

$$\omega_y = 1.745 \times 10^{-3} \text{ r/s}$$

$$M_x = 2.618 \text{ N-m}$$

and the thrust of each thruster must be 1.309 N (0.2943 lb).

(c) The higher-altitude orbits require less torque (and thus less thrusting and propellant) than do the lower-altitude orbits and therefore are preferable.

A secondary problem is the handling of the rotation of antennas and sensors that are directed and pointed along one of the axes of the spacecraft. These devices may be designed to be omnidirectional, so as to be unaffected by the rotation of the spacecraft, or *despun* (rotated) in the opposite direction by an electric motor (requiring energy) so as to remain "fixed." However, a spacecraft with a rotating sensor is no longer a single-spin stabilized spacecraft but rather is a dual spinner; dual spinners are discussed in the next section.

This section will close with a few remarks on the *despinning process* in which the spin rate of a spacecraft is reduced, set equal to zero, or even reversed in direction. Many spacecraft are spun for stability at high rates during launch, orbit insertion, or orbit transfer and enter an operational orbit at higher spin rates than required (three-axis stabilized spacecraft, discussed in Section 10-6, have a zero spin rate). Obviously, despinning can be accomplished with thrusters with an accompanying requirement for fuel. However, passive mass movement techniques in which spin angular momentum is reduced are quite effective, being simple, dependable, and comparatively light.

One such device, known as a *yo-yo*, consists of one or two expendable masses at the end of cords that are initially wrapped about the spin axis and about the spinning body. Considering a two-mass system, after the weights are released, say by explosive bolts, centrifugal force unwraps the cords, as shown in Fig. 10-3-3. As the cords unwind, the spin moment of inertia increases but the spin angular momentum (and total kinetic energy) remains constant, thus decreasing the spin angular velocity of the body and transferring angular momentum to the masses. (Ice skaters use this technique to control their spin velocity by extending their arms to slow down and by tucking them in to increase their spin rate.) When the cords are fully extended, they are released simultaneously, allowing the masses to escape and carry away part or all of the original spin angular momentum.

The masses may be released radially or tangentially. If a single mass is used, it must be wrapped about the spacecraft in the plane containing the center of mass to minimize attitude perturbation during the unwinding process; if such winding is not possible, two masses should be used. If the masses are not released when they reach the end of the cords, they will continue to spin with decreasing cord lengths, transferring angular momentum back into the spacecraft and reversing the spacecraft spin direction.

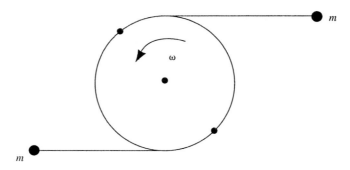

Figure 10-3-3. Despinning using a yo-yo device.

Since angular momentum and kinetic energy are conserved during the despinning process, it is possible to develop relationships among the cord length l, the total mass of the unwinding weight(s) m, the spacecraft radius and spin moment of inertia R and I_{sp}, the initial spin velocities ω_{spo}, the time t, and the spin velocity at any instant of time ω_{sp}. The relationship between the spacecraft spin rate and time is

$$\omega_{sp} = \omega_{spo} \frac{c - \omega_{spo}^2 t^2}{c + \omega_{spo}^2 t^2} \qquad (10\text{-}3\text{-}13)$$

where

$$c = \frac{I_{sp}}{mR^2} + 1 \qquad (10\text{-}3\text{-}13a)$$

The relationship between ω_{sp} (r/s) and the cord length l at any instant of time is given by

$$\omega_{sp} = \omega_{spo} \frac{cR^2 - l^2}{cR^2 + l^2} \qquad (10\text{-}3\text{-}14)$$

If the desired spin velocity is specified, the required cord length can be obtained by rearranging Eq. (10-3-14) to yield

$$l = R\sqrt{c \frac{\omega_{spo} - \omega_{sp}}{\omega_{spo} + \omega_{sp}}} \qquad (10\text{-}3\text{-}15)$$

Let us consider the special case where all of the spin is to be removed; ω_{sp} *is to be zero*. Equation (10-3-15) reduces to

$$l = R\sqrt{c} = \sqrt{R^2 + \frac{I_{sp}}{m}} \qquad (10\text{-}3\text{-}16)$$

Notice that *the cord length for total despinning is independent of the initial spin rate*; however, the time required is not. With $\omega_{sp} = 0$, Eq. (10-3-13) reduces to

$$t = \frac{l}{R\omega_{spo}} = \frac{\sqrt{c}}{\omega_{spo}} = \frac{\sqrt{(I_{sp}/mR^2) + 1}}{\omega_{spo}} \qquad (10\text{-}3\text{-}17)$$

so that the time to despin is inversely proportional to ω_{spo}.

Sec. 10-4   Dual-spin Stabilization

This despinning process lends itself to a simple example in which the initial spin velocity of a spacecraft is reduced to a specified value and is completely despun.

Example 10-3-2

A satellite is spinning about its axis of symmetry with a spin rate of 5 r/s (47.75 rpm). It has a radius of 1.5 m and a moment of inertia about its spin axis of 300 kg-m². Two masses, 3 kg each, are used to despin the satellite.
(a) Find the cord length and deployment time to reduce the spin rate to 1.0 r/s.
(b) Find the cord length and time required to reduce the spin rate to zero.
(c) Do parts (a) and (b) with a single mass of 6 kg.
Solution
(a) From Eq. (10-3-13a),
$$c = \frac{300}{6 \times 1.5^2} + 1 = 23.22$$
Substituting this value of c into Eq. (10-3-15) to find l for $\omega_{sp} = 1.0$ r/s yields
$$l = 1.5\sqrt{23.22\left(\frac{5-1}{5+1}\right)} = 5.902 \text{ m}$$
Now with c and ω_{sp} specified, Eq. (10-3-13) can be written as
$$1 = 5\left(\frac{23.22 - 25t^2}{23.22 + 25t^2}\right)$$
which can be solved to find that it takes 0.787 s for the spin rate to be reduced from 5.0 r/s to 1.0 r/s.
(b) To reduce the spin rate to zero,
$$l = 1.5\sqrt{23.22} = 7.228 \text{ m}$$
and the time to reach this state is
$$t = \frac{7.228}{1.5 \times 5.0} = 0.9637 \text{ s}$$
(c) The results are the same as long as the mass is equal to the sum of the two masses. If only one 3-kg mass were to be used, the results would be different and would be identical to the results using two 1.5-kg masses.

This technique of using an auxiliary moving mass(es) to exchange momentum can be applied to the reduction of uncontrolled tumbling to simple spinning, which can then be adjusted to the desired spin rate or despun using the yo-yo technique above. Such a detumbling system would move an internal control mass in accordance to a control law so selected as to leave the spacecraft in a stable spin state ready to be despun.

10-4 DUAL-SPIN STABILIZATION

Although single-spin stabilization is simple in concept, it has the major drawback that antennas or sensors can be pointed only by moving the entire spacecraft; the spacecraft and its sensors rotate together as a single unit. Furthermore, to be dynamically stable a single-spin spacecraft needs to be oblate, short, and squat, whereas booster considerations (such as drag) favor the slim cylinder, a prolate configuration. Dual spinners evolved to eliminate these drawbacks yet retain the major spin-stabilized advantage of reduced energy expenditure. As might be expected, the benefits of dual spinning are accompanied by associated problems.

The simplest dual spinner is the oblate spinning spacecraft with a despun antenna-sensor platform that was mentioned in the preceding section. The platform and body both rotate about the axis of symmetry (the major axis), although the angular velocity of the platform is that required to keep the platform pointing toward the Earth and therefore much smaller than that of the body. Furthermore, the platform itself is much smaller than the body. Consequently, the stability criteria are essentially those of the single spinner. However, constraints associated with the booster diameter can limit the body (the effective gyroscopic rotor) diameter, which in conjunction with major axis spinning can restrict the spacecraft size.

The limitations on spacecraft size can be bypassed by spinning the body about a minor axis, that is, by using a prolate configuration. Although a spinning prolate configuration is inherently unstable, appropriate energy dissipation (either inherent or introduced) in the two elements of the spacecraft, the platform and the body, can be used to make the spacecraft dynamically stable. It may be necessary to add a nutational damper to the body to handle nutation about the minor axis.

Gyrostat is a term often used to describe an axisymmetric dual-spin configuration in which there is damping in both elements. A simple gyrostat may also be defined as a rigid body that carries an internal axisymmetric rotor whose axis and center of mass are fixed within the body.

Dual-spin spacecraft may have many configurations and do not necessarily need to be absolutely rigid. The requirement for dynamic stability establishes criteria that determine acceptable configurations and the nature of added damping subsystems. Stability analyses to establish these criteria may be rigorous or may use an energy sink approach. In either case, such analyses are beyond the scope of this section and of this book. With the omission of the detailed and complex analyses for the many possible dual-spin configurations, this has been a brief section with the main purpose of introducing the subject and its implications and complications.

Although dual-spin stabilization is essentially passive in nature, the introduction of energy dissipating mechanisms and of pointing mechanisms calls for devices, such as electric motors, to replace energy and to point sensors. In the next section we look at a truly passive stabilization scheme that exploits the gravity gradient.

10-5 GRAVITY-GRADIENT STABILIZATION

Although it is customary to speak of spin stabilization as a passive technique, applied torques are required to compensate for the effects of disturbance torques and to precess the spin axis so as to point the sensors in the desired direction; in other words there are active elements in spin stabilization. Gravity-gradient stabilization, on the other hand, is a purely passive technique that has the theoretical capability of

Sec. 10-5   Gravity-gradient Stabilization

keeping sensors pointed toward the planetary surface, providing a passive solution to the pointing problem. There are, however, limitations to the effectiveness of this technique because the magnitude of the gravity gradient (and thus of the restoring torques) is small, limitations as to acceptable spacecraft configurations and inertia distributions, orbital shape, and orbital altitude. It should be noted that spin stabilization is unaffected by orbital shape and altitude.

In Chapter 2 we saw that the acceleration due to a gravitational attraction (the gravitational force per unit mass) can be expressed as

$$g = \frac{F_g}{m} = \frac{\mu}{r^2} \qquad (10\text{-}5\text{-}1)$$

where μ is the planetary gravitational parameter and r is the distance from the center of the planet; g has the units of m/s². The gravity gradient is simply the derivative of g with respect to r and is

$$\frac{dg}{dr} = -\frac{2\mu}{r^3} \qquad (10\text{-}5\text{-}2)$$

That the magnitude of the gravity gradient, which is a measure of the decrease in the gravitational attraction with an increase in orbital altitude, is small can be seen from the following figures for a geocentric circular satellite:

h (nmi/km)	r (m)	g (m/s²)	dg/dr (m/s²/m)
175/322	6.70×10^6	8.879	-2.651×10^{-6}
300/552	6.93×10^6	8.300	-2.395×10^{-6}
19,490/3586	42.24×10^6	0.223	-0.011×10^{-6}

These figures show just how small the gravity gradient is, how it decreases with altitude, and indicate that gravity-gradient stabilization should be more appropriate for low-altitude (planetary-resource) satellites than for geosynchronous communications satellites, which it is.

A general analysis of the use of gravity-gradient torques to stabilize a rigid body shows that to be stable the body should have one axis longer than the other two so that its moment of inertia about the long axis is less than that about the other two axes; the most effective configurations are the rod or the dumbbell. The stable position is the one in which the long axis lies along the vertical and points toward the planetary surface.

Let us look at the simplest case of a symmetrical dumbbell comprising two equal point masses connected by a massless rigid rod of length L so that the center of mass is halfway between the two masses, as shown in Fig. 10-5-1, where $l = L/2$, and where θ, the angle between the satellite axis and the vertical $\mathbf{r}_0$, is positive in a clockwise direction. Let us further assume that this dumbbell satellite is in a circular geocentric orbit (as described by the motion of the center of mass) and that the altitude is such that the position vectors drawn from each mass and from the center of mass are essentially parallel.

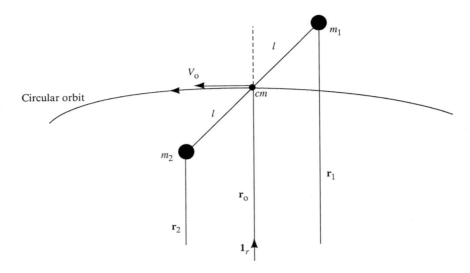

Figure 10-5-1. Gravity-gradient stabilization with a symmetrical dumbbell configuration.

At the center of mass, the sum of the applied forces is zero; that is, the centrifugal force is equal to the gravitational force, so that

$$\Sigma F_0 = \frac{2mV_0^2}{r_0} - 2mg_0 = \frac{2mV_0^2}{r_0} - \frac{2m\mu}{r_0^2} = 0 \qquad (10\text{-}5\text{-}3)$$

where m is the mass of each individual point mass.

At each individual mass, the sum of the forces is not necessarily zero. The resultant forces, F_1 and F_2, at each mass can be written as

$$F_1 = \frac{mV_1^2}{r_1} - mg_1 = m\left[\frac{V_1^2}{r_0 + l\cos\theta} - \frac{\mu}{(r_0 + l\cos\theta)^2}\right] \qquad (10\text{-}5\text{-}4)$$

$$F_2 = \frac{mV_2^2}{r_2} - mg_2 = m\left[\frac{V_2^2}{r_0 - l\cos\theta} - \frac{\mu}{(r_0 - l\cos\theta)^2}\right] \qquad (10\text{-}5\text{-}5)$$

The moment arm for generating a torque is equal to $l\sin\theta$ for each force. With the "small" angle assumption for θ ($\cos\theta \cong 1$ and $\sin\theta \cong \theta$), the moment arms become equal to $l\theta$; $r_1 \cong (r_0 + l)$ and $r_2 \cong (r_0 - l)$. With a positive torque in a clockwise direction, the applied torque about a transverse axis can be written as

$$M = F_2 l\theta - F_1 l\theta = l\theta(F_2 - F_1) \qquad (10\text{-}5\text{-}6)$$

Expressing V_1 as $V_0 + \Delta V$ gives

$$V_1^2 = (V_0 + \Delta V)^2 = V_0^2 + 2V_0\Delta V + (\Delta V)^2 \qquad (10\text{-}5\text{-}7)$$

With the assumption that $\Delta V \ll V_0$, V_1^2 can be simplified with the approximations that $(\Delta V)^2 \cong 0$ and $(2\Delta V/V_0) \ll 1$:

$$V_1^2 \cong V_0^2 + 2V_0\Delta V = V_0^2\left(1 + \frac{2\Delta V}{V_0}\right) \cong V_0^2 \qquad (10\text{-}5\text{-}8)$$

Sec. 10-5   Gravity-gradient Stabilization

Similarly, V_2^2 can be approximated by V_0^2 and the torque equation can now be written as

$$M = ml\theta\left\{V_0^2\left[\frac{1}{r_0-l}-\frac{1}{r_0+l}\right]-\mu\left[\frac{1}{(r_0-l)^2}-\frac{1}{(r_0+l)^2}\right]\right\} \quad (10\text{-}5\text{-}9)$$

Combining the terms within the brackets yields

$$M = ml\theta\left[\frac{2V_0^2 l}{r_0^2-l^2}-\frac{4\mu r_0 l}{(r_0^2-l^2)^2}\right] \quad (10\text{-}5\text{-}10)$$

With the final approximation that $l^2 \ll r_0^2$, Eq. (10-5-10) becomes

$$M = 4m\left(\frac{l}{r_0}\right)^2 \theta\left(\frac{V_0^2}{2}-\frac{\mu}{r_0}\right) \quad (10\text{-}5\text{-}11)$$

Since the bracketed term is the specific energy of the orbit (from the vis-viva integral) and is equal to $-\mu/2a$ ($-\mu/2r_0$ for this circular orbit), an approximate expression for the torque can finally be written as

$$\boxed{M \cong -2m\left(\frac{l}{r_0}\right)^2 \frac{\mu}{r_0}\theta} \quad (10\text{-}5\text{-}12)$$

Note that $(\mu/r_0) = V_{cs}^2$.

Although the development of Eq. (10-5-12) from the nonlinear force expressions is full of approximations and is somewhat tedious, the resulting expression is linear and does provide insight into the nature, magnitude, and effects of the gravity-gradient torque in terms of the steady-state orbital conditions and spacecraft configuration. We first see that the torque is negative (counterclockwise) when the angle θ is positive (clockwise); this is a restoring torque that seeks to drive θ to zero and return the satellite to a vertical orientation along r_0. We also see that the torque is quite small; in fact, the expression contains the term $(l/r_0)^2$, which earlier in the development was taken to be on the order of zero but is retained here so that there might be an expression to examine. Equation (10-5-12) shows the importance of the length and mass of the satellite; the restoring torque is directly proportional to the mass and to the square of the length. The torque is also inversely proportional to the cube of r_0, the semimajor axis of the circular orbit; the higher the orbit, the smaller the torque. Finally, the restoring torque increases as θ increases, leading to the limiting of the deviation from the vertical. The fact that the restoring torque is equal to zero when $\theta = 0°$ and changes direction when passing that point indicates that the vertical is a statically stable position for the long axis; when disturbed, the first movement of this dumbbell-shaped body is back toward its equilibrium position.

With respect to the dynamic stability, in the absence of any external damping, the torque is equal to the product of the moment of inertia and the angular acceleration so that the dynamic equation of motion can be written as

$$\ddot{\theta}(t) + \frac{2m\mu l^2}{r_0^3 I}\theta(t) = \ddot{\theta}(t) + \omega_n^2\,\theta(t) = 0 \quad (10\text{-}5\text{-}13)$$

Equation (10-5-13) is the linear differential equation of a marginally stable (undamped) system (a harmonic oscillator) whose response to a disturbance or applied input is a bounded sinusoid with a circular frequency ω_n, where

$$\omega_n = \sqrt{\frac{2m\mu l^2}{r_0^3 I}} \qquad (10\text{-}5\text{-}14)$$

Since the moment of inertia of a symmetrical dumbbell about a transverse axis is

$$I = 2ml^2$$

Eq. (10-5-14) reduces to

$$\boxed{\omega_n = \sqrt{\frac{\mu}{r_0^3}}} \qquad (10\text{-}5\text{-}15)$$

where ω_n, the *undamped natural frequency* with units of r/s, *is independent of both m and l*, and P, the period of the oscillation, is given by

$$P = \frac{2\pi}{\omega_n} = 2\pi\sqrt{\frac{r_0^3}{\mu}} \qquad (10\text{-}5\text{-}16)$$

But Eq. (10-5-16) is the expression for the period of a circular satellite. Consequently, *the long axis of the spacecraft will complete one oscillation per orbit and remain aligned with the vertical.* Although the frequency (and period) of the oscillation is independent of m and l, the magnitude of the restoring torque is not.

Example 10-5-1

The configuration of a geocentric satellite is that of a symmetrical dumbbell with two masses of 50 kg each and connected by a massless structure that is 40 m long. The satellite is placed into a circular orbit at 175 nmi (322 km).
(a) Find the undamped natural frequency ω_n and the period of the oscillations P. Compare P with the orbital period.
(b) Find the magnitude of the restoring torque if θ is somehow disturbed to $+5°$.
(c) Increase the masses to 100 kg each and the length between them to 60 m. Find the restoring torque for $\theta = +5°$ and compare with that of part (b).

Solution

(a) $l = 20$ m and $r_0 = 6.70 \times 10^6$ m. Just as a matter of interest, the moment of inertia about the transverse axes is

$$I = 2 \times 50 \times 20^2 = 40{,}000 \text{ kg-m}^2$$

and using Eq. (10-5-14) gives

$$\omega_n = \sqrt{\frac{2 \times 50 \times (3.986 \times 10^6) \times 20^2}{(6.70 \times 10^6)^3 \times 40{,}000}} = 1.151 \times 10^{-3} \text{ r/s}$$

The simpler expression of Eq. (10-5-15) yields

$$\omega_n = \sqrt{\frac{3.986 \times 10^{14}}{(6.70 \times 10^6)^3}} = 1.151 \times 10^{-3} \text{ r/s}$$

Sec. 10-5 Gravity-gradient Stabilization

The period of the oscillation is

$$P = \frac{2\pi}{\omega_n} = 5458 \text{ s} = 90.96 \text{ min}$$

The period of the circular orbit is

$$P_{cs} = 2\pi \sqrt{\frac{(6.70 \times 10^6)^3}{3.986 \times 10^{14}}} = 5458 \text{ s} = 90.96 \text{ min}$$

We see that the oscillatory period is indeed equal to the orbital period.

(b) The restoring torque can be found from Eq. (10-5-12), with $l/r_0 = 2.985 \times 10^{-6}$ and $\theta = +0.08726$ rad, to be

$$M = \frac{-2 \times 50(2.985 \times 10^{-6})^2 \times (3.986 \times 10^{14}) \times 0.08726}{6.70 \times 10^6} = -4.626 \times 10^{-3} \text{ N-m}$$

which is not very large, to say the least. However, as θ increases, so does the magnitude of the restoring torque.

(c) Increasing m and l will not affect the frequency or period but will increase the restoring torque. Using a slightly different form of Eq. (10-5-12) yields

$$M = -\frac{2 \times 100 \times 30^2 \times (3.986 \times 10^{14}) \times 0.08726}{(6.70 \times 10^6)^3} = -2.082 \times 10^{-2} \text{ N-m}$$

Doubling the mass and increasing L by 50% has increased the restoring torque by a factor of 4.5.

The following example shows the effects of increasing the orbital altitude on the frequency of the oscillations and on the magnitude of the restoring torque.

Example 10-5-2

Rework Example 10-5-1 with an orbital altitude of 300 nmi (552 km) so that $r_0 = 6.93 \times 10^6$ m.

Solution

(a) The natural frequency will be

$$\omega_n = \sqrt{\frac{3.986 \times 10^{14}}{(6.93 \times 10^6)^3}} = 1.094 \times 10^{-3} \text{ r/s}$$

and the period of the oscillations will be

$$P = \frac{2\pi}{1.094 \times 10^{-3}} = 5741 \text{ s} = 95.69 \text{ min}$$

The orbital period is

$$P_{cs} = 2\pi \sqrt{\frac{(6.93 \times 10^6)^3}{3.986 \times 10^{14}}} = 5741 \text{ s} = 95.69 \text{ min}$$

The frequency has decreased and therefore the period has increased but the latter is still equal to the orbital period.

(b) The restoring torque becomes

$$M = -4.180 \times 10^{-3} \text{ N-m}$$

which is about 9.5% less than the restoring torque at the lower altitude.

(c) The frequency and period of the oscillations will be unchanged, as in Example 10-5-1, but the restoring torque will be increased, becoming

$$M = -1.881 \times 10^{-2} \text{ N-m}$$

which represents an increase by a factor of 4.5, as was the case at the lower altitude.

As mentioned earlier, the only stable position for any long-axis spacecraft configuration[3] is with the long axis pointed toward the planetary surface. The Space Shuttle takes advantage of this characteristic by parking the spacecraft with its long axis vertical (with the attitude-control thrusters turned off) whenever such an attitude is acceptable. Although the gravity-gradient torques are small, they should not be ignored when a long-axis spacecraft uses active control to maintain an unstable equilibrium position, such as with the long-axis perpendicular to the vertical. Not only will the gravity-gradient torques task the active control actuators, but they may even overpower the actuators under certain conditions, as happened with Skylab.

If the *center of gravity (cg) of a body is defined as the point at which the net gravitational force is concentrated*, the presence of gravitational torques about the center of mass (cm) of a body implies that the cg and the cm are not necessarily the same point. Furthermore, although the center of mass is a function only of the mass distribution of the body, the *center of gravity is a function of both the mass distribution and the attitude of the body*. A body for which the cg and cm always coincide is called *centrobaric* and is not subjected to gravitational torques; a homogeneous symmetric body (a sphere) is centrobaric. If the location of the cg is a function of attitude of the body, the angle θ, the resultant torques should also be a function of attitude. That this is true for our long-axis dumbbell can be seen from the torque equation [Eq. (10-5-10)]; if $\theta = 0$, the cg and cm both lie on the vertical, the net gravitational force passes through the cg, and the gravitational torque is zero; when $\theta > 0$, the cg moves away from the vertical. In the vicinity of a planetary surface, the changes in altitude are small, as is the variation of the gravitational force, and the bodies of interest are not in a force-free or torque-free environment where gravitational torques can be significant. Consequently, it is customary to assume that the cg and cm are the same for nonastronautical applications, but this assumption should not be made for space flight without checking its validity.

To summarize this section, gravity-gradient torques are small and to be effective in stabilization (and pointing) call for a spacecraft configuration with a "long" axis of least inertia. The spacecraft is stable when this long axis is aligned with the local vertical. The frequency of any oscillations (the *libration*) of the long axis is such that the period is that of the orbit. Such a spacecraft is stabilized in both pitch and roll (rotation about the transverse axes, the axes with most inertia) but not in yaw (about the long axis). Finally, although gravity-gradient stabilization is theoretically a purely passive scheme, active control is required for stabilization in yaw and may be needed to augment and improve the performance of the gravity-gradient scheme,

---

[3] Any spacecraft can be reconfigured in orbit to take advantage of the gravity gradient by the extension of boom(s).

Sec. 10-6 Active-control Stabilization

particularly with noncircular orbits, nonideal configurations, and nonspherical planets.

10-6 ACTIVE-CONTROL STABILIZATION

Conceptually, active control is a straightforward solution to the stabilization and pointing problems. Sensors determine the attitude of the spacecraft (or payload platform), a controller compares the actual attitude with the desired attitude, and actuators reorient the spacecraft (or payload platform) until it attains the desired attitude. If the payload sensors and antennas are an integral part of the spacecraft, they are pointed by pointing (rotating) the entire spacecraft. It may be simpler, however, and require less energy to maintain a large-inertia spacecraft in a stabilized and fixed attitude and to point a separate payload platform with a much smaller inertia.

Figure 10-6-1 is a schematic block diagram of a typical active-control stabilization and pointing system. This is a negative feedback system in which the plant (whatever is to be controlled) can be either the spacecraft or the payload platform. With a separate payload platform, there would be two control systems, one to stabilize the spacecraft (maintain a desired attitude) and the other to point the platform while isolating it from any spacecraft motion. In the former case, the disturbances shown in Fig. 10-6-1 are any disturbance torques, and in the latter case, the disturbances are any motion of the platform itself. The stabilization of the body of a spacecraft with active control is sometimes referred to as a *three-axis stabilization system*.

At times the sensors used in the control system(s) will be referred to as the *control sensors* to distinguish them from the payload sensors, the sensors used to accomplish the mission of the spacecraft, even though a particular sensor may serve in a dual role. Although it may be possible to determine attitude from ground

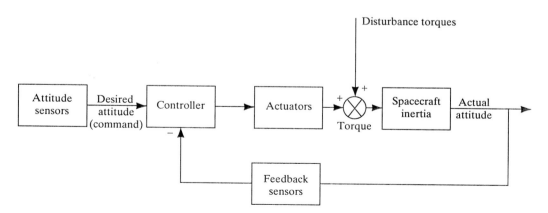

Figure 10-6-1. Typical active-control stabilization and pointing system.

measurements and computer processing, only *autonomous* sensors, sensors aboard the spacecraft, are considered here, inasmuch as they are not subject to communication delays and interruptions and do not require computer processing and smoothing. This does not mean that commands cannot be sent from the ground to the attitude control system, to the controller.

Control sensors used to establish attitude (and position) may be categorized as Earth sensors, Sun trackers, star trackers, and planet trackers. Earth sensors, for example, seek to determine the location of the horizon, from which the local vertical can be determined, by such techniques as infrared scanning, balanced radiation, and edge trackers. Without a knowledge of the horizon, a minimum of two nonparallel position lines (LOPs) are required. Such LOPs can be obtained from Sun and star trackers (sensors), which may be either analog (a solar cell grid) or digital (a slotted grid with photocell receivers). For geocentric satellites, Polaris in the northern sky and Canopus in the southern sky are often used in conjunction with the Sun to provide attitude determination. Planet trackers differ from star trackers in that the planet is close enough to have an extended image and to vary in size and illumination.

The most common control sensors for the maintenance of a specified attitude are gyroscopes of varying types (rotors, ring lasers, and spheres). They are used to sense angular motion resulting from disturbances, providing outputs that are proportional to angular velocity or angular displacement. The gyroscopes can be installed on a platform mounted to the spacecraft or mounted directly on the spacecraft itself. The outputs, after processing, become inputs to actuators that either move only the platform or the entire spacecraft, thus effectively isolating either the platform or the spacecraft from external disturbances. These sensor gyroscopes themselves do not directly stabilize the platform or the spacecraft by brute force, as is the case with spin stabilization, but rather, control the actuators that do the actual stabilization.

These gyroscopes can also accept commands, in the form of torques, that fool the gyroscopes into thinking that the platform or spacecraft is rotating so that the resultant actuating torques "stabilize" the platform or spacecraft in the new desired attitude. The gyroscopes serve dual purposes, as angular velocity sensors and conduits for transmitting command torques to the actuators.

Actuators fall into two major categories: mass ejection devices and momentum exchange devices. Mass ejection devices are commonly referred to as *thrusters* and are characterized by a depletion of mass, mass that must be boosted into space from the surface of the Earth. Thrusters can be further broken down into cold-gas and hot-gas thrusters, comprising propellants, nozzles, and possibly control valves.

The expulsion of a compressed cold gas through an expanding nozzle is simple and flexible in its operation. The commonly used gases, such as nitrogen, helium, and the Freons, are nontoxic, readily available, and inexpensive and pose no thermal or contamination problems to surfaces. However, the specific impulse is low and the pressurized tanks are heavy, but the specific volume is reasonable and the range of possible thrusts is large. Although starting and stopping thrusting requires only the operation of a control valve, missions of long duration require numerous cyclings of this valve as well as a large amount of gas.

Compared with cold-gas thrusters, hot-gas thrusters have the advantages of

Sec. 10-6 Active-control Stabilization

higher specific impulse and lighter weight and the disadvantages of greater complexity, lower minimum thrust, and high-temperature and potentially reactive exhaust gases. Hot-gas thrusters are small, low-thrust rocket motors with the characteristics discussed in Chapter 6. Liquid propellant thrusters use pressure to feed the propellant(s) into the combustion chamber and use control valve(s) for repetitive firings; mono-propellants require a single valve, whereas bi-propellants, which have higher specific impulses, require two valves.

Solid propellant thrusters have problems achieving reproducible thrust profiles and the multiple thrust initiations and terminations associated with repeated short-duration firings. The latter can be partially solved by using segmented charges; however, the duration of each firing is predetermined and cannot be varied. The hybrid propellant thruster, in which the fuel is solid and the oxidizer is liquid, has greater flexibility in that the initiation, duration, and termination of firing can be controlled by the operation of a liquid oxidizer control valve.

Because of the added complexities and energy losses associated with throttling, the thrust of a gas thruster is fixed and only the duration (and direction, if thrust vector control is available) can be controlled. Consequently, gas thrusters operate in an on–off mode (as a relay controller, so to speak) and the resultant motion of the spacecraft will be subjected to limit cycles. Therefore, for this and other reasons, torque motors rather than thrusters are normally used with stable (reference) platforms and with payload platforms carrying mission sensors. Although cold- and hot-gas thrusters are the current choice, other thrusters using mass ejection could be developed to supplement and even replace them.

Whereas thrusters reduce the momentum of a spacecraft by the ejection of mass, momentum exchange actuators leave the total momentum unchanged and merely transfer momentum back and forth between the actuator(s) and the spacecraft itself. Typical moment exchange actuators are flywheels, both solid and fluid; momentum wheels and spheres; and control moment gyros (CMG).

Flywheels are mounted with their individual axes of rotation aligned with the spacecraft body axes; they are at rest when not being used to reorient the spacecraft. A typical solid flywheel unit might consist of a rigid wheel with a relatively large moment of inertia, a drive motor, and possibly a tachometer (for control purposes). Accelerating a flywheel to a specified angular velocity will accelerate the spacecraft in the opposite direction to a velocity determined by the relationship between the moments of inertia of the flywheel and spacecraft. When the spacecraft has reached the desired position, its rotation is stopped by bringing the flywheel to rest. Since the angular momentum of the flywheel–spacecraft combination is conserved, the spin angular momentum of the spacecraft must be the negative of the flywheel spin angular momentum. Equating these momenta yields the relationship

$$(I_{sp}\,\omega_{sp})_{S/C} = (I_{sp}\,\omega_{sp})_{wheel} \qquad (10\text{-}6\text{-}1)$$

so that

$$\omega_{sp,\,S/C} = \frac{I_{sp,\,wheel}}{I_{sp,\,S/C}}\,\omega_{sp,\,wheel} \qquad (10\text{-}6\text{-}2)$$

Therefore, a specified rotation of the spacecraft can be accomplished by driving the flywheel through an appropriate number of revolutions and then stopping it.

The actions of a hard flywheel can be achieved by the circulation of a liquid through a closed tube that may be of any shape. Considering a circular tube containing a flowing liquid, it is easy to visualize the tube as the rim of a rotating flywheel even though the tube itself remains stationary; the inertia of the fluid can be very large in relation to its mass. Conventional pumps can be used to produce the flow of nonconducting fluids and electromagnetic pumps can be used with conducting fluids.

A *momentum wheel* is essentially a flywheel that is designed to rotate through a specified angle rather than attain an angular velocity. Rather than mount individual flywheels (or momentum wheels) on each spacecraft axis, a *momentum sphere* (usually suspended electrostatically or magnetically) can be used to provide torques about all three axes without the cross-axis gyroscopic coupling associated with individual wheels. Introducing these control torques, however, is not as simple with the sphere as with the wheel.

Control moment gyros (CMG) differ from angular sensing gyroscopes in that the spin angular momentum is much larger, so much larger that its precession will precess the spacecraft itself. Control torques are applied to the gyroscope's gimbal(s) by motors and gear trains so that the ratio of the control torque to the resultant vehicle torque is low. CMGs can be used singly or in pairs (back to back) on each axis, or three or more can be grouped with a gimbal configuration whereby the resultant spin angular momentum can be reoriented directly about any axis or axes, thus reducing the weight of the system.

One disadvantage of flywheels and CMGs is that they are susceptible to saturation and "gimbal lock" and may need to be *desaturated* or *reset*. Flywheels saturate when they reach their maximum operating speed and gyroscopes experience gimbal lock when the spin axis and the input axis coincide. (Although gimbal lock can be avoided by adding extra outside gimbals, such gimbals are large and heavy.) Desaturation and reset are accomplished by applying a torque to the vehicle (with thrusters) in such a direction that in order to maintain the spacecraft orientation the flywheel must transfer momentum back to the spacecraft (by slowing down) or the gyroscope must precess back to its original position.

There are external sources that might be used in combination with certain spacecraft characteristics to generate actuating torques, such as planetary magnetic fields (in conjunction with internal magnets) and solar radiation pressure (in conjunction with movable panels). Although they might in theory reduce energy requirements, they can be complicated in themselves and are not necessarily reliable or dependable, nor are they currently in use.

With regard to spacecraft energy, it is apparent that active control is dependent on reliable and adequate sources of energy, sources that may have to function for many years. Two possible sources are solar collectors of various types and isotopic power generators. Fuel cells are another possibility but do require a finite and expendable fuel supply.

Sec. 10-6   Active-control Stabilization

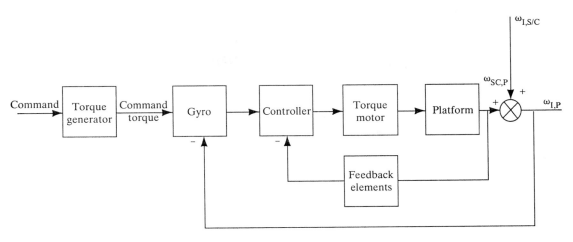

Figure 10-6-2. One axis of a three-axis stabilization and pointing system for a platform.

The usual practice in three-axis stabilization is to stabilize a platform or a spacecraft about each axis individually (three single-degree-of freedom channels), treating the coupling outputs from each channel as disturbance inputs to the other two channels. Let us look at one channel of a three-axis stabilization scheme for a stable platform, which could also be a payload platform, for a star tracker, for example. A combination block diagram and schematic of such a negative-feedback system is shown in Fig. 10-6-2, with inner-loop feedback to provide the desired stability and time response. We see that there are two classes of inputs, disturbances and commands. The objectives of the control system are to minimize the effects of any disturbances on the motion of the platform (filter out the disturbances, the regulator problem) and to follow any commands (the servomechanism problem).

For a platform one disturbance input is the angular velocity of the spacecraft with respect to inertial space, $\omega_{I,S/C}$, about the axis of interest. This angular velocity is sensed by the gyro, usually a SDF integrating gyro, whose output signal is processed by the controller before being sent to the torque motor,[4] which drives the platform in the opposite direction with an angular velocity with respect to the spacecraft, $\omega_{S/C,P}$, which combines with the disturbance angular velocity to yield the angular velocity of the platform with respect to inertial space, $\omega_{I,P}$, namely,

$$\omega_{I,P} = \omega_{S/C,P} + \omega_{I,S/C} \qquad (10\text{-}6\text{-}3)$$

The objective of keeping $\omega_{I,P}$ (and $\theta_{I,P}$) = 0 is accomplished when $\omega_{S/C,P} = -\omega_{I,S/C}$.

This channel also accepts an angular velocity command, $\omega_{I,P,cmd}$, in the form of an input to a torque generator attached to the input axis of the SDF gyro. The generated torque precesses the spin axis as though it were an angular velocity of the spacecraft and the system functions to move the platform in such a way as to make

---
[4] The platform torque motors are not classified as actuators since they do not drive the spacecraft itself.

the difference between the input angular velocity and the actual angular velocity equal to zero, at which time the platform ceases movement. Ignoring any disturbance inputs, the objective of the system in response to a command is to set

$$\omega_{I,P} = \omega_{I,P,\text{cmd}} \tag{10-6-4}$$

The same system with a few changes can be used to stabilize the spacecraft against disturbance torques and to reorient the spacecraft in response to commands. The hardware changes are the replacement of the torque motor by an actuator and the platform by the spacecraft itself. In addition, the disturbances now appear as disturbance torques that combine with the applied torque developed by the actuator to form a resultant torque that acts on the spacecraft. (This is essentially the system of Fig. 10-6-1.) With the system objective of keeping $\omega_{I,S/C} = 0$ in the presence of a disturbance torque, M_d, the immediate objective is to keep the resultant applied torque equal to zero by making the torque developed by the actuator equal to the negative of the disturbance torque. A command to reorient the spacecraft is processed in the same manner as the command to reorient a platform with the objective of making $\omega_{I,S/C} = \omega_{I,P,\text{cmd}}$.

This chapter closes with the comments that nonrigid and flexible structures, such as space stations, pose a severe challenge to the combined efforts of the structural and control system designers and that smaller bodies tethered to larger bodies present interesting dynamic and control problems.

APPENDIX A

Some Useful Vector Operations

Dot Products

$\mathbf{A} \cdot \mathbf{B}$ is a scalar and is the projection of $\mathbf{A}$ along $\mathbf{B}$.

$$\mathbf{A} \cdot \mathbf{B} = AB \cos \theta, \quad \text{where } \theta \text{ is the angle between } \mathbf{A} \text{ and } \mathbf{B}$$

$$\mathbf{B} \cdot \mathbf{A} = \mathbf{A} \cdot \mathbf{B}$$

$$\mathbf{C} \cdot (\mathbf{A} \cdot \mathbf{B}) = \mathbf{A} \cdot (\mathbf{C} \cdot \mathbf{B}) = ABC$$

$$\mathbf{C} \cdot (\mathbf{A} + \mathbf{B}) = \mathbf{C} \cdot \mathbf{A} + \mathbf{C} \cdot \mathbf{B} = AC + BC$$

Cross Products

$\mathbf{A} \times \mathbf{B}$ is a vector that is perpendicular to both $\mathbf{A}$ and $\mathbf{B}$. This vector product is positive in the direction determined by the right-hand rule.

$$|\mathbf{A} \times \mathbf{B}| = AB \sin \theta, \quad \text{where } \theta \text{ is the angle between } \mathbf{A} \text{ and } \mathbf{B}$$

$$\mathbf{B} \times \mathbf{A} = -\mathbf{A} \times \mathbf{B}$$

$$\mathbf{A} \times (\mathbf{B} \times \mathbf{C}) = \mathbf{B}(\mathbf{A} \cdot \mathbf{C}) - \mathbf{C}(\mathbf{A} \cdot \mathbf{B})$$

$$\mathbf{C} \times (\mathbf{A} + \mathbf{B}) = \mathbf{C} \times \mathbf{A} + \mathbf{C} \times \mathbf{B}$$

Differentiation

$\mathbf{A} = A\mathbf{1}_A$, where $\mathbf{1}_A$ is the unit vector of $\mathbf{A}$.

$$\frac{d\mathbf{A}}{dt} = \dot{A}\mathbf{1}_A + A\frac{d\mathbf{1}_A}{dt} \tag{A-1}$$

If $\mathbf{1}_A$ is a unit vector in an inertial reference,

$$\frac{d\mathbf{1}_A}{dt} = 0$$

Some Useful Vector Operations

and

$$\frac{d\mathbf{A}}{dt} = \dot{A}\mathbf{1}_A \tag{A-2}$$

If, however, $\mathbf{1}_A$ is a unit vector in a moving reference frame,

$$\frac{d\mathbf{1}_A}{dt} = \boldsymbol{\omega} \times \mathbf{1}_A$$

where $\boldsymbol{\omega}$ is the vector angular velocity of the moving reference frame with respect to inertial space. Therefore,

$$\frac{d\mathbf{A}}{dt} = \dot{A}\mathbf{1}_A + \boldsymbol{\omega} \times A\mathbf{1}_A \tag{A-3}$$

Theorem of Coriolis

Equation (A-3) is an expression of the theorem of Coriolis, which states that "If **A** is a free vector in a moving reference frame, the time derivative with respect to inertial space is

$$\left.\frac{d\mathbf{A}}{dt}\right|_{\text{inertial space}} = \left.\frac{d\mathbf{A}}{dt}\right|_{\text{moving reference}} + \boldsymbol{\omega} \times \mathbf{A} \tag{A-4}$$

where ω is the angular velocity of the moving reference frame with respect to inertial space.

APPENDIX B

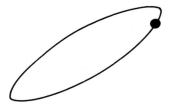
Planetary Values

TABLE B-1 PHYSICAL DATA[a]

Name	Radius ($\times 10^6$ m)	μ (m^3/s^2)	Relative mass	V_{esc} (m/s)
Sun	696.0	1.327×10^{20}	332,900	617,500
Moon	1.738	4.903×10^{12}	0.0123	2,375
Mercury	2.439	2.168×10^{13}	0.0544	4,217
Venus	6.052	3.248×10^{14}	0.8149	10,360
Earth	6.378	3.986×10^{14}	1.000	11,180
Mars	3.393	4.297×10^{13}	0.1078	5,032
Jupiter	71.40	1.267×10^{17}	317.8	59,568
Saturn	60.00	3.791×10^{16}	95.11	35,550
Uranus	25.40	5.788×10^{15}	14.52	21,350
Neptune	24.30	6.832×10^{15}	17.14	23,710
Pluto	1.500	9.965×10^{11}	0.0025 ?	1,153

[a] Mass of the Sun = 1.989×10^{30} kg; mass of the Earth = 5.975×10^{24} kg.

TABLE B-2 ORBITAL DATA

Name	Semimajor axis (AU)[a]	Orbital velocity (m/s)	Period (days)	Eccentricity, ϵ	Inclination (deg)
Mercury	0.3871	47,870	87.97	0.2056	7.004
Venus	0.7233	35,040	224.7	0.0068	3.394
Earth	1.000	29,790	365.2	0.0167	0.000
Mars	1.524	24,140	687.0	0.0934	1.850
Jupiter	5.203	13,060	4,332	0.0482	1.306
Saturn	9.539	9,650	10,760	0.0539	2.489
Uranus	19.18	6,800	30,690	0.0471	0.773
Neptune	30.07	5,490	60,190	0.0050	1.773
Pluto	39.44	4,740	90,460	0.2583	17.14

[a] 1 AU = 1.496×10^{11} m.

APPENDIX C

Additional Illustrations

Additional Illustrations

A. The launch of the Space Shuttle Discovery with the Hubble Space Telescope. (Courtesy of NASA.)

B. The TITAN IV booster. (Courtesy of Pratt & Whitney.)

C. The launch of the Saturn V booster with the Apollo 15 payload. (Courtesy of NASA.)

D. Apollo 15 astronaut (Irwin) on the surface of the Moon along with the Lunar Module and the Rover. (Courtesy of NASA.)

Additional Illustrations

E. The Soviet Soyuz spacecraft, as photographed from an Apollo spacecraft prior to rendezvous and docking. (Courtesy of NASA.)

F. A Delta booster with a business communications satellite to be placed in a stationary geosynchronous orbit. (Courtesy of NASA.)

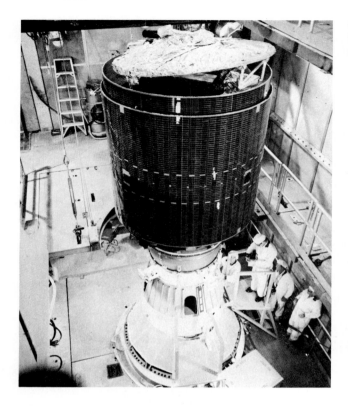

G. A business communications satellite to be placed in a stationary geosynchronous orbit. (Courtesy of NASA.)

H. A SATCOM C-1 communications satellite to be used in a stationary GEO. (Courtesy of GE Astro-Space.)

Additional Illustrations

I. An upper atmosphere research satellite in the integration and testing area. (Courtesy of GE Astro-Space.)

J. Artist's conception of an upper atmosphere research satellite in orbit with boom and panels deployed. (Courtesy of GE Astro-Space.)

K. A tracking and relay satellite (TDRS) leaving the payload bay of Atlantis prior to injection into a GEO. (Courtesy of NASA.)

L. The Gamma Ray Observatory (GRO) leaving the STS payload bay, as seen from the cabin of Atlantis. (Courtesy of NASA.)

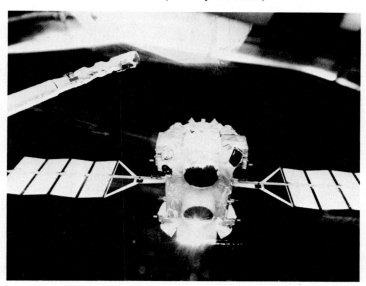

Additional Illustrations

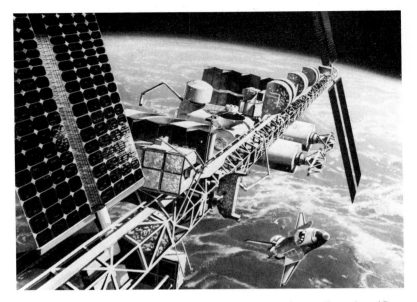

M. Artist's concept of one possible geocentric space station configuration. (Courtesy of NASA.)

N. A pictorial representation of the Solar System. (Courtesy of NASA.)

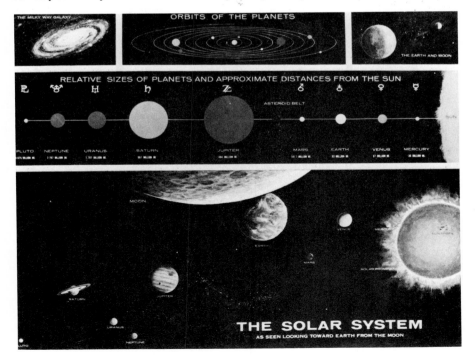

O. Artist's concept of the Galileo spacecraft, a Jupiter visitor, after release from a space shuttle and prior to deployment of antennas and boom. (Courtesy of NASA.)

P. The Magellan spacecraft leaving the payload bay of a space shuttle. (Courtesy of Martin Marietta.)

Additional Illustrations

Q. A computer-generated view of the Voyager 2 spacecraft passing Miranda, a moon of Uranus. (Courtesy of NASA.)

R. Artist's concept of a solar polar exploratory mission, such as Ulysses, in which a Jupiter flyby is used to transfer the spacecraft out of the ecliptic without propulsion. (Courtesy of NASA.)

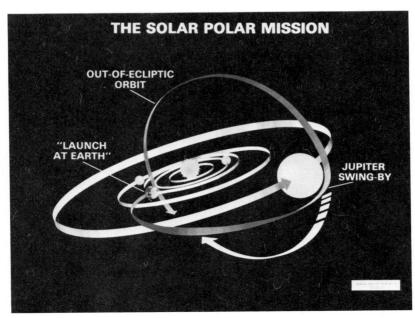

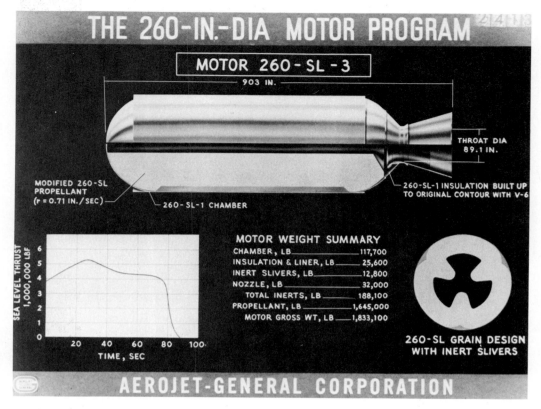

S. An experimental Aerojet-General solid-propellant motor with a thrust of 7317 kN. (Courtesy of NASA.)

T. A schematic of a hybrid rocket motor with liquid fluorine as the oxidizer, a solid fuel grain, and with liquid hydrogen to cool the combustion products prior to entering the nozzle. (Courtesy of NASA.)

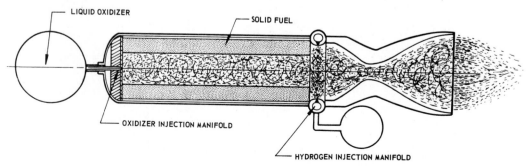

Additional Illustrations

U. An exploded view of an experimental aeroassisted entry vehicle: heat shield with ablative coating; vehicle body; and solid rocket motor for developing desired entry velocity. (Courtesy of NASA.)

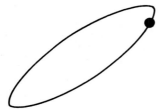

Selected References

BALL, K. J., and OSBORNE, G. F., *Space Vehicle Dynamics*. Oxford: Oxford University Press, 1967.

BATE, R. R., MUELLER, D. D., and WHITE, J. E., *Fundamentals of Astrodynamics*. New York: Dover Publications, Inc., 1971.

BATTIN, R. H., *An Introduction to the Mathematics and Methods of Astrodynamics*. Washington, DC: AIAA, Inc., 1987.

BERMAN, A. I., *The Physical Principals of Astronautics*. New York: John Wiley & Sons, Inc., 1961.

CHAPMAN, D. R., "An Approximate Analytical Method for Studying Entry into Planetary Atmospheres," *NASA Technical Report R-11*, Washington, DC, 1959.

DANBY, J. M. A., *Fundamentals of Celestial Mechanics*. Richmond, VA: Willmann-Bell, Inc., 1988.

EGGERS, A. J., "Atmosphere Entry," *AIAA Selected Reprint Series*, Washington, DC, undated.

ESCOBAL, P. R., *Methods of Orbit Determination*. New York: John Wiley & Sons, Inc., 1965.

FLANDRO, G. A., "Fast Reconnaissance Missions to the Outer Solar System Utilizing Energy Derived from the Gravitational Field of Jupiter," *Astronautica Acta*, Vol. 12, No. 4, pp. 329–337, 1966.

GAZELY, C., "Atmospheric Entry," in *Handbook of Astronautical Engineering*. New York: McGraw-Hill Book Company, 1961.

GREENWOOD, D. T., *Flight Mechanics of Space and Re-entry Vehicles*. Ann Arbor, MI: The University of Michigan Engineering Summer Conferences, 1964.

GRIFFIN, M.D., and FRENCH, J. R., *Space Vehicle Design*. Washington, DC: AIAA, Inc., 1991.

HANKEY, W. L., *Re-entry Aerodynamics*. Washington, DC: AIAA, Inc., 1988.

HESS, W. N., *The Radiation Belt and Magnetosphere*. Waltham, MA: Blaisdell Publishing Co., 1968.

KANE, T. R., LIKINS, P. W., and LEVINSON, D. A., *Spacecraft Dynamics*. New York: McGraw-Hill Book Company, 1983.

KAPLAN, M. H., *Modern Spacecraft Dynamics and Control*. New York: John Wiley & Sons, Inc., 1976.

KOELLE, H. H., *Handbook of Astronautical Engineering*. New York: McGraw-Hill Book Company, 1961.

LOH, W. H. T., *Dynamics and Thermodynamics of Planetary Entry*. Englewood Cliffs, NJ: Prentice Hall, 1963.

MARTIN, J. J., *Atmospheric Reentry*. Englewood Cliffs, NJ: Prentice Hall, 1966.

NEWELL, H. E., *Beyond the Atmosphere*. Washington, DC: NASA, 1980.

OSMAN, TONY, *Space History*. New York: St. Martin's Press, Inc., 1983.

PRATT, T., and BOSTIAN, C. W., *Satellite Communications*. New York: John Wiley & Sons, Inc., 1986.

REGAN, F. J., *Re-entry Vehicle Dynamics*. Washington, DC: AIAA, Inc., 1984.

REPIC, E. M., and EYMAN, J. R., "Approach Navigation and Entry Corridors for Manned Planetary Atmosphere Braking," *Journal of Spacecraft*, Vol. 6, No. 5, pp. 532–537, May 1969.

RIMROTT, F. P. J., *Introductory Attitude Dynamics*. New York: Springer-Verlag, 1989.

ROY, A. E., *The Foundations of Astrodynamics*. New York: Macmillan Publishing Company, 1965.

RUPPE, H. O., *Introduction to Astronautics*, Vol. 1. New York: Academic Press, Inc., 1966.

SEIFERT, H. S., *Space Technology*. New York: John Wiley & Sons, Inc., 1959.

SEIFERT, H. S., and BROWN, K., *Ballistic Missile and Space Vehicle Systems*. New York: John Wiley & Sons, Inc., 1961.

SHEVELL, R. S., *Fundamentals of Flight*, 2nd ed. Englewood Cliffs, NJ: Prentice Hall, 1989.

SINGER, S. F., *Torques and Attitude Sensing in Earth Satellites*. New York: Academic Press, Inc., 1964.

SZEBEHELY, V. G., *Adventures in Celestial Mechanics*. Austin, TX: University of Texas Press, 1989.

WALBERG, G. D., "Aeroassisted Orbit Transfer Window Opens on Missions," *Aeronautics & Astronautics*, pp. 36–43, November 1983.

WIESEL, W. E., *Spaceflight Dynamics*. New York: McGraw-Hill Book Company, 1989.

WOESTEMEYER, F. B., "General Considerations in the Selection of Attitude Control Systems," *Aerospace Vehicle Flight Control Conference*. New York: Society of Automotive Engineers, 1965.

WOODCOCK, G. R., *Space Stations and Platforms*. Malabar, FL: Orbit Book Company, 1986.

Index

A

Ablation, 237, 241
Acceleration:
  angular, 313
  of gravity, 10
  centripetal, 313
  Coriolis, 313
  inertial, 311–13
Active control, 311, 336
Activity sphere (*see* Sphere of influence), 86
Actuators, 334
  mass ejection, 334–35
  momentum exchange, 334–36
Aerobraking, 200, 242
  aeroassist, 200, 203, 205, 242
  aerocapture, 200, 203, 242
Aldrin, Buzz, 3
Angle:
  of attack, 282
  bank (roll), 316
  Euler, 314, 316
  pitch, 316
  roll, 316
  yaw, 316

Angular momentum, 13, 313–14, 319
  of point mass (external), 314
  of rigid body (internal), 314
  spin, 319
Anomaly:
  eccentric, 63, 69
  mean, 63, 65
  true (*see* Position angle), 63, 248, 251
Angular velocity, 313–17
Aphelion, 17
Apoapsis, 17
Apogee, 17, 33
Apollo program, 2–3, 291, 239, 258
Approach distance, 89, 108, 110
Apsides, 17, 19
  rotation of, 275
Argument of latitude, 251
Argument of the periapsis, 251
Armstrong, Neil, 3
Asteroids, 85
Astronomical Almanac, 85
Astronomical unit AU, 87, 135, 343
Asymmetrical trajectory, 281
Atmosphere, 197
  density ratio, 210
  of Earth, 210, 212
  planetary, 210, 221

361

Atmosphere (*cont.*)
  sensible, 188, 197, 206, 233
Atmospheric drag, 200, 242, 258
Attitude control, 311
Axisymmetric body, 317

B

Ballistic coefficient (BC), 209–10
Ballistic factor, 214
Ballistic missile, 133, 280–92
  time of flight, 290–91
  trajectory, 281
    long–way, 288, 303
    maximum–range, 297
    minimum–energy, 297
    short–way, 288, 303
Ballute, 242
Bielliptical transfer, 50
Booster, 160, 185, 280
Burnout, 280–82, 286
Burn time, 145, 153, 160

C

Caliber, 224
Canonical units, 83, 134–37
Celestial mechanics, 7
Celestial sphere, 249
Center of gravity (cg), 332
Center of mass (cm), 9, 10, 332
Central force field, 7, 11, 85
Centrobaric, 332
Circle, 17
Circular error probable (CEP), 302
Circular orbit, 22, 30–33
Circular satellite, 22
Collision (capture) cross section, 90
Conic sections, 17–21
  circle, 17
  ellipse, 17–20
  hyperbola, 17, 20
  parabola, 17, 21
Control moment gyro (CMG), 336
Coordinate transformation, 256
Coriolis:
  acceleration, 313
  theorem of, 314, 341
Cowell's method of perturbations, 257
Cranking, 124
Cross–range:
  distance: 232
  error, 301
Cruise boosters, 192–93

D

de Bergerac, Cyrano, 2, 142
Declination, 249, 257
Density specific impulse, 154, 184
Departure trajectory, 93
Despinning, 323
Direct entry (*see* Entry, ballistic)
Double–r iteration method, 257
Down–range:
  distance, 231
  error, 301
Drag, 188, 199, 224
  atmospheric, 200, 224, 258
  loss, 185, 186, 188–89, 199, 242
Dynamic stability, 233
DynaSoar, 241

E

Earth–surface satellite, 23, 30, 31
Eccentric anomaly, 63, 69
  hyperbolic, 69
Eccentricity, 16–17, 248, 289–90
  vector, 253
Ecliptic, 84, 112, 248
Effective structural factor, 159, 163, 168
Elevation angle, 14, 19, 24–25
  burnout 283–84, 286, 292–95, 303
  entry, 209, 211
Ellipse, 17
Elliptical orbits, 33–36
Encke's method of perturbations, 257
Energy, specific:
  kinetic, 13, 197
  mechanical, 12, 18, 54, 198
  potential, 13
Entry, 197–98
  angle, 209, 211
  atmospheric, 198, 206
  ballistic, 198, 209–12, 243
  corridor, 207–08, 225, 240
  lifting, 198, 225–34, 243
  orbital decay, 224, 234
  problem, 198–208
    deceleration, 214–22, 234
    heating, 234–42
  skip, 233–34
  vehicle (E/V), 197

Ephemeris, 85, 248
Epoch, 85, 248, 251
Equation(s) of motion, 11, 208, 211, 225, 247–48, 311–17
Error coefficients, 299
Escape, 13, 22, 39, 87, 91, 101, 185
  direct, 42
  Oberth maneuver, 41, 89, 92
  from parking orbit, 42
  trajectories, 39–42, 98
  velocity, 23
Euler:
  angles, 314, 316
  axes, 314
  rates, 317
Euler's moment equations, 314
Excess velocity, 23
Exhaust velocity, 147, 148
  characteristic, 153
  effective, 150

F

Fast transfer, 48–49, 59, 83, 107
Field of view, 258
Flight path angle (*see* Elevation angle), 14
Flyby, 82, 89, 199
  hyperbolic, 107, 112, 123
Flywheel, 335–36
Free-fall range, 281

G

Gagarian, Yuri, 3
Gauss:
  method of, 257
  problem, 59, 76, 257
Geocentric–equatorial coordinate system, 248
Geocentric orbits, 29
  geostationary, 33
  geosynchronous Earth (GEO), 32, 258, 265
  high Earth (HEO), 33
  low Earth (LEO), 33
Gimballing, 154
Glide Trajectories:
  equilibrium, 225, 233, 240
  negative L/D, 234
Goddard, Robert H., 143

Gravitational:
  constant, 8, 83
  force, 8, 85, 327
  parameter, 9
Gravity, 10
  loss, 185–87, 189
  turn, 190–91, 199
Gravity–gradient:
  stabilization, 326–33
  torque, 329, 332
Gravity assist, 83, 112
Grazing trajectory, 200
  multipass, 200, 203
Gross lift–off weight (GLOW), 162, 172
Ground tracks, 30, 54, 247, 257, 276
  geosynchronous Earth orbit (GEO), 265, 270
  inclination effects, 263
  longitudinal displacement, 260, 263
Gyroscope, 317, 334
  equations, 319
  torque, 321
Gyrostat, 319, 326

H

Half–cone angle, 212, 222
Hansen's method of perturbations, 257
Heat sink, 237
Heating, aerodynamic, 234–42
  ballistic, 238–39
  lifting, 239–41
  protection, 237
Heliocentric–ecliptic coordinate system, 248
Heliocentric orbits, 40
Hinge symmetry, 270, 279
Hohmann, Walter, 42
Hohmann transfer, 42–50, 83, 101, 117, 120, 123, 191
Hubble Space Telescope, 55, 258
Hyperbola, 17, 20
Hypergolic ignition, 156

I

Impact parameter, 87–89, 110
Impulsive thrusting, 38, 133, 145, 153
Inclination, 250
Inertial reference, 14, 29, 98, 312, 315
Inertial upper stage (IUS), 155
Influence coefficients, 299, 301

Intercept, 128–31, 134
Intersection angle, 107, 114, 123

K

Kepler, Johannes, 7
Kepler's equation, 66
Kepler's Laws of planetary motion, 7, 60, 62, 84
Keplerian orbit, 197, 198, 205, 280

L

Lambert, Johann Heinrich, 72
Lambert's theorem, 59, 72–78
Landing footprint, 225, 234, 243
Laplace, method of, 257
Launch velocity, 32, 45
Launch window, 83, 126, 277
Law(s):
 of cosines, 49, 51, 115
  Kepler's, 7, 60, 62, 84
  Newton's universal gravitation, 8, 85
Lead angle, 83, 124–31, 280
Leading-edge flyby, 123
Libration, 310, 332
Lift, 192–93, 199, 234
 negative, 208
Lifting body, 241–42
Lift-to-drag ratio (L/D), 198, 226–34
Limit cycle, 335
Linear momentum, 312
Light-second, 2
Light-year, 2, 83
Line of nodes, 251
Line symmetry, 270, 275
Load factor, aerodynamic, 232
Local horizon, 14, 315
Longitude:
 of ascending node, 251
 celestial, 251
 at epoch, 252
 geographic, 251
 of periapsis, 251

M

Mars, 3, 100–101, 108, 119, 129, 210, 221
Mass ratio, 157–58, 179, 183
Mean motion, 66
Mixture ratio, 155
Molyniya, 276

Moments of inertia, 314
 matrix, 314
 principal, 315
Momentum wheel, 336
Moon, 3, 137

N

NASA (National Aeronautics & Space Administration), 2
NASP (National Aerospace Plane), 183
n–body problem, 4
Newton, Isaac, 7, 11
Newton's law of universal gravitation, 8
Node(s):
 ascending, 251, 252
 descending, 251
 line of, 251
 regression of, 275
Nozzle, 147
 expansion ratio, 149
 ideal, 150
 plugged, 154
Nutation, 318–19
 damper, 319

O

Oberth, Hermann, 41
Oberth escape maneuver, 41
Oblate rigid body, 317
Optimum burnout angle, 295
Orbit determination, 76–79, 247, 252–57
Orbit shaping, 29, 36–38, 42
Orbit trimming, 42, 258
Orbital constants:
 specific angular momentum, 13
 specific mechanical energy, 12
Orbital elements, 54, 137, 247–53, 276
Orbital mechanics, 7
Orbital transfers, 42–50
Orbital transfer vehicle (OTV), 44
Orbits:
 circular, 22–23
 elliptical, 21–22
 hyperbolic, 23
 parabolic, 23

P

Parabola, 17, 21
Parking orbit, 42, 99, 108, 185, 191, 277

Patched conic, 82, 91–104, 112
Payload ratio, 158, 159, 174–76
Periapsis, 17
Perifocal coordinate system, 249
Perigee, 17, 33, 280
Perihelion, 17
Period, orbital, 22, 31, 33, 62
Perturbations, 8, 54, 257
  periodic, 257
  secular, 257
  variation of parameters, 257
Plane change, 50–53, 112, 124
  aerodynamic turn, 225
Planetary capture, 82, 104–12
Planetary passage, 82, 104, 112–24
Planetocentric encounter, 101
Planetocentric orbits, 30
Poe, Edgar Allen, 142
Point mass, 8, 310
Pointing, 311, 321
Polar equation, 7, 16–18, 283
Polar orbit, 263
Position angle (see True anomaly), 11, 16
Powered boost, 143, 185–92
Precession, 318–19, 321
Pressure ratio, 149
Prograde (direct) orbit, 251
Prolate rigid body, 317
Propellant mass fraction, 152, 158
Propulsion system, 144
  bipropellant, 153
  hybrid, 154–55
  liquid, 153
  monopropellant, 154
  solid, 154, 156
Pumping, 124

Q

Qbo parameter, 283–88

R

Radiation, 237, 241
Range equation, 282–85, 298
  rotating Earth, 302–4
Range sensitivities, 292–98
Reentry (see Entry), 280
  point, 282
Rendezvous, 131–34
Regression of the nodes, 275
Residual velocity, 23–24, 92

Retrograde orbit, 270
Right ascension, 249, 257
Rigid body, 311, 313
Rocket engines, 154–56
  chemical, 145, 154
  electric, 156
  nuclear, 156

S

Scale height, 210
Schuler period, 23, 31
Semilatus rectum, 16, 20
  of parabola, 77
Semimajor axis, 17, 248
Sensors, 333–34
  autonomous, 334
  control, 333, 334
  payload, 333
Sidereal day, 232
Single–stage–to–orbit (STO), 192
Solar day, 252
Solar system, 83–85
Space, 2–4, 82
Space Defense Initiative (SDI), 3
Space Shuttle (STS), 3, 170, 233, 240, 317
Space station, 3
Spacecraft, 160, 185
Specific impulse, 145, 155, 184
  density, 154, 184
Speed of light, 83
Sphere of influence (SDI), 82, 85–91, 104, 200, 233
Sputnik, 2
Stability:
  dynamic, 233, 310, 329
  static, 329
Stabilization:
  active, 311, 333–38
  gravity–gradient, 326–33
  passive, 311
  spin, 317, 325
  three–axis, 333
Stable platform, 337
Stage, 169
Staging, 143, 169–70, 179–85
Star Wars, 3
Station keeping, 54, 258
Step, 159, 169
  structural factor, 159, 163, 168
Sun–synchronous satellite, 275
Synodic period, 125

T

Thermal conversion fraction, 235–236
Thermal protection, 243
Three–body problem, 4, 91, 137
Throttling, 154
Thrust, 146–47, 150
   coefficient, 153
Thrusters, 317, 334
Thrust–to–weight ratio, 133, 152, 160
Time of flight (TOF), 59–60, 73
   ballistic missile, 290–91, 304
   ellipse, 62–68
   hyperbola, 68–72, 78
   parabola, 60–62, 77
Time of periapsis passage, 250, 276
Topocentric–horizon coordinates, 256
Trailing–edge flyby, 123
Tropos, 256
Tsiolkowski, Konstantin, 142–43
Tumbling, 325
Turning angle, 20–21, 113, 115
Turns, aerodynamic, 225, 232, 242
Two–body problem:
   equations, 25–26
   problem, 8–11

U

Units, 4–5, 143
   canonical, 83, 134–37, 253
   conversions, 5
Universal gravitation, 8
Universal variable, 79

V

Van Allen radiation belt, 258
Velocity budget, 82, 101, 142, 186
Venus 210, 221
Vernal equinox, 248
Verne, Jules, 2, 142
Vertical launch, 189
Vis–viva integral, 13, 283
von Neumann, John, 83

W

Wait time, 126
Weight, 10, 159
Wells, H. G., 142

Y

Yo–yo device, 323–24